Global Navigation Satellite Systems: Signal, Theory and Applications

Global Navigation Satellite Systems: Signal, Theory and Applications

Editor

Gustaf Kron

scitus
academics

Global Navigation Satellite Systems: Signal, Theory and Applications
Edited by **Gustaf Kron**

ISBN: 978-1-68117-412-9
Library of Congress Control Number: 2016937477

SCITUS Academics LLC,
www.scitusacademics.com
Box No. 4766, 616 Corporate Way,
Suite 2, Valley Cottage,
NY 10989

Notice

Printed in United States of America on Acid Free Paper

Preface

Satellite Navigation is a global coverage system for autonomous geospatial positioning. It allows small electronic receivers to determine their location at high precision (within a few meters) by means of time signals transmitted by satellites over a sight via radio. The signals also enable electronic receivers to calculate the current local time with high precision, allowing synchronization of time. The Global Navigation Satellite System (GNSS) can be called a global satellite navigation system. GNSS is a satellite system used geographically to locate the recipient of a user anywhere in the world. GNSS recipients are used with GPS, GLONASS, Galileo or Beidou systems in many applications. Automobiles can be equipped with GNSS receivers in the factory or as aftermarket equipment. Units often show moving maps and location, speed, direction and nearby roads and attractive places. Air navigation systems usually show a moving map and are often connected to the autopilot for enroute navigation. Boats and ships can use GNSS to navigate the world's entire lakes, seas and oceans. Maritime GNSS units include water-friendly functions such as "man overboard" (MOB) functions, which instantly mark the location where a person fell overboard and facilitate rescue efforts. GNSS can be used for heavy construction, mining and precision agriculture equipment. The blades and seats of the construction equipment are controlled automatically in GNSS-based machine control systems. Agricultural equipment can use GNSS to automatically control the driver or as a visual aid on the screen. Walkers, climbers and even ordinary pedestrians can use GNSS to determine their position with or without separate maps in urban or rural areas.

Spacecraft begins to use the GNSS as a navigation instrument. The addition of a GNSS receiver to a spacecraft enables precise determination of the orbit without tracking the ground. On the other hand, this enables autonomous navigation, flying formation and autonomous meeting of spacecraft. GNSS plays an important role in high-precision navigation, positioning, timing and scientific positioning. This is a very precise, constant, all-time and real-time technique.

The book aims to present recent results and developments in GNSS theory and real-world cases. This book is an important tool for GNSS designers, engineers, scientists and the user market.

Table of Contents

1

CHAPTER

AN SDR-BASED REAL-TIME TESTBED FOR GNSS ADAPTIVE ARRAY ANTI-JAMMING ALGORITHMS ACCELERATED BY GPU

Hailong Xu, Xiaowei Cui * and Mingquan Lu

Department of Electronic Engineering, Tsinghua University, Weiqing Building, Tsinghua, Beijing 100084, China

ABSTRACT

Nowadays, software-defined radio (SDR) has become a common approach to evaluate new algorithms. However, in the field of Global Navigation Satellite System (GNSS) adaptive array anti-jamming, previous work has been limited due to the high computational power demanded by adaptive algorithms, and often lack flexibility and configurability. In this paper, the design and implementation of an SDR-based real-time testbed for GNSS adaptive array anti-jamming accelerated by a Graphics Processing Unit (GPU) are documented. This testbed highlights itself as a feature-rich and extendible platform with great flexibility and configurability, as well as high computational performance. Both Space-Time Adaptive Processing (STAP) and Space-Frequency Adaptive Processing (SFAP) are implemented with a wide range of parameters. Raw data from as many as eight antenna elements can be processed in real-time in either an adaptive nulling or beamforming mode. To fully take advantage of the parallelism resource provided by the GPU, a batched method in programming is proposed. Tests and experiments are conducted to evaluate both the computational and anti-jamming performance. This platform can be used for research and prototyping, as well as a real product in certain applications.

Keywords: Global Navigation Satellite System (GNSS) anti-jamming; adaptive array; software-define radio (SDR); Space-Time Adaptive Processing (STAP); Space-Frequency Adaptive Processing (SFAP); nulling; beamforming

INTRODUCTION

Due to the vulnerability of Global Navigation Satellite System (GNSS) signals to radio frequency (RF) interferences, anti-jamming design has always been an important issue in both military and civilian applications. In order to increase the resistance of GNSS receivers to intentional and unintentional interferences, adaptive array processing techniques such as space-time adaptive processing (STAP) and space-frequency adaptive processing (SFAP) are widely used. Compared to space adaptive processing (SAP), they exploit both temporal (frequency) and spatial domain, providing more degrees of freedom (DOF) [1]. Previous literature shows that wide-band and narrow-band interference can be mitigated effectively through these algorithms.

Jamming and anti-jamming are best considered together. An anti-jamming algorithm with certain parameters may perform quite differently while the jamming pattern varies. It is necessary to study the relation, so that we can choose the most appropriate anti-jamming algorithm with the most appropriate parameters against specific interference scenarios. Although software tools such as MATLAB can be used to simulate on the principle level, it is quite far from the actual physical environment, reducing the validity of the results. It is hoped that different adaptive array algorithms can be tested in real-time in real interference scenarios, while the parameters can be flexibly configured in the field. In order to meet this demand, a GNSS adaptive array anti-jamming testbed with a high level of flexibility and configurability is designed and implemented in this paper.

Adaptive array algorithms are very computational costly, so traditionally they are implemented on hardware such as Field Programmable Gate Arrays (FPGAs) or Application Specific Integrated Circuits (ASICs) to meet the real-time requirement. However, this hardware implementation approach has some drawbacks and is not suitable for our testbed. Firstly, implementing on hardware requires a long development cycle, and is difficult to update and customize. Secondly, once the program is compiled and burned into the hardware, algorithms and parameters are difficult to be configured. Thirdly, it is hard to observe any intermediate states while the algorithm is running. On the contrary, a software-defined approach makes up all the short-comings above. It has greater flexibility, and can provide rich functionality. Using high-level programming languages such as C++ and C#, many libraries can be used, making the development process much faster. Different algorithms and parameters can be integrated into one implementation, making it a fertile ground for researchers to explore more possibilities. The real-time display of intermediate states by the graphic user interface (GUI) also makes assessing and verifying the algorithms more direct and timely. In summary, a software defined approach is preferred in the design and implementation of our testbed.

While a software implementation exceeds a hardware implementation in many ways such as flexibility, configurability and cost, special attention should be paid to satisfy the real-time requirement, especially in applications with a high data throughput. Adaptive array anti-jamming algorithms such as STAP and SFAP include a large number of matrix multiplication, linear equations solving and other complex operations. The amount of calculation is so high that even a modern Central Processing Unit (CPU) with many cores has difficulties to run them in real time. However, Graphics Processing Units (GPUs) have achieved great success in general-purpose computing in the last ten years. Unlike a CPU which has 4 cores or 8 cores with most transistors devoted to data caching and flow controlling, a GPU generally has hundreds of cores with most transistors devoted to data processing. Tens of thousands of threads can run concurrently on a single GPU, providing great speed-up against CPU in computational intensive applications. So in this project a GPU is selected as an accelerator.

Using a GPU as an accelerator for GNSS software-defined receivers has been studied intensively in the past ten years [2,3,4,5]. One of the most representatives is the STARx developed by a group of researchers in Tsinghua University, China [6]. By putting the acquisition search and correlation operations on the GPU, the receiver is able to process all civil signals from all GNSS systems in real time. Despite that software receivers have become a common way to evaluate signals and test new algorithms nowadays, work in GNSS adaptive array anti-jamming field to employ a software-defined approach is still limited, due to the high data throughput and computational complexity of adaptive array algorithms. As far as the authors know, the two most updated works in open literature are contributed by a group of researchers in Stanford University [7,8]. In [7] a real-time GPS receiver with adaptive beam-steering capability using a software-defined approach based on a quad-core CPU coupled with a GPU is developed. This receiver offers sufficient computational capability to support a four-element antenna array to form 12 beam-steering channels for all tracking satellites in view using SAP, and processes digital samples of GPS L1 Coarse/Acquisition (C/A) and L5 signals at a 40 MSPS rate with 14-bit resolution. After that, a CPU-based real-time beamforming receiver was implemented by the same group of researchers [8]. This receiver extends the function of the previous one by implementing the STAP algorithm. Single Instruction Multiple Data (SIMD) instructions assembly coding and multithreaded programming are used to reduce computational complexity. However, both of the two receivers are especially developed for beamforming and are limited in flexibility and configurability. Furthermore, they both exploit the iterative Least Mean Square (LMS) method to reduce computational complexity by avoiding the matrix inversion operation. Distinguished from the above two works, our testbed in this paper is for anti-jamming algorithms including not only SAP and STAP, but also SFAP, towards both adaptive nulling and beamforming. The Sample Matrix Inversion (SMI) method which requires matrix inversion operations is implemented. Special efforts are made to provide a high configurability of algorithms and parameter, as well as a high computational performance, which makes our testbed to exceed work in previous literature. A comparison between the previous works and our testbed is summarized in Table 1.

The above capabilities of our testbed are attributed to the GPU programs to a great extent. Although massive parallelism can be provided by GPU, it is the programmers' duty to map algorithms properly to the hardware architecture in order to fully take advantage of the parallel computational resources. This is often a tricky task, which contributes to the fact that until now, GPU programming is still highly application-relevant. In consideration of this, the GPU programming details are fully documented for reference. In particular, a batched method which allocates the computing tasks more efficiently on GPU is highlighted. It aims to calculate many small problems concurrently rather than serially. The batched method has been proposed in previous work [9,10], but has not been used in GNSS anti-jamming applications. In addition, Compute Unified Device Architecture (CUDA) codes are carefully tuned to achieve the largest possible speedup. Programming techniques are introduced with great detail.

Table 1. Overview of previous work and testbed presented in this paper.

Work	Jiwon Seo (Stanford) [7]	Yu-Hsuan Chen (Stanford) [8]	This Paper
Anti-jamming mechanism	Adaptive beamforming only	Adaptive beamforming only	Both adaptive nulling and beamforming
Algorithm types	SAP	STAP	SAP, STAP and SFAP
Configurability	Limited	Limited	Parameters can be configured in a wide range
Weights calculation method	A propagating (iterative) approach. No inverse calculation of the covariance matrix	A propagating (iterative) approach. No inverse calculation of the covariance matrix	The SMI approach
Antenna element number	4 antenna elements	4 antenna elements	Configured antenna element number, which is 8 in maximum
Process samples	14-bit resolution samples at a rate of 40 MSPS (20 MSPS inphase and quadrature samples)	16-bit resolution samples at a rate of 40 MSPS (20 MSPS inphase and quadrature samples)	14-bit resolution samples at a rate of 40 MSPS (20 MSPS inphase and quadrature samples)
Processing Core	Intel Core i7 950 CPU and a NVDIA GTX 480 GPU	An x86-based multi-core CPU	An x86-based multi-core CPU and a NVDIA GTX 770 GPU
Application	For anti-jamming reference station receivers	For anti-jamming stationary receivers	For researching, prototype testing, as well as anti-jamming stationary receivers

The rest of this paper is organized as follows: first, the system design is outlined, with emphasis on the anti-jamming processing unit on which the software runs. Then, Section 3 gives a brief review of the STAP and SFAP algorithms, which is a requisite for the following contents. After that, the software architecture and GPU programming techniques are introduced in detail in Section 4. To show that the software implementation meets the real-time requirement very well, test results of computational performance are given in Section 5. In Section 6, experiments are conducted in real environment with interferences to validate the algorithms, while Section 7concludes this paper.

2. SYSTEM DESIGN

In this section, the overall architecture and data flow path of the platform are outlined, including the hardware modules as well as the general purpose processors in which the software are embedded. In the system design, special attention is paid to adapt the high data throughput and computational complexity.

2.1. Platform Architecture and Data Flow

As Figure 1 shows, the anti-jamming testbed performs the signal transforming and processing from the antenna array all the way to the GNSS receiver. In this paper we mainly focus on signals of GPS L1 C/A and China's Beidou System (BDS) B3, which are two representatives of a low-chip-rate signal with 2 MHz bandwidth and a high-chip-signal with 20 MHz bandwidth respectively. At the very beginning, the antenna elements transform the radio signals in L-band to voltage signals, after which the low noise amplifiers (LNAs) amplify the signals with a gain of about 30 dB. Figure 2 shows the geometry of the specially designed planar antenna array with 8 elements. The objective of the design is that we can select different elements to form multiple types of array geometry with the number of antenna elements varying from 4 to 8. The square patch antennas of the array are dual-band, with the frequency centered at GPS L1 (1575.42 MHz) and BDS B3 (1268.52 MHz), respectively. Alternatively, a third-party antenna array and LNAs can also be used, as long as the element number is 8 or less.

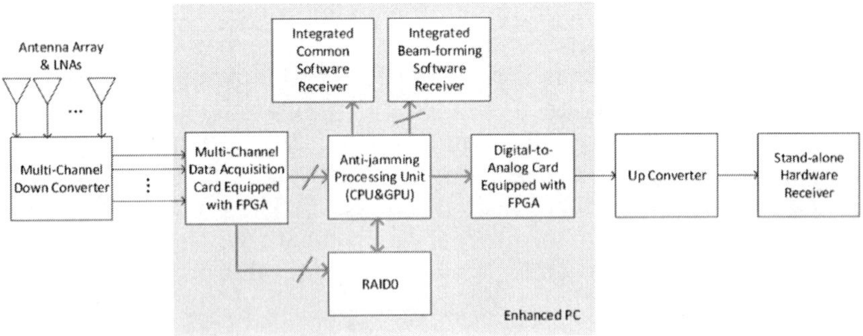

Figure 1. System architecture and data flow path of the adaptive array anti-jamming testbed. The thin black lines and arrows indicate the analog signal stream, while the bold blue lines and arrows indicate the digital data stream. The short slashes upon the lines indicate that the stream contains multiple channels.

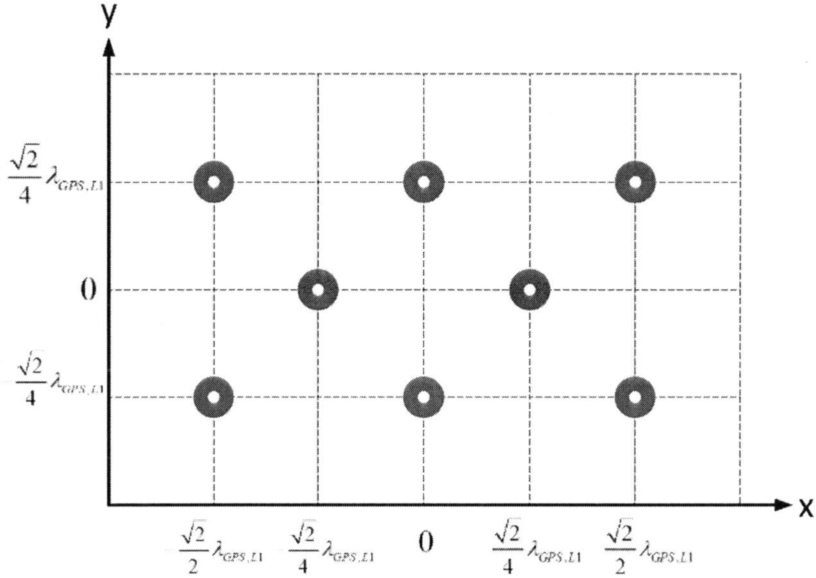

Figure 2. Geometry of the dual-frequency antenna array for GPS L1 and BDS B3 signals. The distance between the two nearest antenna elements is one half of the GPS L1 carrier wavelength.

Afterwards, the multi-channel down converter transforms the L-band signal to intermediate-frequency (IF) of about 45 MHz. The bandwidth of the RF front-ends is 2 MHz for GPS L1 C/A signal and 20 MHz for BDS B3 signal, respectively. The multi-antenna signals are then digitalized by a data acquisition card, which consists of 8 analog-to-digital converters (ADCs) and an FPGA whose framework logic can be modified. The card operates at a sampling rate of 60 MSPS to avoid spectrum aliases. The quantization resolution is 14 bits, thus enough ADC dynamic range is provided to support intensive jamming scenarios, although 2 bits are enough for a common receiver [11]. For the conversion to in-phase and quadrature (I/Q) signals as well as further reducing the data throughput, the 60 MSPS real data are then processed by the FPGA logic and transformed to 20 MSPS complex data. Performing the conversion to I/Q digitally also avoids the imbalance of analog I/Q circuits [12]. Then the output 20 MSPS multi-antenna complex data are transferred to the CPU and GPU for anti-jamming processing.

The anti-jamming processing unit is the computing core of the whole system and can work in two modes, namely the nulling mode and the beamforming mode. In the nulling mode, the multi-antenna digital complex data from the ADC card are transferred to the GPU for adaptive nulling processing. After the interferences are mitigated, the output single-channel complex data are transferred to the digital-to-analog converting (DAC) card, in which they are converted to IF analog signal. The signal is then up-converted to L-band before transferred to a stand-alone hardware receiver. As an alternative, the output data of the anti-jamming processing unit can also be directly transferred through the GPU memory to an integrated common software receiver, which is a modified version of STARx [6]. The word "common" is used to distinguish the single-input-channel receiver from a beamforming receiver which has multiple input channels.

In the beamforming mode, the anti-jamming processing unit uses the elevation and azimuth information of each visible satellites from the software receiver to form a dedicated beam in the antenna radio pattern towards each satellite. The multi-beam complex data are then transferred to the integrated beamforming software receiver. The feedback path of the elevation and azimuth information is not depicted in the figure, for its data bandwidth is small compared with the data stream.

Besides working in real-time mode, the testbed can also run in a post-processing mode, in which the input multi-antenna data of the anti-jamming processing unit are read from a file. These data are previously saved to the hard disk from the data acquisition card in a data fetching and logging process. If an eight-element antenna array is used, the data throughput to the hard disk can be as high as 80 × 8 = 640 MB/s, making data storage bandwidth the bottleneck of the whole procedure. Thus, a RAID hard disk array is used, as Figure 1 shows [13].

2.2. Anti-Jamming Processing Unit

A hybrid computing architecture is used in the anti-jamming processing unit, which comprises an Intel i7 CPU with four 2.5 GHz cores, and an NVidia GeForce GTX 770 GPU. The CPU is responsible for light sequence tasks, while the GPU is responsible for heavy parallel tasks.

GeForce 770 is NVidia's consumer level product designed for gamers, providing a high computing capacity at a reasonably low cost. It has 1536 CUDA cores with a base clock of 1046 MHz. The memory is as much as 2 GB memory with a speed of 7 Gbps. This GPU is not so advanced compared to the following products such as GTX 980 and TITAN X, but it will be seen that with the programming optimization techniques in this paper, the capability of this GPU is enough to satisfy the real-time requirement.

In practice, the data acquisition card, the DAC card, the disk array, and the GPU are all integrated into an enhanced PC and transfers data through the PCI-E bus. The common software receiver and the beamforming software receiver runs on the anti-jamming processing unit.

3. ADAPTIVE ARRAY ANTI-JAMMING ALGORITHMS

Both STAP and SFAP have been well studied and documented in [14,15,16], so we just give a brief review of them for reference. Meanwhile, the parameters and their range of values which we implemented in this project are emphasized. Implementations of the algorithms will be introduced in detail in the next section.

Throughout this paper, the following notation is adopted: small bold letters stand for vectors and capital bold letters stand for matrices, while superscripts H and T denote conjugate transpose and transpose respectively.

3.1. Spatial-Time Adaptive Processor (STAP)

3.1.1. Algorithm Architecture

A space-time adaptive processor (STAP) extends a space adaptive processor (SAP) by utilizing a finite impulse response (FIR) filter behind each element, thus introducing resolution in the temporal (frequency) domain and obtaining more degrees of freedom (DOF) [11]. Figure 3 shows the architecture of STAP, where M denotes the number of array elements, and N denotes the filter tap length. xm(n) denotes the complex sample data of element m at the time n, and y(n) denotes the complex output.

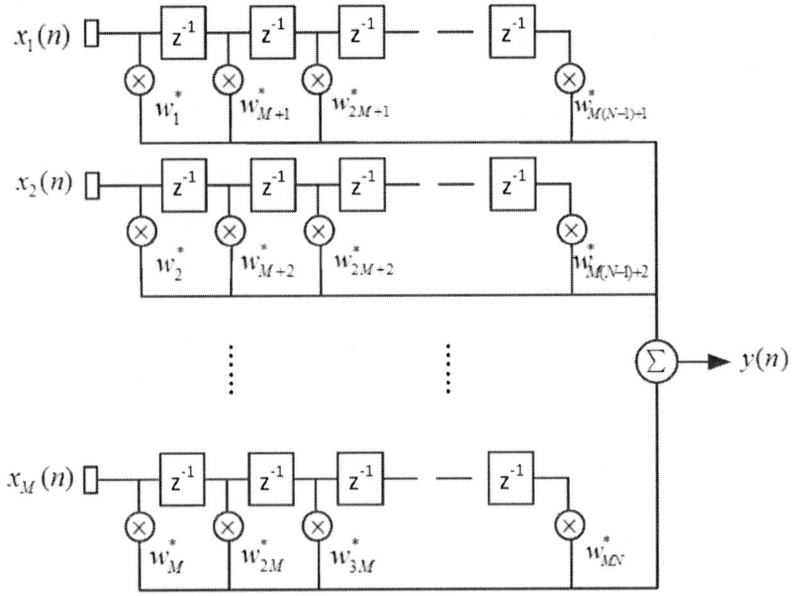

Figure 3. The architecture of a spatial-time adaptive processor (STAP).

The input data vector x(n) and the weight vector w can be written as

$$\mathbf{x}(n) = [x_1(n), x_2(n), \ldots, x_M(n), x_1(n-1), x_2(n-1), \ldots, x_M(n-1), \ldots, \\ x_1(n-N+1), x_2(n-N+1), \ldots, x_M(n-N+1)]^T \tag{1}$$

$$\mathbf{w} = [w_1, w_2, \ldots, w_{MN}]^T \tag{2}$$

Then y(n) can be written in the dot product form:

$$y(n) = \mathbf{w}^H \mathbf{x}(n) \tag{3}$$

In GNSS anti-jamming applications, the weight vector is usually solved according to the linearly-constrained-minimum-variance (LCMV) criterion, in which case the optimized w can be written as [14]:

$$\mathbf{w}_{opt} = \frac{\mathbf{R}^{-1}\mathbf{s}}{\mathbf{s}^{\mathrm{H}}\mathbf{R}^{-1}\mathbf{s}} \tag{4}$$

where s is the constraint vector and R is the correlation matrix of the input data which is defined as follows:

$$\mathbf{R} = E\left\{\mathbf{x}\mathbf{x}^{H}\right\} \tag{5}$$

Assuming that the input data have a zero mean, then R is also the covariance matrix. In practice, it is usually estimated by

$$\hat{\mathbf{R}} = \frac{1}{K}\sum_{k=0}^{K-1}\mathbf{x}(n-kN)\mathbf{x}(n-kN)^{H} \tag{6}$$

where parameter K denotes the number of snapshots taken to estimate the covariance matrix. The parameter kis multiplied by N to ensure that the successive input data vectors do not overlap with each other, thus can be regarded as independent.

To implement adaptive beamforming, s can be set to a steering vector of one visible satellite which can be written as

$$\mathbf{s} = \left[\mathbf{0}^{T},\ldots\mathbf{0}^{T},\mathbf{s}_1{}^{T},\mathbf{0}^{T},\ldots,\mathbf{0}^{T}\right]^{T} \tag{7}$$

where 0 is the zero vector of M×1 and s1 can be written as

$$\mathbf{s}_1 = \left[e^{j\theta_1},e^{j\theta_2},e^{j\theta_3},\cdots e^{j\theta_M}\right]^{T} \tag{8}$$

where θi(i=1,2,···M) is the carrier phase shift of antenna element i due to the geometrical relationship between the antenna array and the satellite. In practice, the steering vector can be calculated by the direction of the satellite and the orientation of the antenna array measured by an Inertial Measurement Unit (IMU). It is necessary to note that, s can be written as the simple form because some non-ideal factors of the antenna and RF front-ends are ignored. These factors include the radio pattern's anisotropy of each antenna element, inaccuracy of the array geometry, the mutual coupling effect, the finite ground-plane effect, mismatches between RF channels, different delays of cable lines, *etc.* [7,17,18].

In adaptive nulling, s is usually set to

$$\mathbf{s} = [0, 0, \ldots, 0, 1, 0, 0, \ldots, 0]^T \quad (9)$$

in which the values are all zero except one. In this case, the approach is also called the power inversion (PI) method [14].

3.1.2. Configurable Parameters

The antenna element number M and the filter tap length N are two basic parameters of STAP. In this project, M can vary from 4 to 8, while N can be equal to or less than 10. This makes that the size of R vary from 4 × 4 to 80 × 80. As we will see, this range of the covariance matrix scale will have an important influence on the GPU-based implementation.

The number of snapshots K is another important parameter. On one hand, it determines how accurate the estimation of R is. An inadequate estimation may result in a distorted antenna pattern far from the expected. Fante has stated in [15] that effective nulling can be achieved when K=4MN, which is a relatively small number. However, larger K surely helps to get a more accurate estimation of R in a stationary process. On the other hand, K decides the updating period of w together with N in our implementation. When N is fixed, the smaller K is, the more frequent the weight vector updates. Normally, we hope the weight vector has a short updating period in order to adapt to the fast changing environment. In summary, the choice of K is a balance between the estimating accuracy of R and the adapting capability. In this paper, K is set to vary from 100 to 1000, while larger K is possible but will not likely bring any obvious benefits.

Sometimes an inadequate estimation of the covariance matrix is inevitable, especially when the environment statistical characters are time-variant. One way to ameliorate the side effects brought by inadequate estimation of the covariance matrix is diagonal loading [19], by which we mean that R^ is added by a diagonal matrix as follows:

$$\hat{\mathbf{R}}_{dl} = \hat{\mathbf{R}} + \sigma_{dl}\mathbf{I} \quad (10)$$

where σdl is defined as the diagonal loading factor and I is the identity matrix. In [20] σdl is set to be 3 or 10 dB greater than the thermal noise power. In this paper, σ_{dl} can be configured to a fixed number, or more reasonably, can be set to be proportional to the trace of the covariance matrix, meaning that

$$\sigma_{dl} = \alpha_{dl} \times \text{trace}(\hat{\mathbf{R}}) \tag{11}$$

where αdl is a small number. In our experiments, it is also found that diagonal loading has a compensating effect when the interference is very strong, in which case ill-condition of the covariance matrix causes that the eigenvalues representing the noise are too small compared to the biggest eigenvalue, thus cannot be expressed precisely due to the precision of floating point numbers. The drawback of diagonal loading is that it can reduce the depth of nulls in the antenna pattern. In our implementation, the way how diagonal loading is conducted is configurable.

In both beamforming and nulling, selection of a reference tap is required. The reference tap corresponds to the non-zero elements in s. Moore has stated in [21] that reference tap selection has some kind of relationship with the output signal distortion. Simulation results also showed that selecting the center tap as the reference offers the best overall performance [21]. In spite of this, we leave the reference tap index as a parameter to be configured.

To conclude this subsection, the configurable parameters in STAP and their value ranges are summarized inTable 2.

Table 2. Configuration parameters of the Spatial-time adaptive processor (STAP) and their value range.

Parameter	Meaning	Value Range
M	Antenna element number	4, 5, 6, 7, 8
N	Filter tap number	1 to 10
K	Number of snapshots taken to estimate the covariance matrix	100 to 1000
N_{ref}	Reference tap index	1 to N
σ_{dl} or α_{dl}	Diagonal loading factor	Positive small number

3.2. Spatial-Frequency Adaptive Processor (SFAP)

3.2.1. Algorithm Architecture

Spatial-frequency adaptive processor (SFAP) is proposed as an alternative of STAP to reduce the size of the covariance matrix to be inversed [16]. Figure 4 depicts the schematic overview of the spatial-frequency adaptive processor. Here, $x_m(n)$ ($m=1,2\cdots M$, $n=0,1,\cdots N-1$) denotes the complex sample data of element m at time, and $\tilde{x}_m(l)$ denotes the frequency sample of antenna m in frequency bin l after the discrete Fourier transform (DFT) module. In this paper, we use the same letter N to represent both the filter tap length in STAP and the DFT length in SFAP. The meaning should be obvious in the specific context.

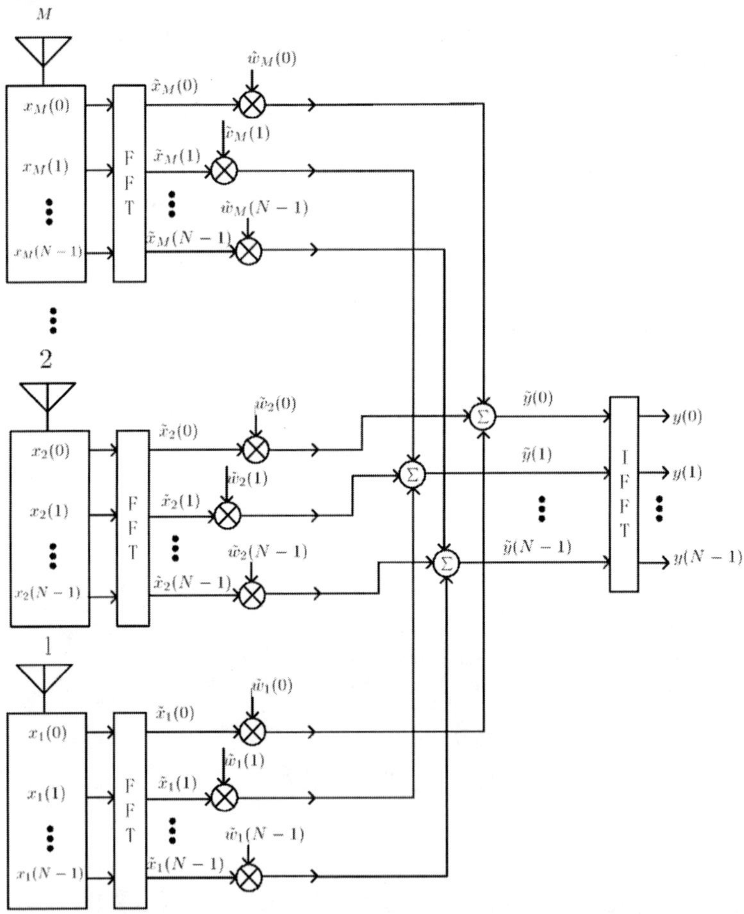

Figure 4. Architecture of the spatial-frequency adaptive processor (SFAP) with M antenna elements and a DFT length of N.

In each frequency bin the output y˜(l) is calculated by

$$\tilde{y}(l) = \tilde{w}(l)^H \tilde{x}(l) \tag{12}$$

where w˜(l) is the weight vector and x˜(l) is the input vector, both in the frequency domain.

The weight vector is given by

$$\tilde{w}(l) = \frac{\tilde{R}(l)^{-1}\mathbf{s}}{\mathbf{s}^H \tilde{R}(l)^{-1}\mathbf{s}} \tag{13}$$

where R˜(l) is the covariance matrix of frequency bin l which is defined as

$$\tilde{R}(l) = \mathrm{E}\left\{\tilde{x}(l)\tilde{x}(l)^H\right\} \tag{14}$$

In practice R˜(l) is usually estimated by

$$\hat{\tilde{R}}(l) = \sum_{i=0}^{K-1} \tilde{x}(l,i)\tilde{x}(l,i)^H \tag{15}$$

where x˜(l,i) denotes the input vector in frequency bin l at snapshot i, and K is the snapshot number used to estimate R˜(l).

In nulling, s can be set to

$$\mathbf{s} = [1,0,0,\cdots,0]^T \tag{16}$$

In beamforming, s is set to be the steering vector.

3.2.2. Configurable Parameters

The three primary parameters of SFAP are M, N and K. For this study, M varies from 4 to 8, while N is set to be a power of 2 for the computational efficiency of FFT. In this project, N can be 1024 at maximum. The parameter K determines the weight updating period together with N, and influences the accuracy of the estimation of R˜(l). Besides, diagonal loading is also used to desensitize the covariance matrix estimation, just like in the situation of STAP.

In addition to the four parameters above, Gupta has shown in [16] that when the interference scenario consists of narrow-band as well as wide-band sources, windowing the time domain samples leads to improved performance in SFAP. He also demonstrated that 20% to 25% sample overlapping is needed to achieve the optimum performance. Thus in this project, we leave the window type and the overlapping ratio as two configurable parameters.

To conclude this subsection, all the configurable parameters in SFAP are summarized in Table 3.

Table 3. Configuration parameters of the Spatial-frequency adaptive processor (SFAP) and their value range.

Parameter	Meaning	Value Range
M	Antenna element number	4, 5, 6, 7, 8
N	DFT length	64, 128, 256, 512, 1024
K	Number of snapshots to estimate the covariance matrix	32 to 500
σ_{dl} or α_{dl}	Diagonal loading factor	Positive small number
Window type	Function to window the time domain samples	Rectangle, Triangle, Bartlett, Hann, Hamming, Blackman, *etc.*[22]
overlap	Overlapping ratio	Non negative number less than 1

4. A SOFTWARE-DEFINED APPROACH TO THE ANTI-JAMMING PLATFORM

4.1. Software Architecture

The software is the part of the testbed that needs the most development efforts. In this project, we use Visual Studio 2010 (Microsoft, Redmond, WA, USA) and CUDA® 7.0 (NVidia, SantaClara, CA, USA) on the operating system of Windows 7 as the developing environment. Strictly speaking, Windows 7 is not regarded as a real-time operating system, but real-time applications developed on it are still possible due to the strong processing power of the processors. The software is developed in C++.

Figure 5 shows the architecture of our software implementation. The software is mainly divided into four modules, namely the data stream controlling module, the anti-jamming processing module, the software receiver processing module and the GUI module, each of which can be further divided into several function units. These modules are driven by four threads, which are for data fetching, data delivering, data processing and GUI display, respectively. Some units are allocated to the GPU, while others are allocated to the CPU, just as shown in the figure. This allocation is mainly based on the data intensity of each task.

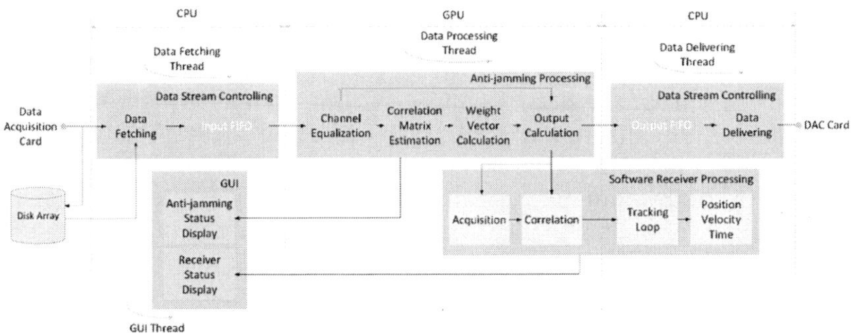

Figure 5. Software architecture.

4.1.1. Data Fetching and Delivering

In the data fetching thread, the data fetching unit receives raw data from the data acquisition card in packet form, and then send them to the input first-in-first-out (FIFO) buffer unit after the packet headers are parsed and stripped. The input FIFO buffer is allocated on CPU memory and aims to guarantees the continuity of the data flow. In the real-time mode, the raw data are extracted at the output end of the buffer by the data processing thread for following processing. The occupancy of the buffer should stay at a low level, which indicates that the data processing speed is faster than the data arriving speed. Instead of transferring the raw data to the input FIFO buffer, the data fetching unit can also log the raw data to the disk array directly for post-processing. Thus, in the post-process mode, the data fetching unit extracts the raw data from a file on the disk array.

If a standalone hardware receiver is used, then the data delivering thread is created to receive the output data from the anti-jamming processing module. The data are first transferred to the output FIFO buffer allocated in CPU memory, then extracted by the data delivering unit and packetized. Once an interrupt signal from the DAC card arises requiring for new data, a data packet is send to the card. Under normal conditions, occupancy of the output buffer should be neither full nor empty, and stay stable. The input and output FIFOs are the main contributions of the total latency of this system.

4.1.2. Data Processing

The data processing thread is responsible for anti-jamming and software receiver processing. It has been well known that mismatches of the RF channels and different cable delays can reduce the interference mitigation performance seriously [23,24,25], so in this project, the multi-element raw data are first equalized by the channel equalization unit. The channel equalization is realized by filtering the raw data of each channel by a FIR filter. The filter coefficients are obtained beforehand through a calibration process, in which a strong white Gauss noise (WGN) signal is split into eight paths, each of which is injected to a RF channel through the LNA input. Then the output data of each ADC channel are used to train the coefficients using the Wiener filtering method. The calibration process may be conducted over time due to aging of the RF devices.

Besides channel equalization, covariance matrix estimation, weight vector calculation and output calculation are the three basic steps in both STAP and SFAP when the SMI method is used. These three units are the most computational intensive parts of the whole software, so must be carefully programmed and tuned. The implementation details will be introduced with emphasis in following subsections. For SFAP, a DFT unit and an IDFT unit should also be added, but they are ignored in Figure 5 for brevity.

If the integrated software receiver is used, then the output data after the interference suppression are transferred to the software receiver processing module for acquisition, tracking and position-velocity-time (PVT) calculation [26,27]. Of particular importance is that for beamforming, the input data to the receiver include multiple channels, each of which corresponds to a particular beam. In this project, both the common software receiver and the beamforming software receiver are modified versions of STARx, the implementation of which has been introduced very well in [6], thus is ignored in this paper.

4.2. GPU Architecture and Programming Model

The rest of this section focuses on the implementation of STAP and SFAP on the GPU. Firstly, it is necessary to give a brief review of the GPU architecture and its programming model.

4.2.1. Thread Hierarchy

CUDA uses a single-program-multiple-data (SPMD) paradigm, which means that different data are processed by the same program concurrently. This is based on the hardware architecture of the GPU. In general, one single GPU contains a number of Streaming Multiprocessors (SMs), each of which is an array of Streaming Processors (SPs). Each SP is a fully pipelined, in-order microprocessor that executes a set of instructions serially, while many SPs in the same or different SMs can execute at the same time. For instance, GeForce GTX 980 which uses GM204 GPU contains 16 SMs, each of which contains 128 SPs, so that as many as 2048 threads can be executed at the same time.

In CUDA a kernel function executes a specific task, and generates a large number of threads to exploit data parallelism after launched. On programming logic level, these threads are further divided into blocks within the same grid. Each thread is distinguished from others by its block index in the grid (denoted by blockIdx) and thread index in the block (denoted by theadIdx). When mapping the programming logic architecture to the hardware architecture, each block must reside on one single SM, while each SM can contains more than one block at the same time. This determinates that the number of threads in one block must not exceed a certain value due to limited resources per SM, which include the number of SPs, the amount of shared memory as well as the register file size. However, the number of blocks within one grid can be very large, which is as much as 65,535 generally. Details about thread hierarchy in CUDA can be found in [28,29,30].

4.2.2. Memory Hierarchy

Besides the execution model described above, data memory also plays an important role in GPU programming. Memory on the CPU side is called host memory in GPU programming context, whereas within the GPU, global memory, shared memory and registers are the main concerns. Global memory is off-chip and public to all threads within the grid. Accessing it is very slow compared to shared memory and registers.

Shared memory resides on SM and are used as L1 cache. It is visible to all threads within the same block and can be accessed much faster than global memory, but the amount of it is limited (generally 48 KB per SM). Registers have the smallest accessible domain which is restricted to within one single thread, but they have the smallest access latency that is ignorable. The register file size is limited, and often it is the complier's role to decide how much of it is used.

It is well known that data transfer between global memory on GPU and host memory on CPU is very costly, so should be minimized as much as possible. Besides, it is always preferable to let the global accesses to coalesce, which is perhaps the most important performance consideration in CUDA programming [31].

Another key point of memory optimization is to use as much fast-access memory and as little slow-access memory as possible to maximize bandwidth [10]. Caching the frequently-used data to shared memory and registers can significantly lower memory access cost. However the amount of them is limited, thus they should not be overused. Details about memory in GPU programming can also be found in textbooks and tutorials such as [28,29,30].

4.2.3. The Occupancy Metric

Memory access is often the bottleneck of performance, so it is important to launch as many active threads onto the hardware as possible to keep the SMs busy and hide latencies through context-switching. Occupancy is a metric to measure how busy the hardware is, and is defined as the ratio of the number of actual active threads to the maximum number of possible [31]. It is determined by the kernel function and the resource offered by the SM, which includes the amount of registers and shared memory, as well as the maximum thread number per SM. Low occupancy can result in performance degradation in most cases, so programmers are encouraged to prevent this from happening. In order to reach this goal, the block size, the used registers per thread, as well as the used shared memory per block should be reduced. However, Volkov has showed in [32,33] that under certain circumstances, faster codes may run at a lower occupancy. The key point is to use more registers and do more parallel work per thread.

4.3. Batched STAP Implementation

This subsection introduces the implementation of STAP, in which a batched method is used to make the best use of the GPU parallelism.

The data going to the anti-jamming processing module are in a stream form. To reduce the data transfer count as well as to take advantage of the GPU parallelism, the data stream is divided into chunks in host memory, then the whole data in one chunk are transferred onto global memory at a time. In this project, the time length of one chunk is 50 ms. This is the result of a balance between data transferring time and memory cost. Thus, the sampling period number within one chunk is

$$L_{\text{chunk}} = 20\ \text{MSPS}\ \times 50\ \text{ms}\ = 10^6\ \text{samples} \quad (17)$$

Figure 6 depicts the batched implementation of STAP. After one chunk of input data has been transferred onto global memory, the chunk is further divided into Z batches, while each batch contains K non-overlap input vectors of MN×1. Thus, Z can be calculated by

$$Z = \text{ceil}\left(\frac{L_{\text{chunk}}}{NK}\right) \quad (18)$$

where "ceil" denotes the function of rounding up. In the input data stream shown in Figure 6, rows corresponds to different antenna elements, while columns corresponds to different sampling period. For each batch, one covariance matrix of MN×MN is estimated, and one weight vector of MN×1 is calculated using the SMI method. Then KN outputs are calculated from the input data using the same weight vector. Thus, each batch corresponds to a weight updating period, which is calculated by

$$T_{update} = KNT_s \quad (19)$$

where Ts is the sampling period and N is the filter tap length.

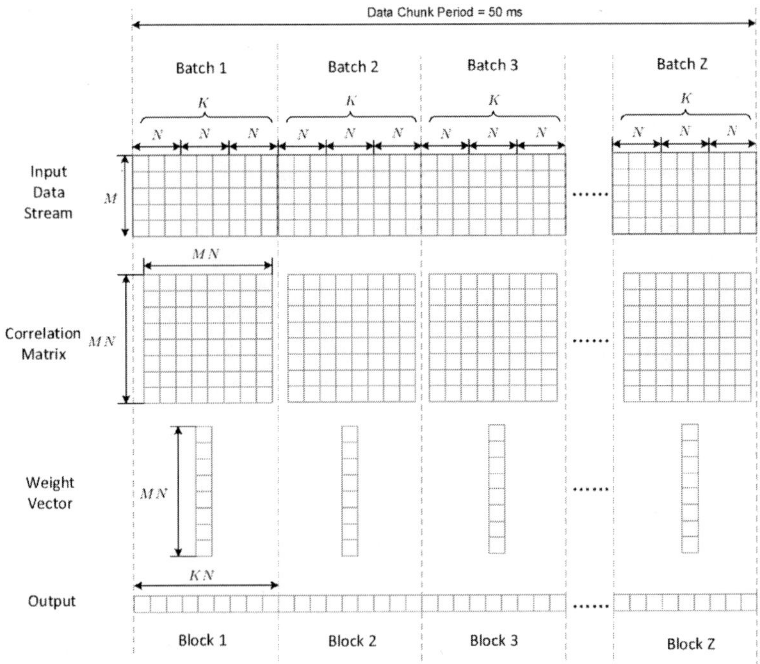

Figure 6. The batched implementation of STAP.

It can be seen that each batch can be regarded as a small "problem" independent with others. One way to calculate these problems is to process them one by one. However, this is usually inefficient since the scale of each problem is not big enough to obtain a high occupancy. Furthermore, the kernel function invoking expense is costly when the batch number per chunk is large. In order to address this issue, a batched method is proposed in this paper, in which all the batches within one chunk are processed concurrently by a single kernel launch.

In the rest of this subsection, the implementation of the first two calculation steps in STAP are introduced in detail, namely the covariance matrix estimation and the weight vector calculation, respectively. The output calculation is implemented by a kernel function which is called *gpu_output_batched*. This calculation step contains mainly matrix-vector multiplications, thus the implementation details are ignored here.

4.3.1. Batched Covariance Matrix Estimation

The covariance matrix estimation is one of the computational hotspots and should be fully optimized to meet the real-time requirement. In practice, the covariance matrix is estimated by a time average method given by Equation (6). As an alternative, this equation can be rewritten in matrix form as follows.

$$\hat{\mathbf{R}} = \frac{1}{K}\mathbf{X}\mathbf{X}^H \tag{20}$$

where X is the input data matrix written by

$$\mathbf{X} = \left[\mathbf{x}^{\mathrm{T}}(n), \mathbf{x}^{\mathrm{T}}(n-N), \ldots, \mathbf{x}^{\mathrm{T}}(n-KN+N)\right] \tag{21}$$

A specialized kernel function called *gpu_Rxx_batched* is developed to calculate the Z correlation matrices within one chunk concurrently. In the kernel, each covariance matrix of one batch is mapped to a single block, while each element to be calculate in the covariance matrix is mapped to a single thread within the block. Only elements on and below the diagonal are calculated for the conjugate symmetry of the covariance matrix.

Each element in the input data matrix is used many times, so is cached onto shared memory to eliminate reluctant accesses to global memory. This also helps the global memory accesses to coalesce. However, special attention should be paid not to overuse shared memory, because the amount of it is limited. For GTX 770, the amount of shared memory is 48 kB per SM at maximum. This results that when MN and K are relatively large, the size of the input data matrix can be quite big, which makes caching the whole of it onto shared memory within one single transfer inappropriate or impossible. To address this issue, the input data matrix is further divided into multiple successive slices, as shown in Figure 7 on the left side. Each slice contains Kp input vectors, and is cached and processed one after another. The parameter Kp needs to be carefully tuned to obtain the best performance.

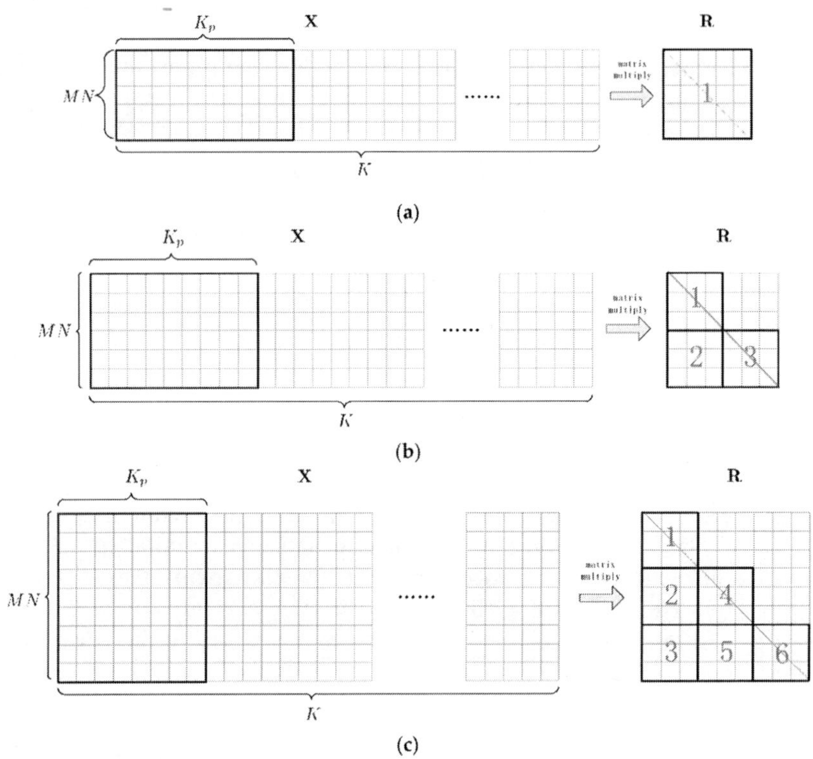

Figure 7. Schematic of the covariance matrix estimation. (**a**) Covariance matrix estimation of size smaller than 30 × 30; (**b**) Covariance matrix estimation of size from 30 × 30 to 59 × 59; (**c**) Covariance matrix estimation of size from 60 × 60 to 80 × 80.

Each element of the covariance matrix is stored in the on-chip register file while being calculated by the corresponding thread. In this project the size of the covariance matrix varies from 4 × 4 to 80 × 80, so*gpu_Rxx_batched* should adapt all the various sizes. On the other hand, the thread number per block is limited. For GTX 770, the maximum thread number per block is 1024. For a covariance matrix of 80 × 80, if one thread calculates only one element on or below the matrix diagonal, then there needs to be 3240 threads per block, which exceeds 1024. To address this issue, the covariance matrix is portioned into tiles, as shown in Figure 7 on the right side. For a covariance matrix whose size is smaller than 30 × 30, one tile occupies the entire matrix. For a covariance matrix whose size varies from 30 × 30 to 59 × 59, three tiles cover all the elements on and below the diagonal. For a covariance matrix whose size varies from 60 × 60 to 80 × 80, the number of tiles increases to 6. For each covariance matrix, the tiles are processed serially by the block, which means that if the covariance matrix is divided into v tiles of t×t, then each block should contains t×t threads and each thread should calculate vmatrix elements. Table 4 shows the block size corresponding to different covariance matrix sizes. Thus, the number of threads per block is decreased to the size of a tile, which is less than 901, as shown in Table 4. At last, the resulted covariance matrix in each batch is copied to global memory.

Table 4. Thread number of *gpu_Rxx_batched* per block for covariance matrices of different sizes.

Matrix Size	Tile Count	Thread Number
4 × 4 to 29 × 29	1	16 to 841
30 × 30 to 59 × 59	3	225 to 900
60 × 60 to 80 × 80	6	400 to 729

4.3.2. Batched Cholesky Solving

The calculation of the weight vector can be viewed as a problem of solving linear equations, such avoiding the computational cost of matrix inversion. This solving process can be divided into three steps. Firstly, the covariance matrix can be factorized by the Cholesky method into a multiplication form which is given by

$$\mathbf{R} = \mathbf{L}\mathbf{L}^{\mathrm{H}} \tag{22}$$

where L is a lower triangular matrix. Then the vector b=R−1s is obtained by two iteration processes, which are Ld=s and LHb=d in turn. Finally, b is normalized to obtain the desired weight vector, which is

$$\mathbf{w} = \frac{\mathbf{b}}{\mathbf{s}^{\mathrm{H}}\mathbf{b}} \tag{23}$$

Parallel implementation of Cholesky solver has been investigated in previous works. In [9] Molero proposed a batched Cholesky solver for local RX anomaly detection in hyper spectral images. The method calculates hundreds of different and independent small problems simultaneously, such reaching a high degree of parallelism on a GPU. In this work we adopt a similar implementation method by developing a kernel called*gpu_cholesky_batched*. In this kernel, each block processes one problem and contains MN threads, while each thread corresponds to one row of the covariance matrix, as shown in Figure 8. The covariance matrix and the constraint vector are cached onto shared memory at the beginning of the kernel launch. The algorithm of the Cholesky solver for each block is as shown in Algorithm 1 in Appendix A.

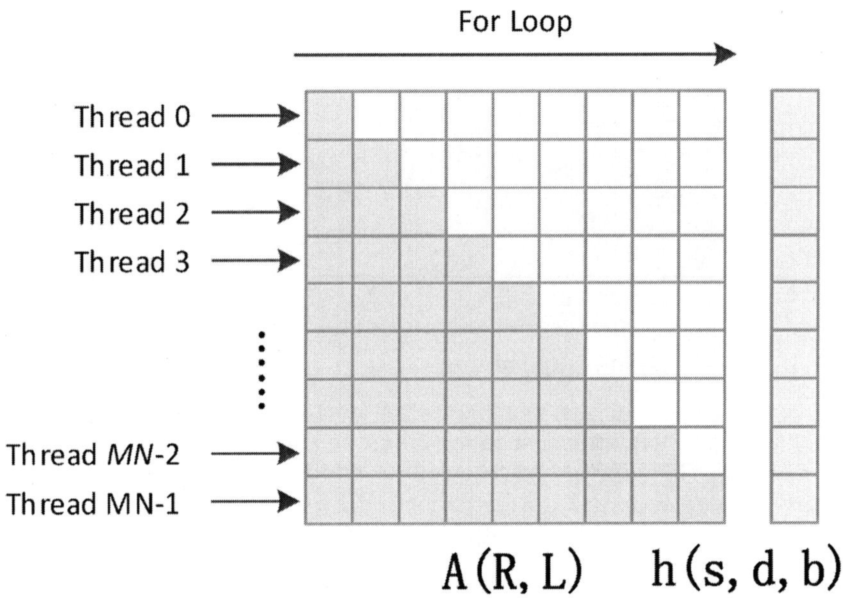

Figure 8. Architecture of the Cholesky solver.

From Figure 8 and Algorithm 1, it can be seen that the covariance matrix R and the lower triangle matrix Lshare the same memory in the Cholesky factorization. The same is for vector s, d, and b in the forward and backward iterations. Thus, the used shared memory is reduced. It can also be seen that there are two "for" loops in each block, and every iteration in the loop depends on results of the former iteration. This means that the operations in each block are not fully parallelized. This is also why the costly function *syncthreads* which acts as a barrier where all threads within one block must wait before any is allowed to proceed is repeatedly used. However, this lack of parallelism is balanced by the fact that all the batched are processed in parallel.

4.4. Batched SFAP Implementation

Similar to STAP, one chunk of data is also divided into batches. However, one difference in the batched implementation of SFAP with STAP presented in this paper is that, since the key point of SFAP is to do SAP in each frequency bin, DFT and IDFT should be carried out at the beginning and end of the whole process, respectively. Thus, after DFT, each batch of data is further divided into N small batches corresponding to Nfrequency bins. Here, N denotes the DFT length. The DFT and IDFT on a GPU can be done by the off-the-shelf library CUFFT® which is well tuned, so we use functions in it directly instead of reinventing the wheel.

The batched implementation of SFAP is depicted in Figure 9. For brevity, only calculation steps in frequency domain are showed. For each frequency bin of each batch, one covariance matrix estimate is calculated, then one weight vector is generated. After that, outputs in frequency domain are calculated, ready to be transformed back to time domain by IDFT. The data of each batch in time domain may overlap with adjacent batches at the two ends, so the weight updating period is calculated by

$$T_{update} = (1 - overlap)KNT_s \tag{24}$$

where overlap is the overlapping ratio.

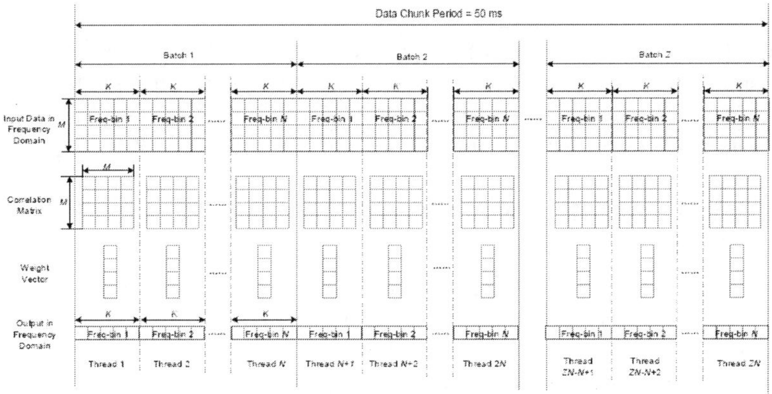

Figure 9. Data layout in the spatial-frequency adaptive processor (SFAP) after the FFT.

The key point in designing the kernels *gpu_Rxx_batched* and *gpu_cholesky_batched* in STAP is that they should adapt to the different sizes of the covariance matrices, which vary from 4 × 4 to 80 × 80. Different from this, the covariance matrix size in frequency domain is relatively small, which varies from 4 × 4 to 8 × 8. This makes it possible to further improve the performance by certain measures.

4.4.1. Kernel Function Design Strategy

It is generally recommend that in order to hide latencies, one should run more threads per multiprocessor [10], which means that a high occupancy is preferred. However, Volkov has stated in [32,33,34] that it is possible to gain high speed up at low occupancy. The key is to calculate more outputs per thread and hide latency by instruction-level optimization (ILP). Our batched implementation of SFAP is based on this strategy.

The kernel we developed is called *gpu_Rxx_cholesky_weight_output_sfap*, which indicates that all the calculation steps in frequency domain are completed in one single kernel function. In this kernel, each small batch (*i.e.*, one frequency bin of one batch) is assigned to a single thread. The covariance matrix and the constraint vector s are allocated as registers instead of shared memory. These registers are reused as much as possible to avoid more register allocation, thus more threads can be loaded to SM at the same time. Registers are further saved by only loading and calculating the matrix elements on and below the diagonal. The process of the kernel is as follows: firstly, the input data of one small batch are loaded to the register file by one thread, then multiplied and accumulated to estimate the covariance matrix, just as depicted in Equation (15). Secondly, Cholesky factorization is conducted and the weight vector is generated after the two-phase iteration. All the operations are strictly serial within one thread, so the costly function *syncthreads* does not need to be called repeatedly. Finally, the input data are read from global memory again to be weighted by the weight vector and summed on registers, and then written back to global memory.

Performance can benefit from at least two aspects. Firstly, all the three calculation steps are integrated into one single kernel, reducing both accessing to global memory repeatedly and the multiple kernel launch overhead. Secondly, in memory hierarchy of GPU, registers are nearest to arithmetic logical units (ALUs) so have the minimum latency. Furthermore, it does not have the problem of bank conflicts and thread synchronization which shared memory suffers from [30]. Although the input data reading operation from global memory is time costly, but once it is completed, following operations can be accomplished on-chip with nearly no memory latency.

To improve latency within one thread, IPL is used, which means overlapping independent operations so that one can run while the others are waiting for memory accesses to be finished [33]. Besides, loops are unrolled to gain better performance [34]. Since CUDA does not support dynamic allocation of registers, five versions of the kernel are developed, respectively, for 4 to 8 antenna elements.

4.4.2. Performance Preview

In order to get a deeper insight into the problem, we use the GPU occupancy calculator provided by NVIDIA to give a performance preview [35]. The kernel tested is the 8-elements version which has the highest computational quantity. In the calculator, the Compute Capability (CC) is set to 3.0, which is the same with GTX 770, and block size is set to 128. Results are presented in Figure 10. It can be seen that 63 registers are allocated by the complier per thread, which is the maximum number of available registers in a GPU of CC 3.0. This results in that 32 warps are active per SM. In a GPU every 32 threads are bound to be a warp, so the total number of active threads per SM is 1024. The maximum warp number is 64 per SM, so the occupancy we get is 50%, which is low. This is due to that the allocated registers per thread are too many, which can be seen in the right graph presented in Figure 10. However, we will show in the next section that a considerably good performance can still be obtained.

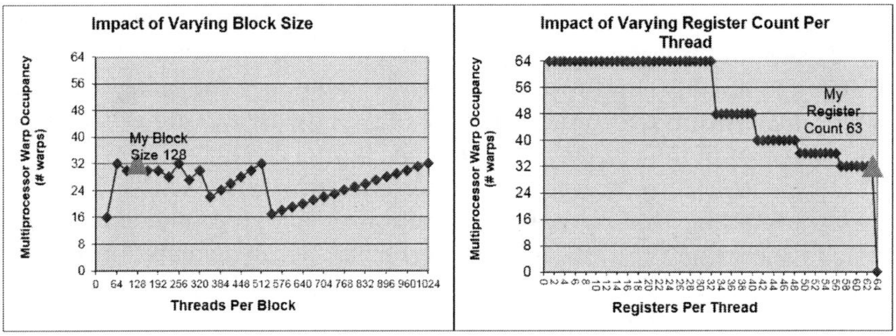

Figure 10. Results of occupancy calculations. (**Left**) Impact of varying block size on occupancy; (**Right**) Impact of varying register count per thread on occupancy.

4.5. Precision Consideration

Precision is an important factor influencing the anti-jamming performance in both signal digitalizing and calculation. In this platform, the ADC has a resolution of 14 bits, resulting in an effective dynamic range of more than 65 dB which is needed to avoid saturation in the digital domain while sampling the received signal containing interference [36]. A 14-bit resolution is also considered to be enough to limit the side effect on interference cancellation brought by quantization errors [37].

After being digitalized, the sampled raw data are transferred to processors for calculation. In a hardware implementation using a FPGA, data are usually processed in a fixed point form, and the precision usually depends on the truncated bits after multiplication or other operations. However, in a general-purpose processor like CPU and GPU, data are usually manipulated in floating point type to get higher precision, because fixed point operations will not bring extra benefit in general. In particular for GPU, 32-bit single-precision operations are much faster than 64-bit double precision operations. For example, on a GPU of Compute Capability 3.0, the number of single-precision operations per clock cycle per SM is 192, while that of double precision operations is only 8 [30]. Thus, the fixed point raw data are transformed to single-precision type before being calculated by kernel functions.

It is known that the relative precision of 32-bit single-precision operations is about 1×10^{-7} [38]. This implies an upper limit of the eliminable jamming power, which corresponds to the situation when the ill-condition degree of the covariance matrix is so large that the minimum eigenvalue cannot be represented with enough precision. In practice, this effect can be compensated by diagonal loading.

In situations when accuracy is of lower priority than the performance, the fast math library can be used instead of the standard math library in CUDA for further optimization. This is done by switching on the –*use_fast_math*compiler option. Test results show that using this method, the time cost of the *gpu_cholesky_batched* kernel can be reduced by about 15% to 25% according to different problem scales. In our implementation, this optimization option is adopted, and experiment results show that the anti-jamming performance can be maintained.

4.6. Whole Picture of the Testbed

Until now, both the hardware and software of this testbed have been introduced in detail, so now it is time to give a whole picture of the platform, which is shown in Figure 11 below.

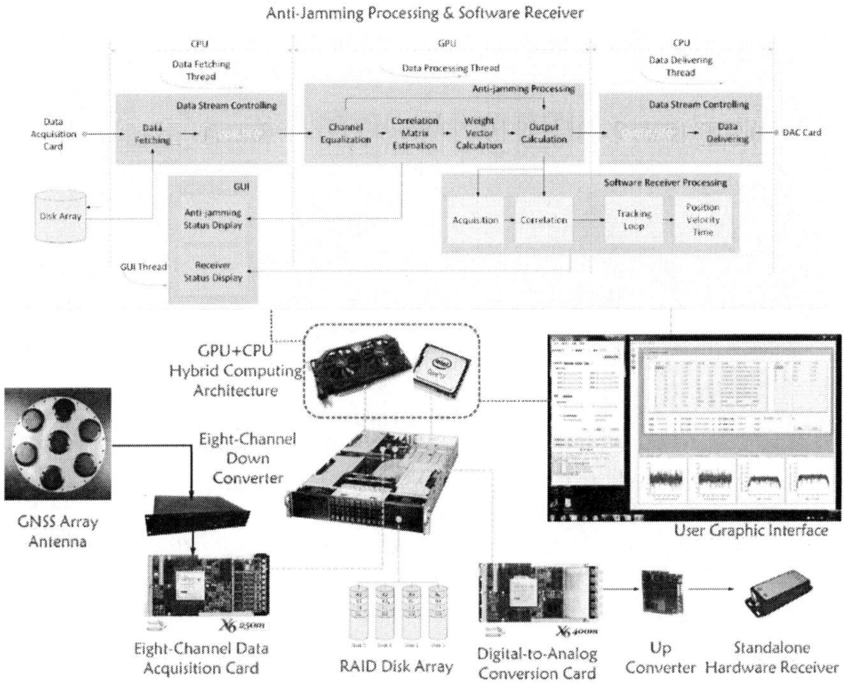

Figure 11. The complete schematic picture of the real-time testbed for adaptive GNSS array anti-jamming.

5. COMPUTATIONAL PERFORMANCE

The focus of the software-defined approach in the last section is to create programs efficient enough to meet the real-time requirement. In this section, the computational performance of both the STAP and SFAP implementation is measured by inserting CPU clock timers to the codes and measuring time costs of different calculation steps. It will be seen that, with our programming techniques, the real-time requirement can be satisfied very well for a wide range of the configurable parameters.

All the results in this section are obtained using GTX 770, which is a not so advanced GPU compared to the newly released GTX 980 or GTX TITAN X. It can be expected that better computational performance can be gained with more advanced GPU without modifying the program, so results in this section are conservative.

5.1. Computational Performance of STAP

Figure 12 depicts the computational performance measurement results of our batched STAP implementation. In the test, time costs of the three calculation steps, namely the covariance matrix estimation, the Cholesky solving, and the output calculation are measured for one chunk of input data, which is 50 ms long. The number of antenna elements (*i.e.*, M) is set to 8, which is the maximum in this project and represents the situation with the highest computational quantity. The filter tap length (*i.e.*, N) is set to 1, 3, 7, 10 respectively, while the number of snapshots taken to estimate the covariance matrix (*i.e.*, K) ranges from 100 to 1000 with a step size of 100. According to the last section, the number of batches (*i.e.*, the number of blocks, or grid size) is determined by Nand K together. Block size of all the three kernels, which are *gpu_Rxx_batched*, *gpu_cholesky_batched*, and*gpu_ output_ batched*, is set to 256 as a result of carefully tuning.

From the figure it can be seen that, given particular values of M and N, the time costs of the covariance matrix estimation and the output calculation stay roughly consistent with K varying. This result makes sense because the arithmetic operations in the corresponding two kernels are relatively simple, thus global memory access may become the bottleneck of the performance. However, the time costs of Cholesky solving decreased with Kincreasing. This is because smaller K corresponds to more batches to be processed. The comparison between the four graphs also makes clear that the time costs of the Cholesky solving increase with N much faster than the other two calculation steps, which is partly because for *gpu_cholesky_batched*, arithmetic operations play the central role of performance, rather than memory accesses.

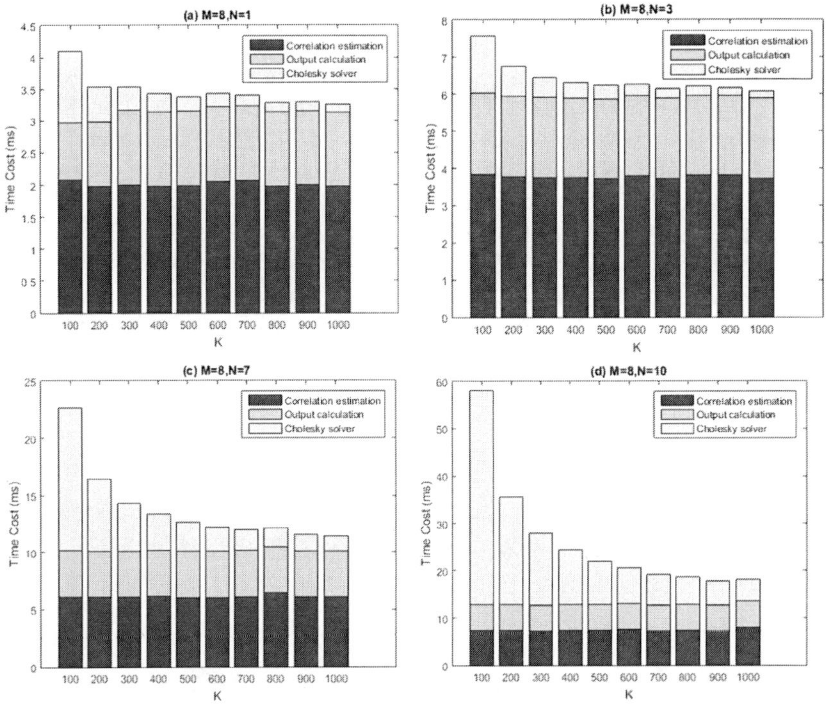

Figure 12. Performance of the batched STAP implementation. M denotes the number of antenna elements and N denotes the filter tap length. The real-time threshold is 50 ms. (**a**) Results of time cost when M is 8 and N is 1; (**b**) Results of time cost when M is 8 and N is 3; (**c**) Results of time cost when M is 8 and N is 7; (**d**) Results of time cost when M is 8 and N is 10.

The real-time requirement demands that, given a data chunk of 50 ms, we must process it within 50 ms. This is achieved for most cases in Figure 12, except for the situation when M=8, N=10 and K=100. However, this set of parameters is not practical, since the relation K≥4MN is generally required to be satisfied for effectively interference mitigation. Thus, we can say that the real-time requirement is satisfied in a practical sense.

5.2. Computational Performance of SFAP

Before exploring the performance of the whole process, we compare the two implementation methods introduced in Section 4.3 and Section

4.4 respectively in the context of SFAP. In the recommended method introduced in Section 4.4, all the three calculation steps in the frequency domain are accomplished by the kernel*gpu_Rxx_ cholesky_weight_ output_sfap*, while in the not recommended method which is introduced in Section4.3, they are accomplished by *gpu_Rxx_ batched*, *gpu_ cholesky_batched*, and *gpu_output_batched* respectively. In the test, one chunk of input data is used, which is 50 ms long. The number of antenna elements (*i.e.*, M) is set to 8, the DFT length (*i.e.*, N) is set to 256, and the parameter overlap is set to 0. K ranges from 10 to 100 with a step size of 10. Only time costs of the calculation in the frequency domain are measured, meaning that DFT and IDFT are not included. Figure 13 depicts the result, from which it can be seen that the recommended method outperforms the not recommended method a lot, and the smaller K is, the more obvious the improvement is.

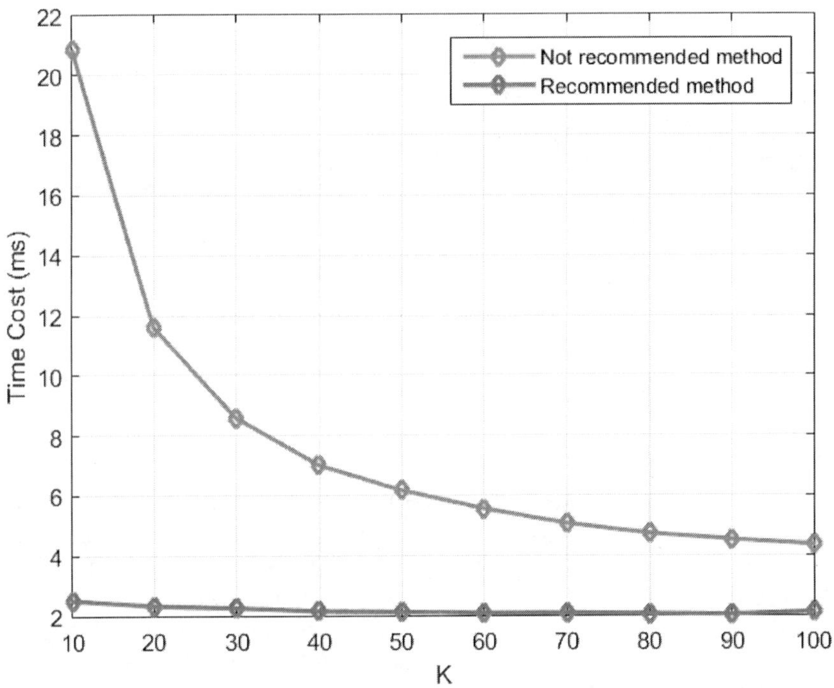

Figure 13. Comparison between the two implementation methods introduced in Section4.3 and Section4.4 in the context of SFAP. The parameters M and N were 4 and 256, respectively.

Figure 14 depicts the computational performance measurement results of our batched SFAP implementation using the recommended method. Here, M is set to 8, while N is set to 128, 256, 512 and 1024 respectively. The parameter K ranges from 20 to 200 with a step size of 20. The parameter overlap is set to 0.4, which is a conservative value. In the figure, the "data preparation" denotes the data arrangement before the DFT. It can be seen that the two most computationally costly parts of the whole process is FFT and the calculation in frequency domain (*i.e., gpu_Rxx_ cholesky_weight_ output_sfap*). The total time cost per data chunk is less than 10 ms, leaving more than 40 ms for other operations, meaning that the real-time requirement is satisfied very well.

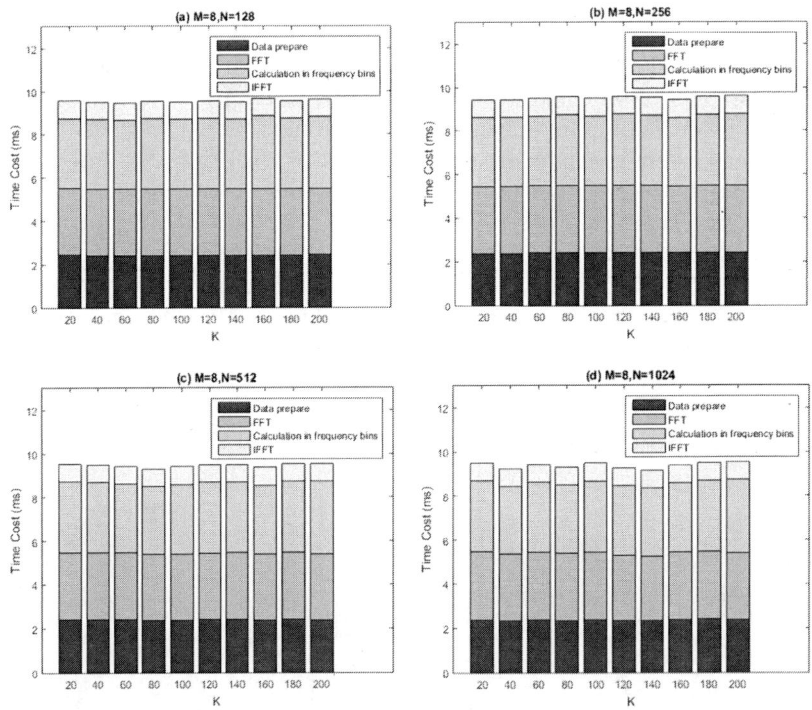

Figure 14. Performance of the batched SFAP implementation. M denotes the number of antenna elements and N denotes the FFT length. The real-time threshold is 50 ms. (**a**) Results of time cost when M is 8 and N is 128; (**b**) Results of time cost when M is 8 and N is 256; (**c**) Results of time cost when M is 8 and N is 512; (**d**) Results of time cost when M is 8 and N is 1024.

5.3. Real-Time Processing Time Distribution

The total time cost of the software includes not only the anti-jamming processing on the GPU mentioned above, but also the data transfer between device and host, as well as the software receiver processing. In this subsection we measure the time cost distribution among the three steps.

Firstly the real-time processing time distribution for STAP is measured. In the test, the number of antenna elements (*i.e.*, M) is set to 8 and the filter tap length (*i.e.*, N) is set to 10, thus the size of the covariance matrix is 80 × 80, which is the maximum in our implementation. The number of snapshots taken to estimate the covariance matrix (*i.e.*, K) is set to 500, resulting in the weight updating period to be 0.25 ms according to Equation (19). According to Section 5.1, this set of parameters represents computational complexity near to the maximum required by this project. The measurement results are shown in Figure 15. The *X*-axis in the graph corresponds to continuous time, segmented into processing periods of 50 ms length. The *Y*-axis corresponds to the time cost of each processing period. Shown as stacked bars are the data transfer (black), anti-jamming processing (dark brown) and software receiver processing (light brown). The software receiver processes data every 100 ms, so the bar representing the time cost of the receiver processing appears every two periods. It can be seen that for this set of parameters, more than 15 ms of idle time on average are left for operations other than the three kinds in the figure, so the real-time requirement is satisfied very well.

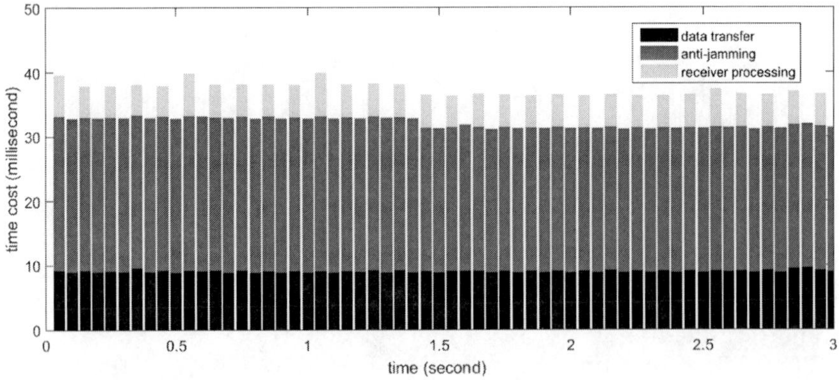

Figure 15. Time cost distribution for real-time processing with the space-time adaptive processor (STAP).

The real-time processing time distribution for SFAP is measured in a similar manner with STAP. The number of antenna elements (*i.e.*, M) is set to 8 and the DFT length (*i.e.*, N) is set to 256. The number of snapshots taken to estimate the covariance matrix (*i.e.*, K) is set to 200 and the parameter overlap is set to 0.4, so the weight updating period is 0.768 ms according to Equation (24). According to Section 5.2, this set of parameters is representative, for the computational complexity of SFAP is almost constant with varying M, N and K. The results are presented in Figure 16. It can be seen that about 30 ms of idle time on average are left for other operations per 50-millisecond-long period, meaning that the real-time requirement is satisfied very well. Besides, the data transfer time is slightly less than STAP. This is due to that the data type of the input data in SFAP is INT16 whose size is 2 Bytes, while the data type of the input data in STAP is single precision floating point, whose size is 4 Bytes, resulting in that the total amount of the input data to be transferred to the global memory in SFAP is less than that in STAP.

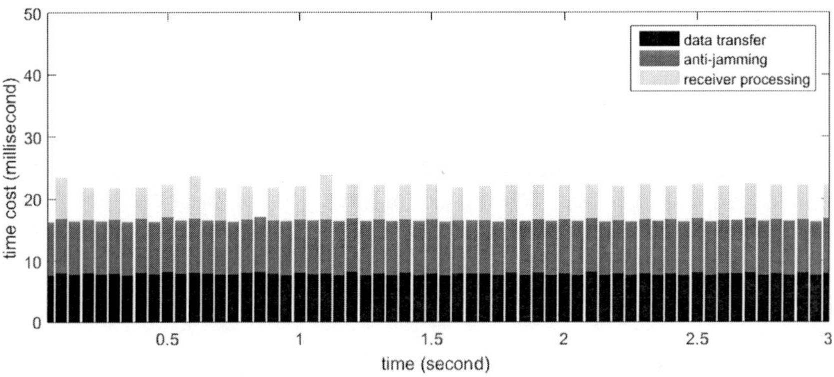

Figure 16. Time cost distribution for real-time processing with the space-frequency adaptive processor (SFAP).

6. ANTI-JAMMING EXPERIMENTS

A major advantage of the platform is that the intermediate states of both the anti-jamming processing and the software receiver processing can be observed in the field or analyzed afterwards. These intermediate states include eigenvalues of the covariance matrix, updating weight vectors, C/N0 of each tracking channel, *etc.* More insights to the anti-jamming performance can be obtained based on these states. In this section, examples of adaptive nulling both in STAP and SFAP are given, showing how anti-jamming algorithms with different parameters react to the changing of the interference environment. Besides, the performance of adaptive beamforming is measured. Experiments for adaptive nulling are conducted using the BDS B3 signal, whose bandwidth is 20 MHz, while the experiment for beamforming is conducted using the GPS L1 C/A signal.

6.1. Jamming Scenarios

In this section, the adaptive array anti-jamming performance is measured in two different jamming scenarios, as shown in Figure 17. In the first scenario, there is one wideband jammer in the north direction of the antenna array, covering all the bandwidth of the BDS B3 signal which is 20 MHz. The jammer stays silent at the beginning to let the receiver capture signals from visible satellites and is turned on at about second 66 with an interference-signal-ratio (ISR) being about 60 dB. Then every more than 10 s, the jammer is strengthened gradually by 6 dB, increasing the ISR range from 60 dB to more than 90 dB. This scenario lasts for about 180 s. In the second scenario, four wideband jammers are located at the four ends of a square. All jammers stay silent at first, and then are turned on one by one, with an ISR of each jammer being about 45 dB. Information of the 4 jammers is presented in Table 5. This scenario lasts for about 100 s. Figure 18 is a photo of the experiment setup.

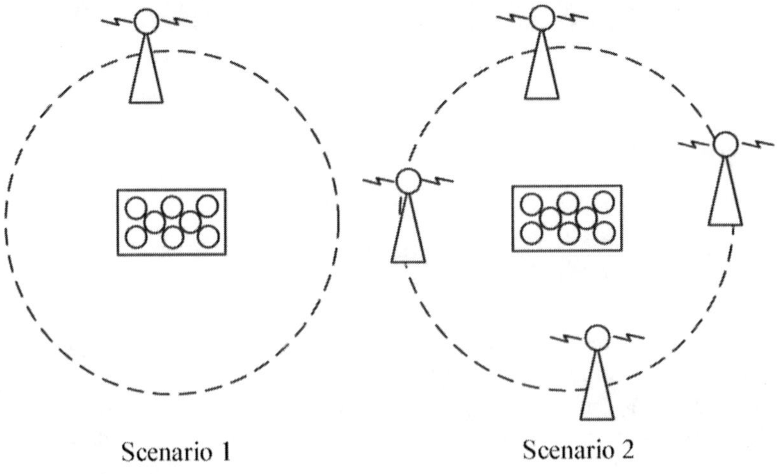

Figure 17. The two jamming scenarios used in the following experiments.

Figure 18. A photo of the experiment setup.

Table 5. Information of the 4 jammers in Scenario 2.

Jammer	Bandwidth	Start Time
Jammer 1	10 MHz	Second 32
Jammer 2	20 MHz	Second 51
Jammer 3	20 MHz	Second 69
Jammer 4	20 MHz	Second 84

In order to compare the anti-jamming performance of different algorithms and parameters against the same interference environment, instead of running the testbed in a real-time mode, the multi-element signals received in the two jamming scenarios are digitalized and logged to the disk array respectively, after which the logged file on the disk array are processed in a post-processing mode. The results obtained in this way are the same with the real-time mode.

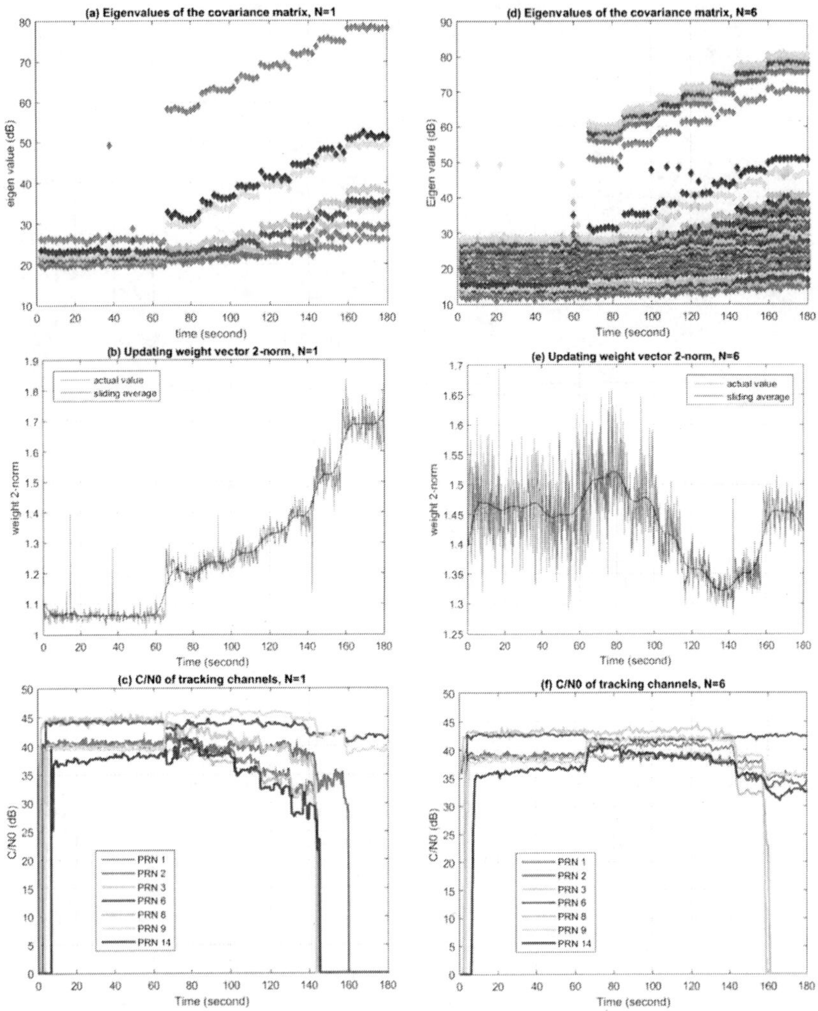

Figure 19. Performance of the STAP nulling during Jamming Scenario 1. (**a**) Eigenvalues of the corvariance matrix when N is 1; (**b**) 2-norm of the updating weight vector when N is 1; (**c**) C/N0 of the tracking channels whenN is 1; (**d**) Eigenvalues of the corvariance matrix when N is 6; (**e**) 2-norm of the updating weight vector when N is 6; (**f**) C/N0 of the tracking channels when N is 6.

6.2. STAP Nulling Experiments

Firstly, the STAP nulling performance in Scenario 1 was measured. In the experiment, the antenna element number (*i.e.*, M) was set to 8, while the number of snapshots taken to estimate the covariance matrix (*i.e.*, K) was set to 200. In order to desensitize the side effect brought by inadequate estimation of the covariance matrix, diagonal loading was conducted according to Equations (10) and (11), where αdl was set to 1×10^{-6}. Figure 19shows the changes of covariance matrix eigenvalues, the Eucidian 2-norm of the updating weight vector, and C/N0 with time, respectively. The left three graphs correspond to the situation when N=1. In this case, the STAP algorithm degrades to SAP, while the weight updating period is 10 microseconds. The right three graphs correspond to the situation when N=6, thus the weight updating period is 60 microseconds.

From Figure 19a,d, it can be seen that the eigenvalues increase with the jamming signal power. When N=1, the biggest eigenvalue is a lot larger than the second biggest eigenvalue, while the biggest five eigenvalues are close to each other when N=6. This indicates that while STAP provides more degrees of freedom than SAP, more degrees of freedom are also consumed in STAP by the wideband interference. In the ideal case where one wideband jammer exits in the environment, the largest eigenvalue in SAP represents the jamming signal, while other eigenvalues represent the noise and are more or less the same [39]. However, it can be seen that several large eigenvalues also appears in the actual experiment. This may be due to the antenna element mutual coupling effect and RF channel mismatches. It is also worth noting that the smallest eigenvalue increases with the jamming signal power because of the diagonal loading operation.

Figure 19b,e depict the changes of the Euclidian 2-norm of the updating weight vector. Although this metric does not have an obvious physical meaning, it provides a simple way to indicate the value changes of the weight vector along with the environment. In the case of N=1, the norm increases with the jamming signal power, while in the case of N=6, the norm changes with the environment, but the relation is not intuitive. This result also confirms that by observing the value or other characteristics of the weight vector, changes in the interference environment can be observed.

Figure 19c,f depict the C/N0 changes of several tracking channels during processing. Since the jammer is very strong, if no anti-jamming algorithms are used, all the tracking channels will lose tracking (*i.e.*,

C/N0 = 0) immediately after the jammer is turned on. But with the help of the anti-jamming algorithms, channels can still be tracked, as presented in the graphs. It can be seen that when the jammer is turned on, there is a sudden change in C/N0 for most tracking channels, while some of them increase and some of them decrease. This is due to the sudden distortion of the antenna array radio pattern, resulting changes of the gain in different directions, since in the PI method the array pattern toward the satellites are not constrained. In the case of N=1, after the jammer is turned on, C/N0 of each tracking channel decrease with the increasing jamming power, and finally lose tracking when the jamming power exceeds a certain threshold. In the case of N=6, the trend of the change is similar, but the jamming power threshold after which C/N0 begins to decrease is higher than the case of N=1, and in the end more tracking channels remain locked. This proves that more filter taps in STAP can lead to better performance in interference similar to Scenario 1.

Similar experiments are conducted for Scenario 2 using the same parameters as for Scenario 1. Results are given in Figure 20. From Figure 20a it can be seen that in the case of N=1, with the number of jammers increasing, the number of "big" eigenvalues increases and keeps the same with the number of jammers. (Here, the word "big" denotes the situation when the eigenvalues are far larger than the ones representing the noise. Specifically in this experiment, it means the eigenvalues are more than 40 dB.) As shown in Figure 20d, in the case of N=6, the number of "big" eigenvalues is 3 for the first jammer and 6 for each of the next three jammers, corresponding exactly to the filter tap length. The number of "big" eigenvalues for the first jammer is less than the other three because the bandwidth of the last three jammers occupies the whole bandwidth of the B3 signal, while the bandwidth of the first jammer is half of the other three. This verifies that the DOFs consumed have a relationship with the bandwidth of the interference. Figure 20b,e show that the changes of the updating weight vector's Eucidian-2 norm are in a relation to the number of jammers in the environment. The 2-norm of the weight vector increases with the number of jammers in SAP, while in STAP the relation is not that obvious. Figure 20c,f show that C/N0 of each tracking channel has a possibility of both increasing and decreasing when a new jammer is turned on. This is due to the distortion of the antenna pattern. Since each jammer is much weaker as compared to Scenario 1, the wideband nature of the jammers is less detrimental to SAP, making the SAP and STAP comparison more similar.

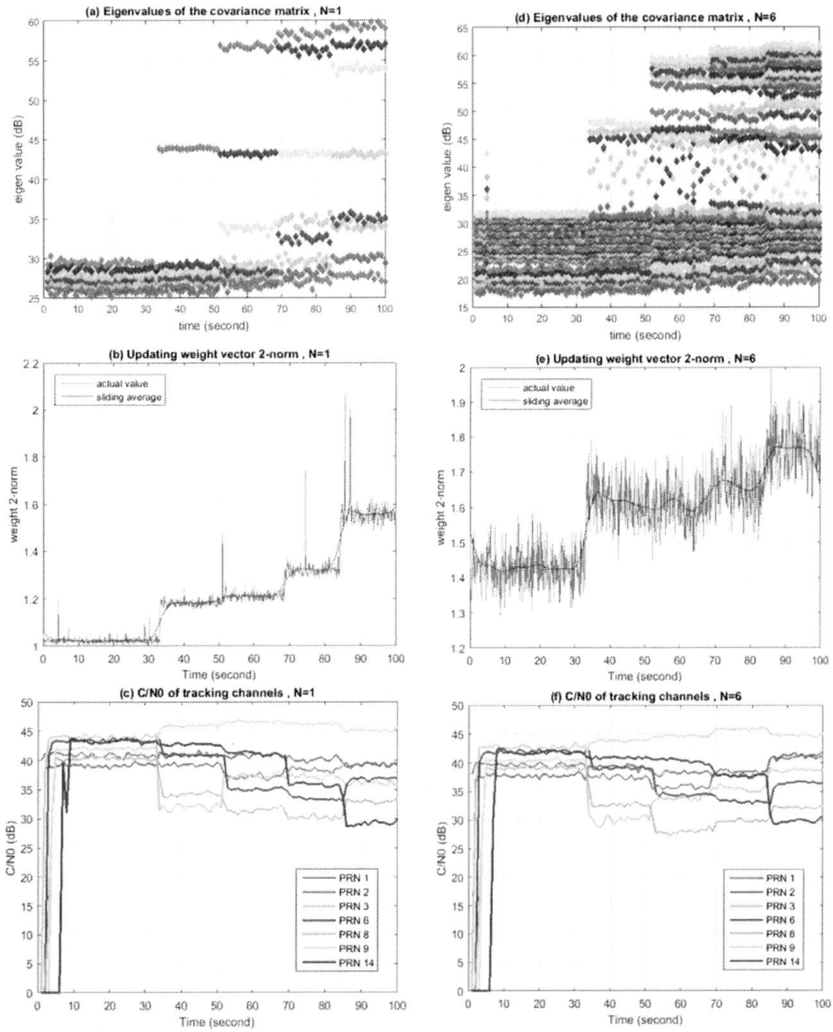

Figure 20. Performance of the STAP nulling during Jamming Scenario 2. (**a**) Eigenvalues of the corvariance matrix when N is 1; (**b**) 2-norm of the updating weight vector when N is 1; (**c**) C/N0 of the tracking channels whenN is 1; (**d**) Eigenvalues of the corvariance matrix when N is 6; (**e**) 2-norm of the updating weight vector when N is 6; (**f**) C/N0 of the tracking channels when N is 6.

Besides, given the weight vector and the geometry of the array, the antenna radio pattern can also be plotted roughly. Figure 21 gives two examples when SAP is used with the presence of one and four jammers, respectively, in Jamming Scenario 2. It can be seen that the pattern is very rough, because all the non-ideal factors of the antenna array and RF channels are ignored, including position inaccuracy of the array geometry, mutual coupling between antenna elements, finite ground-plane effect, mismatches between RF channels, different delays of cable lines, *etc.* [17,18]. Thus the plot is not so accurate and can only be used as auxiliary reference.

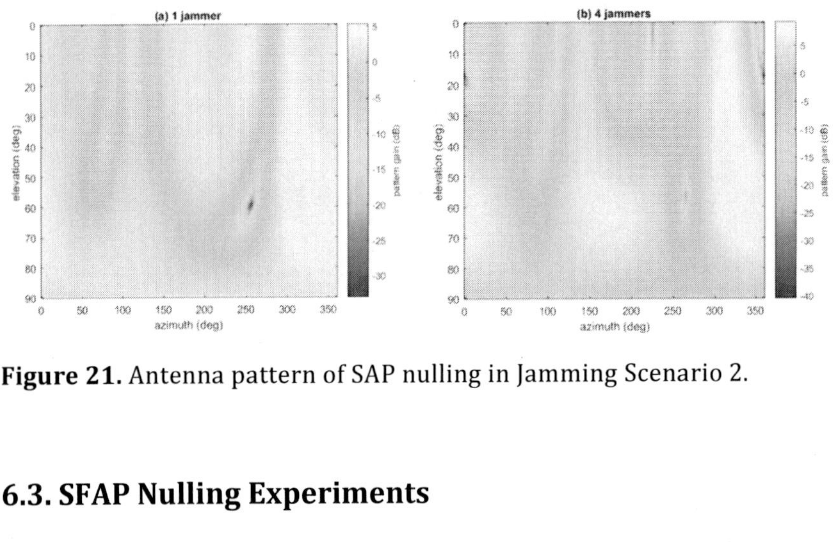

Figure 21. Antenna pattern of SAP nulling in Jamming Scenario 2.

6.3. SFAP Nulling Experiments

For SFAP, similar experiments as presented in Section 6.2 can be conducted, in which eigenvalues of the covariance matrix and the updating weight vector Euclidian 2-norm can be analyzed in a similar manner for each frequency bin. The results and interpretations are similar to those in STAP, so these are not presented here in detail. However, as an example, the C/N0 of each tracking channel with different DFT lengths in Scenario 1 is presented in Figure 22. The antenna element number was set to 8, the overlap parameter was set to 0.4, while the number of snapshots taken to estimate the covariance matrix was set to 100. Diagonal loading was performed, where αdl was set to 1×10^{-6}. Two tests were carried out, with the DFT length set to 128 and 512, respectively. It can be seen that the anti-jamming performance with N=512 is slightly better than that with N=128, since at about second 150 the tracking channel of PRN 3 is still locked in the right figure, while it has already lost tracking in the left figure. However, this performance improvement is not obvious.

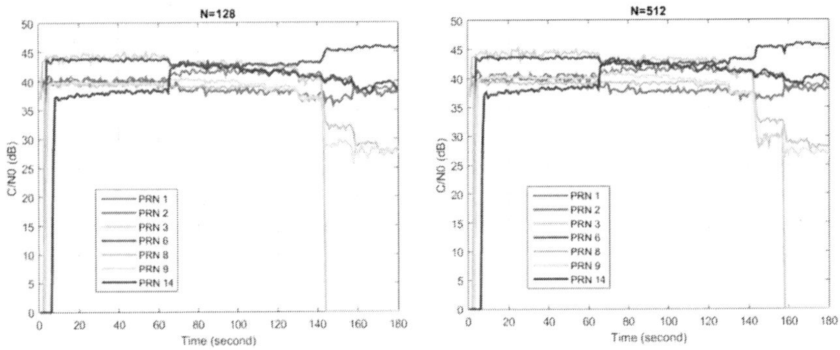

Figure 22. Performance of the SFAP nulling during Jamming Scenario 1. (**Left**) Results of C/N0 when N is 128; (**Right**) Results of C/N0 when N is 512.

Comparing the results presented in Figure 19 and Figure 22, it can be seen that more satellites are lost in the STAP than in SFAP approach. The performance gain is due to that the spectrum resolution in the filtering behind each antenna in SFAP (corresponding to the filter tap length) is better than that in STAP (corresponding to the FFT length). According to [40], when the filter tap length in STAP and the FFT length in SFAP are equal, the signal-to-interference-plus-noise ratios (SINRs) obtained in the two approaches are the same. Since in our implementation, the FFT length in SFAP is significantly larger than the filter tap length in STAP, the performance gain is reasonable.

6.4. Beamforming Experiments

In the last part of this section, results of a STAP beamforming experiment using the GPS L1 C/A signal are briefly introduced. Similar results can be obtained using the BDS B3 signal. In our implementation of beamforming, as many as 12 beams can be pointed to the visible satellites in the sky, but C/N0 of only 4 satellite signals are given in Figure 23 for brevity. C/N0 results without beamforming are also given as a comparison. It can be seen that considerable gains in C/N0 can be obtained by beamforming.

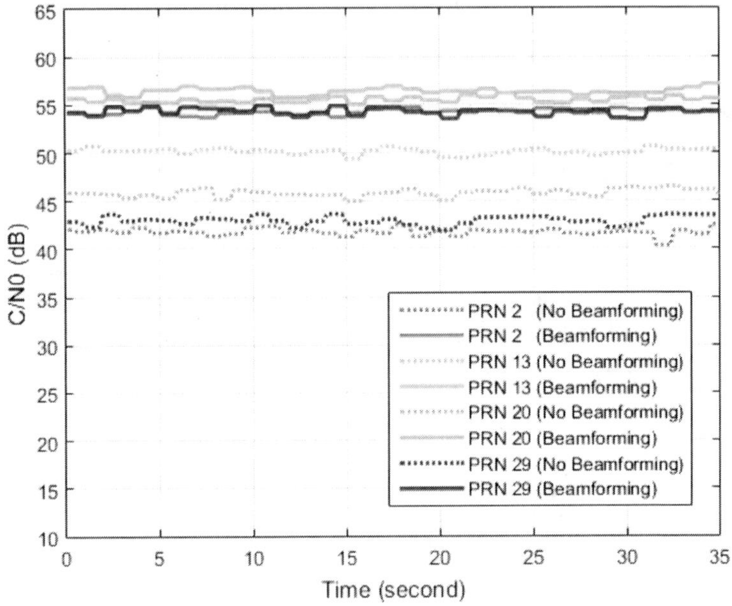

Figure 23. Results of C/N0 with beamforming (solid lines) and without beamforming (dotted lines).

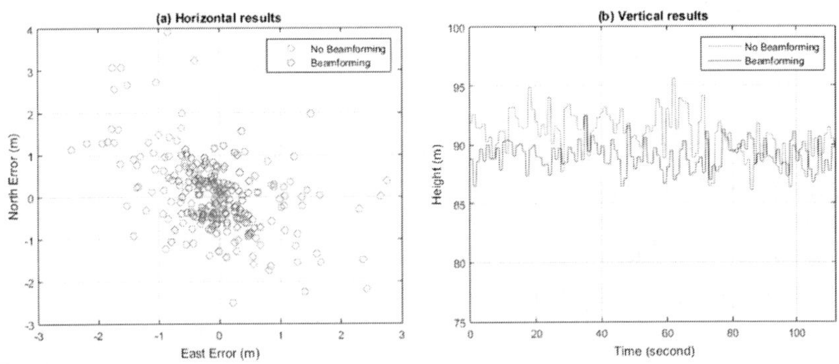

Figure 24. Positioning results with beamforming (red circles in graph-a) and red lines in graph-b) and without beamforming (blue circles in graph-a and blue lines in graph-b). The north error and south error are with respect to the average values, both with beamforming and without beamforming, respectively. The results are updated once per second. (**a**) Horizontal results; (**b**) Vertical results.

Table 6. The average values and standard deviations of the positioning results.

Statistic Character		Beamforming	No Beamforming
Average positioning results	Longitude (°)	116.330340	116.330347
	Latitude (°)	40.001468	40.001470
	Height (m)	89.03	90.52
The standard deviations of the positioning results with respect to the average values	East error (m)	0.32	1.05
	North error (m)	0.57	1.13
	Height (m)	1.17	1.86

Figure 24 depicts the positioning results both with and without beamforming. The position results were calculated once per second. Table 6 gives the average values and standard deviations of the positioning results. It can be seen that with beamforming, more accurate positioning results can be obtained, thanks to the gains in C/N0.

7. CONCLUSIONS

This paper documents the design and implementation of an SDR-based real-time testbed for GNSS array anti-jamming. This platform is able to perform both STAP and SFAP algorithms to multi-element raw data at a sampling frequency of 20 MSPS, in either an adaptive nulling or a beamforming mode. When calculating the updating weight vector, either an SMI method in which the inversion of the covariance matrix needs to be calculated, or an iterative method such as LMS can be used. A wide range of parameters can be configured to explore their relation to the anti-jamming performance in specific interference scenarios. As examples, anti-jamming performances and the intermediate states of varied algorithms and parameters in certain jamming scenarios are given by experiments in real environments.

In order to meet the real-time requirement, a not so advanced GPU, *i.e.*, GTX 770, is used as an accelerator. Batched methods are proposed in the calculation of covariance matrices, the weight vectors as well as the outputs to make the best use of the parallelism of the computational resources provided by the GPU. Special attention is paid to provide a high level of configuration possibilities in the implementation. In the software architecture, besides the anti-jamming processing module, software receiver processing is also provided. Tests showed that the computational performance satisfies the real-time requirement very well.

This testbed significantly outperforms previous similar systems in function, flexibility, configurability as well as computational performance. Besides, as far as the authors know, this is the first time a real-time implementation of SFAP and the SMI method has been realized using a software-defined approach in GNSS array anti-jamming field in open literature. It is hoped that this testbed can be used as a fertile ground for researchers to test and validate different adaptive array anti-jamming algorithms in various jamming scenarios. It can also be used for product prototyping, and even plays a role in certain applications like signal monitoring in challenging environments.

ACKNOWLEDGMENTS

This work is supported by the National Science Foundation of China (GrantU1333203) and the Civil Aviation Administration of China (Grant MHRD20140102). Special thanks are given to Jiannan Shen of Tsinghua University for his help in implementing the software receiver. The authors would also like to thank the anonymous reviewers for their valuable comments and suggestions.

AUTHOR CONTRIBUTIONS

Xiaowei Cui and Mingquan Lu designed the testbed and the experiments; Hailong Xu implemented the software and performed the experiments; Hailong Xu and Xiaowei Cui analyzed the data; Mingquan Lu contributed analysis tools; Hailong Xu wrote the paper.

APPENDIX A

In the implementation of STAP on the GPU, many weight vectors are calculated using the Choleksy factorization method concurrently by the blocks in one kernel. The following table presents the algorithm of the Choleksy solver for each block.

REFERENCES

1. Sklar, J.R. Interference mitigation approaches for the global positioning system. *Linc. Lab. J.* **2003**, *14*, 167–179.

2. Marcel, B.F.; Grégoire, W.; Cyril, B.; Pierre-André, F. Real-Time GNSS Software Receiver: Challenges, Status and Perspectives. In Proceedings of the European Navigation Conference—Global Navigation Satellite Systems, Naples, Italy, 3–6 May 2009.

3. Stöber, C.; Anghileri, M.; Ayaz, A.S.; Dötterböck, D.; Kramer, I.; Kropp, V. ipexSR: A real-time multi-frequency software GNSS receiver. In Proceedings of the 2010 ELMAR, Zadar, Yugoslavia, 15–17 September 2010; pp. 407–416.

4. Petovello, M.G.; O'Driscoll, C.; Lachapelle, G.; Borio, D.; Murtaza, H. Architecture and Benefits of an Advanced GNSS Software Receiver. *J. Glob. Position. Syst.* **2008**, *7*, 156–168.

5. Pany, T.; Falk, N.; Riedl, B.; Hartmann, T.; Stangl, G.; Stöber, C. Software GNSS Receiver. *GPS World.* **2012**,*23*, 60–66.

6. Huang, B.; Yao, Z.; Guo, F.; Deng, S.; Cui, X.; Lu, M. STARx—A GPU Based Multi-System Full-Band Real-Time GNSS Software Receiver. In Proceedings of International Technical Meeting of the Satellite Division of the Institute of Navigation, San Diego, CA, USA, 28–30 January 2013; pp. 1549–1559.

7. Seo, J.; Chen, Y.; de Lorenzo, D.S.; Enge, P.; Akos, D.; Lee, J. A Real-Time Capable Software-Defined Receiver Using GPU for Adaptive Anti-Jam GPS Sensors. *Sensors* **2011**, *11*, 8966–8991.

8. Chen, Y.H.; Juang, J.C.; Seo, J.; Lo, S.; Akos, D.M.; de Lorenzo, D.S.; Enge, P. Design and implementation of real-time software radio for anti-interference GPS/WAAS sensors. *Sensors* **2012**, *12*, 13417–13440.

9. Molero, J.; Garzn, E.; Garca, I.; Quintana-Ort, E.; Plaza, A. *Poster: A Batched Cholesky Solver for Local RX Anomaly Detection on GPUs*; PUMPS: Moscow, Russia, 2013.

10. NVIDIA. *Cublas Library*; NVIDIA Corporation: SantaClara, CA, USA, 2015.

11. Misra, P.; Enge, P. *Global Positioning System: Signals, Measurements, and Performance*; Ganga-Jamuna Press: Lincoln, MA, USA, 2011.

12. Rader, C. A simple method for sampling in-phase and quadrature components. *IEEE Trans. Aerosp. Electron. Syst.* **1984**, *AES-20*, 821–824.

13. Shirriff, K.W. *Sawmill: A logging File System for A High-Performance Raid Disk Array*; University of California: Berkeley, CA, USA, 1995.

14. Trees, V.; Harry, L. *Optimum Array Processing: Part IV of Detection, Estimation, and Modulation*; Wiley: New York, NY, USA, 2002.

15. Fante, R.L.; Vaccaro, J. Wideband cancellation of interference in a GPS receive array. *IEEE Trans. Aerosp. Electron. Syst.* **2000**, *36*, 549–564.

16. Gupta, I.J.; Moore, T.D. Space-frequency adaptive processing (SFAP) for radio frequency interference mitigation in spread-spectrum receivers. *IEEE Trans. Antennas Propag.* **2004**, *52*, 1611–1615.

17. Drefer, A; Niklasch, N.; Klefenz, F.; Schroth, A. Antenna and receiver system with digital beamforming for satellite navigation and communications. *IEEE Trans. Microw. Theory Tech.* **2003**, *51*, 1815–1821.

18. Daneshmand, S. Precise Calibration of a GNSS Antenna Array for Adaptive Beamforming Applications.*Sensors* **2014**, *14*, 9669–9691.

19. Carlson, B.D. Covariance matrix estimation errors and diagonal loading in adaptive arrays. *IEEE Trans. Aerosp. Electron. Syst.* **1988**, *24*, 397–401.

20. Carlson, B.D.; Goodman, L.M.; Austin, J.; Ganz, M.W.; Upton, L.O. An ultralow-sidelobe adaptive array antenna. *Linc. Lab. J.* **1990**, *3*, 291–310.

21. Moore, T.D. Analytic Study of Space-Time and Space-Frequency Adaptive Processing for Radio Frequency Interference Suppression. Ph.D. Thesis, The Ohio State University, Columbus, OH, USA, 2002.

22. Oppenheim, A.V.; Schafer, R.W.; Buck, J.R. *Discrete-Time Signal Processing*; Prentice-Hall Englewood Cliffs: Upper Saddle River, NJ, USA, 1989.

23. Farina, A. Digital equalisation in adaptive spatial filtering for radar systems: A survey. *Signal Process.* **2003**,*83*, 1–29.

24. Farina, A.; Sanzullo, R. Performance limitations in adaptive spatial filtering. *Signal Process.* **2001**, *81*, 2155–2170.

25. Johnson, J.R.; Fenn, A.J.; Aumann, H.M.; Willwerth, F.G. An experimental adaptive nulling receiver utilizing the sample matrix inversion algorithm with channel equalization. *IEEE Trans. Microw. Theory Tech.* **1991**, *39*, 798–808.

26. Kai, B.; Akos, D. A Software-Defined GPS and Galileo Receiver: Single-Frequency Approach. In Proceedings of the International Technical Meeting of the Satellite Division of the Institute of Navigation, Fort Worth, TX, USA, 25–28 September 2007; pp. 1632–1637.

27. Tsui, B.J. *Fundamentals of Global Positioning System Receivers*; Wiley-Interscience: New York, NY, USA, 2000.

28. Nicholas, W. *The Cuda Handbook: A Comprehensive Guide to GPU Programming*; Pearson Education: Upper Saddle River, NJ, USA, 2013.

29. Kirk, D.B.; Wen-Mei, W.H. *Programming Massively Parallel Processors: A Hands-on Approach*; Newnes: Burlington, MA, USA, 2012.

30. NVIDIA. *Nvidia Cuda c Programming Guide*; NVIDIA Corporation: SantaClara, CA, USA, 2015.

31. NVIDIA. *Cuda c Best Practices Guide*; NVIDIA Corporation: SantaClara, CA, USA, 2014.

32. Vasily, V. Better performance at lower occupancy. In Proceedings of the GPU Technology Conference, GTC, San Jose, CA, USA, 20–23 September 2010.

33. Vasily, V. Use registers and multiple outputs per thread on GPU. In Proceedings of the International Workshop on Parallel Matrix Algorithms and Applications (PMAA'10), Basel, Switzerland, 29 June–2 July 2010.

34. Vasily, V. *Unrolling Parallel Loops*; University of California: Berkeley, CA, USA, 2011.

35. NVIDIA. GPU Occupancy Calculator. In *CUDA SDK*; NVIDIA Corporation: SantaClara, CA, USA, 2010.

36. Vicario, J.L.; Antreich, F.; Barcelo, M.; Basta, N.; Cebrian, J.; Cuntz, M.; Gago, O.; Gonzalez, L.; Heckler, M.V.T.; Lavin, C.; *et al.* ADIBEAM: Adaptive digital beamforming for Galileo reference ground stations. In Proceedings of the 23rd International Technical Meeting of The Satellite Division of the Institute of Navigation (ION GNSS 2010), Portland, OR, USA, 21–24 September 2010.

37. Farina, A.; Ortenzi, L. Effect of ADC and receiver saturation on adaptive spatial filtering of directional interference. *Signal Process.* **2003**, *83*, 1065–1078.

38. Goldberg, D. What every computer scientist should know about floating-point arithmetic. *ACM Comput. Surv. (CSUR)* **1991**, *23*, 5–48.

39. Griffiths, J. Adaptive array processing: A tutorial. *IEE Proc. F Commun. Radar Signal Process.* **1983**, *130*, 3–10.

40. Compton, R.T. The relationship between tapped delay-line and FFT processing in adaptive arrays. *IEEE Trans. Antennas Propag.* **1988**, *36*, 15–26.

2

CHAPTER

A REAL-TIME INTERFERENCE MONITORING TECHNIQUE FOR GNSS BASED ON A TWIN SUPPORT VECTOR MACHINE METHOD

Wutao Li [1], Zhigang Huang [1], Rongling Lang [1], Honglei Qin [1,*], Kai Zhou [2] and Yongbin Cao [3]

[1]*School of Electronics and Information Engineering, Beihang University, Beijing 100191, China*

[2]*Academy of Opto-Electronics, Chinese Academy of Sciences, Beijing 100094, China*

[3]*Realsil Microelectronics Inc., Suzhou 215021, China*

ABSTRACT

Interferences can severely degrade the performance of Global Navigation Satellite System (GNSS) receivers. As the first step of GNSS any anti-interference measures, interference monitoring for GNSS is extremely essential and necessary. Since interference monitoring can be considered as a classification problem, a real-time interference monitoring technique based on Twin Support Vector Machine (TWSVM) is proposed in this paper. A TWSVM model is established, and TWSVM is solved by the Least Squares Twin Support Vector Machine (LSTWSVM) algorithm. The interference monitoring indicators are analyzed to extract features from the interfered GNSS signals. The experimental results show that the chosen observations can be used as the interference monitoring indicators. The interference monitoring performance of the proposed method is verified by using GPS L1 C/A code signal and being compared with that of standard SVM. The experimental results indicate that the TWSVM-based interference monitoring is much faster than the conventional SVM. Furthermore, the training time of TWSVM is on millisecond (ms) level and the monitoring time is on microsecond (μs) level, which make the proposed approach usable in practical interference monitoring applications.

Keywords: global navigation satellite system; interference monitoring; twin support vector machine

1. INTRODUCTION

The Global Navigation Satellite System (GNSS), which is a crucial means for navigation and positioning for both civil and military applications, is playing an increasingly significant role in modern society. However, the GNSS signals are extremely fragile and vulnerable, and can easily be influenced by interferences [1]. Therefore, interference monitoring techniques are exceedingly essential for improving the anti-interference performance of GNSS receivers. GNSS interference monitoring must be accurate and real-time. There are plenty of common methods for interference monitoring, such as the hypothesis testing method, method based on characteristic parameter threshold, integrated navigation method, and smart antenna technology.

1.1. Hypothesis Testing Method for Interference Monitoring

The hypothesis testing method establishes hypothesis testing statistics with monitoring indicators, and obtains the threshold of the monitoring indicators with the given significance level. In [2], Balaei and Dempster proposed a statistical inference technique for GPS interference detection, where the statistical parameters of the signal without interference were obtained and then the presence of interference was tested by comparing these parameters as a null hypothesis with the parameters extracted from a later window of data. The indicators utilized to establish hypothesis testing statistics include power spectral density, the correlator output, code phase,*etc.* [2,3]. The hypothesis testing method requires the predictable interference characteristics when making a hypothesis in order to ensure the rationality of the corresponding statistics.

1.2. Interference Monitoring Method Based on Characteristic Parameter Threshold

The method based on characteristic parameter threshold compares the actually measured indicators or monitoring quantities with a preset threshold in order to determine whether there are interferences. The monitoring indicators include the quantization weight gain of the A/D converter [4], power spectral, AGC (automatic gain control) [5] carrier to noise ratio [6], the correlator output [7], *etc.* This method requires a monitoring threshold, which is usually derived from the corresponding characteristic of the GNSS signals without interference. However, the GNSS signal environment is always complex, and especially during transfer, the GNSS signal inevitably suffers from intentional or unintentional interferences. Therefore, utilizing the characteristic without interference as the threshold will increase the false alarm rate and missing report rate.

1.3. Integrated Navigation Method for Interference Monitoring

In 2011 Faurie and Giremus proposed an interference monitoring method based on GNSS and Inertial Navigation System (INS) [8], where GNSS is usually integrated with INS using an extended Kalman filter. The least squares estimated values of the potential variance jump are calculated by studying the impact on the covariance of the Kalman filter outputs induced by the increase of the GNSS noise. With these estimated values as the inputs of continuous Bayesian testing experiments, the presence of interference in the received GNSS signals can be determined according to the experimental results. This method requires the GNSS and INS to work collaboratively, and superior stability of the system as well. Nevertheless, the existence of interferences will inevitably affect the stability of receivers.

1.4. Smart Antenna Technology for Interference Monitoring

Wilson and Ganguly proposed in 2006 a GPS interference monitoring method based on 3D Controlled Reception Pattern Antennas (CRPA) [9], where an implementation of Space-Time Adaptive Processing (STAP) algorithm using FFT was presented. The experimental results indicated that this algorithm could achieve a preferable monitoring effect under lower interference power conditions.

Among all these methods, the hypothesis testing method and the method based on characteristic parameter thresholds require theoretical values or models without interferences. However, due to the complicated environment, the time-varying and uncertain GNSS signal can make it difficult to precisely establish models and determine the theoretical values. As a result, we cannot just rely on the traditional methods to establish the necessary precise models to monitor the interferences. In addition, the integrated navigation method and smart antenna technology require superior stability of systems or superior parameter deployment.

On the other hand, applying the GNSS signals will accumulate huge amounts of data, which contain the influences of interferences on the signal quality and receiver performance. We can monitor the interferences without the precise physical models if the information and knowledge can be mined from these data. This method based on GNSS signals, seeks out regularities from these data, and then utilizes these regularities to monitor the unknown or unobservable data, without any need to know the precise analytic models of the systems in advance, thus making this method a more intelligent and flexible one. At present, the data-driven method has achieved successful applications in fields such as fault diagnosis and fault prediction. In this paper, the data-driven method is introduced into the field of interference monitoring, transforming interference monitoring problems into artificial intelligence classification problems, then used to monitor the interferences.

This paper presents an interference monitoring method, where the parameters manifesting the characteristics of interferences, such as the power spectrum, carrier to noise ratio, the correlator output, *etc.*, are extracted from the GNSS signals, and then are imported into an artificial intelligence classifier to determine whether there are interferences, and to acquire the information such as the type and strength of the interferences.

The selection of the classifier has a crucial impact on the accuracy and efficiency of monitoring. Support Vector Machine (SVM) is chosen as the classifier in this paper. SVM is one of the state-of-the-art classification techniques. Benchmarking studies reveal that the SVM performs the best among current classification techniques [10].

Solving of SVM involves quadratic programming, and the computational complexity rapidly increases as the training samples increase in number. In this paper, the Twin Support Vector Machine (TWSVM) technique is utilized to ensure the interference monitoring is real-time. TWSVM has outstanding classification precision, faster than SVM, and it satisfies the accuracy requirement of the GNSS interference monitoring. Therefore, this paper will focus on the GNSS interference monitoring with the TWSVM technique.

The rest of this paper is organized as follows: in the following section an introduction to SVM is presented. The model of TWSVM and the LSTWSVM algorithm are presented in Section 3. In Section 4, the features used for the interference monitoring are introduced. In Section 5, the algorithm is tested and verified with GPS L1 band C/A signals. Finally, the conclusions are made in Section 6.

2. SUPPORT VECTOR MACHINE

We assume that the training sample set of the interference monitoring is T={(x1,y1),...,(xl,yl)}∈(χ×γ)l, where (xi,yi), i=1,...,l are the n-dimension vectors constituted by the characteristic parameters of the interference monitoring, in which xi∈χ=Rn, and yi∈γ=(1,−1). It is defined that, when yi=1, xi is the characteristic parameter with interferences and when yi=−1, correspondingly xi is the characteristic parameter without interferences.

The SVM aims to build a classification hyperplane wx+b=0, making xiw≥1−b for the sample of yi=1, and xiw≤−1−b for the sample of yi=−1. Therefore, the interference monitoring problem is transformed into the optimization problem by SVM [11] as follows:

$$\min_{w,b,\xi} \frac{1}{2}||w||^2 + C\sum_{i=1}^{l} \xi_i$$
$$s.t.\ y_i(wx_i + b) \geqslant 1 - \xi_i, \xi_i \geqslant 0, i = 1, 2..., l \tag{1}$$

where ξ=(ξ1,ξ2···ξn) is the relaxation factor and C is the penalty parameter. The Wolfe dual of Equation (1) is:

$$\min \frac{1}{2}\sum_{i=1}^{l}\sum_{j=1}^{l} \alpha_i\alpha_j y_i y_j K(x_i, x_j) - \sum_{i=1}^{l} \alpha_i$$
$$s.t.\, 0 \leqslant \alpha_i \leqslant C, \sum_{i=1}^{l} \alpha_i y_i = 0 \tag{2}$$

where each ai is a Lagrange multiplier, which corresponds to the sample (xi,yi); K(xi,xj) is Gaussian kernel function, which is defined as:

$$K(x_i, x_j) = exp(-\frac{||x_i - x_j||^2}{\sigma^2}) \tag{3}$$

Let a1,···,al is the optimum solution of Equation (2), the classification criterion of GNSS signals to be monitored is:

$$f(x) = \text{sgn}\left\{\sum_{i=1}^{l} \alpha_i y_i K(x_i, x) + b\right\} \tag{4}$$

As is shown, the classifier decided by Equation (4) only relies on the samples (xi,yi) when ai≠0, and their input xi is called the Support Vector. The b in Equation (4) can be solved by:

$$b = y_j - \sum_{i=1}^{l} y_i \alpha_i K(x_i x_j) \tag{5}$$

where the aj corresponding to yj should meet the condition of 0≤αj≤C.

The common method to solve Equation (2) is the Sequential Minimal Optimization (SMO) algorithm [12]. However, the training time will rapidly increase as the number of training samples increases, making it hard for interference monitoring to be in real-time. Therefore, in this paper the TWSVM technique is adopted to improve the training speed.

3. TWIN SUPPORT VECTOR MACHINE

3.1. The Model of TWSVM

Unlike the standard SVM, TWSVM constructs two hyperplanes xw1+b1=0 and xw2+b2=0, and then makes the sample points with interferences as close as possible to the hyperplane xw1+b1=0, and as far as possible from the hyperplane xw2+b2=0. Likewise, the sample points without interferences are made as close as possible to the hyperplane xw2+b2=0 and as far as possible from the hyperplane xw1+b1=0.

Supposing that a n-dimension sample set T={(x1,y1),...,(xl,yl)}∈(χ×γ)l with m1 training points with interferences, that is, +1 class of sample, and m2 training points without interferences, that is, −1 class of sample, the +1 class of training points are denoted by the matrix A∈Rm1×n, and the −1 class of training points are denoted by the matrix B∈Rm2×n. The corresponding optimization problems of TWSVM are:

$$\begin{aligned} \min_{w_1,b_1} \quad & \frac{1}{2}(Aw_1 + e_1 b_1)^T (Aw_1 + e_1 b_1) + c_1 e_2^T \xi \\ s.t \quad & -(Bw_1 + e_2 b_1) + \xi \geqslant e_2, \quad \xi \geqslant 0 \end{aligned} \tag{6}$$

$$\begin{aligned} \min_{w_2,b_2} \quad & \frac{1}{2}(Bw_2 + e_2 b_2)^T (Bw_2 + e_2 b_2) + c_2 e_1^T \xi \\ s.t \quad & (Aw_2 + e_1 b_2) + \xi \geqslant e_1, \quad \xi \geqslant 0 \end{aligned} \tag{7}$$

where c1,c2>0 are the penalty parameters which can be selected by the heuristic algorithm proposed in [13], and e1,e2 are the column vectors with all of elements being one.

The computational complexity of the standard SVM is O(n3), and each complexity of the two optimization problems Equations (6) and (7) solved by TWSVM is nearly O(n/2), therefore, the computational speed of TWSVM is about four times faster than that of the standard SVM.

3.2. The Solution of TWSVM

Currently, there are three methods to solve the TWSVM: the Sequential Minimal Optimization (SMO) algorithm [12], the Successive over Relaxation (SOR) iterative method [14,15] and the Least Squares Twin Support Vector Machine (LSTWSVM) algorithm [16]. The LSTWSVM algorithm is the fastest one, as verified by the experimental analysis, so it is utilized as the solution of TWSVM in this paper.

Based on the fact that the Support Vectors are almost distributed around the classification hyperplanes, LSTWSVM narrows the constraint of the optimization problem near to the classification hyperplane, and modifies the original optimization models Equations (6) and (7) to Equations (8) and (9), respectively.

$$\begin{aligned} \min_{w_1,b_1} \quad & \frac{1}{2}(Aw_1+e_1b_1)^T(Aw_1+e_1b_1)+\frac{c_1}{2}\xi^T\xi \\ s.t \quad & -(Bw_1+e_2b_1)+\xi=e_2 \end{aligned} \tag{8}$$

$$\begin{aligned} \min_{w_2,b_2} \quad & \frac{1}{2}(Bw_2+e_2b_2)^T(Bw_2+e_2b_2)+c_2e_1^T\xi \\ s.t \quad & (Aw_2+e_1b_2)+\xi=e_1, \quad \xi\geqslant 0 \end{aligned} \tag{9}$$

Because of the relaxation factor ξ, the area confirmed by the constraints of Equations (8) and (9) is a rectangle.

Substituting the constraint into the objective function, Equation (8) becomes:

$$\min_{w_1,b_1} \quad \frac{1}{2}||Aw_1+e_1b_1||^2+\frac{c_1}{2}||Bw_1+e_2b_1+e_2||^2 \tag{10}$$

Differentiating Equation (10) for w1 and b1, respectively, and making the derivatives zero, we can obtain:

$$A^T(Aw_1+e_1b_1)+c_1B^T(Bw_1+e_2b_1+e_2)=0 \tag{11}$$

$$e_1^T(Aw_1+e_1b_1)+c_1e_2^T(Bw_1+e_2b_1+e_2)=0 \tag{12}$$

Combining Equation (11) with Equation (12), and adapting it into matrix form, then we have:

$$\begin{bmatrix} B^T B & B^T e_2 \\ e_2 B & m_2 \end{bmatrix} \begin{bmatrix} w_1 \\ b_1 \end{bmatrix} + \frac{1}{c_1} \begin{bmatrix} A^T A & A^T e_1 \\ e^T A & m_1 \end{bmatrix} + \begin{bmatrix} B^T e_2 \\ m_2 \end{bmatrix} = 0 \quad (13)$$

$$\begin{bmatrix} w_1 \\ b_1 \end{bmatrix} = \begin{bmatrix} B^T B + \frac{1}{c_1} A^T A & B^T e_2 + \frac{1}{c_1} A^T e_1 \\ e_2^T B + \frac{1}{c_1} e_1^T A & m_2 + \frac{1}{c_1} m_1 \end{bmatrix}^{-1} \begin{bmatrix} -B^T e_2 \\ -m_2 \end{bmatrix} \quad (14)$$

$$\begin{bmatrix} w_1 \\ b_1 \end{bmatrix} = \begin{bmatrix} \begin{bmatrix} B^T \\ e_2^T \end{bmatrix} \begin{bmatrix} B^T & e_2^T \end{bmatrix} + \\ \frac{1}{c_1} \begin{bmatrix} A^T \\ e_1^T \end{bmatrix} \begin{bmatrix} A^T & e_1^T \end{bmatrix} \end{bmatrix}^{-1} \begin{bmatrix} -B^T e_2 \\ -m_2 \end{bmatrix} \quad (15)$$

Defining E=[Ae1] and F=[Be2], Equation (15) becomes:

$$\begin{bmatrix} w_1 \\ b_1 \end{bmatrix} = -\left(F^T F + \frac{1}{c_1} E^T E\right)^{-1} F^T e_2 \quad (16)$$

Similarly:

$$\begin{bmatrix} w_2 \\ b_2 \end{bmatrix} = \left(E^T E + \frac{1}{c_2} F^T F\right)^{-1} E^T e_1 \quad (17)$$

4. INTERFERENCE MONITORING

The process of the interference monitoring based on TWSVM is shown in Figure 1. Firstly, the training samples are extracted from GNSS signals and then imported into the TWSVM to get TWSVM model; secondly the corresponding characteristic parameters are extracted from GNSS signals to be monitored, and then imported into the trained TWSVM model to get the monitoring results.

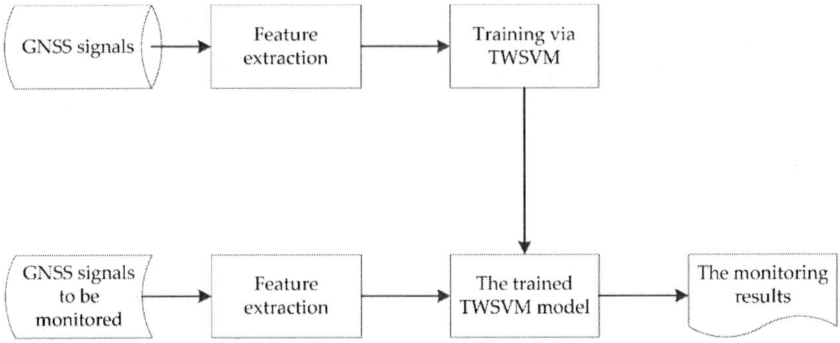

Figure 1. The process chart of the TWSVM-based interference monitoring.

From Figure 1, we can see that it is necessary to extract the features for interference monitoring. Therefore, we firstly study feature extraction from GNSS signals. According to the position of interference monitoring indicators relative to the correlator, the monitoring indicators are divided into pre-correlation interference monitoring indicators and post-correlation interference monitoring indicators. The pre-correlation interference monitoring is mainly implemented through the extraction of the observations of receivers, including antenna, AGC gain, ADC, *etc.* The post-correlation interference monitoring is implemented by observing the carrier to noise ratio, correlator output, *etc.*

4.1. Power Spectral Density

Frequency characteristic is the most direct reflection of signal characteristics. Through observing the frequency characteristic of GNSS signals, a variety of information can be obtained, including signal bandwidth, power spectral density, in-band stray, and interference performance. Therefore, the analysis of frequency characteristic is firstly carried out in interference monitoring. Power Spectral Density (PSD) reflects the signal power distribution within the signal bandwidth. With the measurement of the PSD, whether the power of every frequency is well-distributed or there is abnormal power can be seen. PSD is also an effective means to monitor the GNSS signal interference.

4.2. Carrier to Noise Ratio

Carrier to noise ratio (C/N0) is the ratio of received signal power to noise power spectral density. Carrier to noise ratio affects the performance such as the acquisition, tracking and data demodulation of receivers, which will ultimately affect the positioning precision of receivers. When the inference exists, it superimposes noise on the GNSS signals, thus reducing the carrier to noise ratio of receivers and increasing the measurement errors of pseudo range and pseudo range rate. This will result in the increase of the error of navigation positioning. When the interference power is too high, the carrier to noise ratio will fall below the tracking threshold, making receivers lose their capabilities of obtaining the navigation information from satellite signals. Therefore, carrier to noise ratio can be used as an interference monitoring indicator.

When inferences exist, the carrier to noise ratio is given by:

$$\left(\frac{C_S}{N_0}\right)_{eff} = \frac{\frac{C_S}{N_0}\int_{-\beta_r/2}^{\beta_r/2} G_s(f)df}{\int_{-\beta_r/2}^{\beta_r/2} G_s(f)df + \frac{C_I}{N_0}\int_{-\beta_r/2}^{\beta_r/2} G_I(f)G_s(f)df} \quad (18)$$

where βr is the bandwidth of receiver front-end and processing filter, N0 is the noise PSD, Cs is the carrier power of navigation signals, Gs(f) is the normalized PSD of navigation signals, CI is carrier power of the interference signals, and GI(f) is the normalized PSD of interference signals.

4.3. Correlator Output

When no interferences exist, the correlator output is basically constant, which is decided by pseudo-code autocorrelation, correlation length of the correlator and integral time. However, interferences can make the correlator output vary. Therefore, the correlator output can be also used as an interference monitoring indicator.

4.3.1. Correlator Output Power

Correlator Output Power (COP) is defined as the result of the correlator output power when the former and later correlator are balanced divided by the expected thermal noise threshold, which is given by:

$$COP = \frac{I^2 + Q^2}{N} \tag{19}$$

where I and Q are in-phase and quadrature signal of the correlator output, respectively. N is the expected noise level, varying from different receivers, which is a statistical expected value. Usually, the correlator output signal is:

$$\begin{aligned} I &= \frac{A}{\sqrt{2}} C_w C_{rw} X_w \cos(\phi_w - \phi_{rw}) \\ Q &= \frac{A}{\sqrt{2}} C_w C_{rw} X_w \sin(\phi_w - \phi_{rw}) \end{aligned} \tag{20}$$

where, A is the amplitude of input signal, *Cw* is the C/A code in the sampling rate of w, Crw is the local pseudo-code, Xw is the data code in the sampling rate of w, ϕw is the sampling signal phase, and ϕrw is the local pseudo-code phase.

When there are interferences through receivers, distortion, delay or peak reduction will arise in the receiver correlation peak curve, which directly results in COP variation. Therefore, the quality of the received signals can be judged by COP and the preliminary analysis can determine whether there are interferences.

4.3.2. Correlator Output Power Standard Deviation

The standard deviation of correlator output power (COPσ) can be also utilized to analyze the extent of impact on navigation signals from interferences. The COPσ is the measures of dispersion of COP, which is defined by:

$$COP\sigma = \sqrt{E[COP - E(COP)]^2} = \sqrt{E(COP^2) - [E(COP)]^2} \quad (21)$$

When the input signals are constant, the peak value of *COP* is relatively stable. Thus the COPσ is relatively small. However, when the unstable interferences exist, the correlation peak jitters increase and the value of COPσ gets larger.

5. EXPERIMENTS

5.1. Experimental Analyses of Monitoring Indicators

In this section, the GPS L1 C/A code signal is tested, and the experimental scheme is shown in Figure 2. The GPS signals used in the experiment are generated by simulation, and the power ratio between GPS signals and the additional white Gaussian noise is −18 dB. Broadband Gaussian interference covers the whole frequency range of GPS signals, and narrowband Gaussian interference only covers the frequency range with the center frequency of 3.563 MHz and the bandwidth of 80 KHz. The receiver bandwidth is 2 MHz.

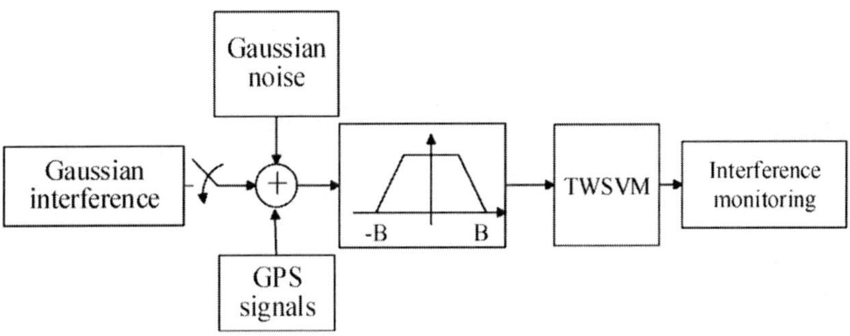

Figure 2. Experimental scheme of the interference monitoring.

5.1.1. Influence of Interference on C/N0

When noise and RFIs are superimposed on the GPS digital intermediate frequency signal, with different RFIs and the increase of interference signal power, the output C/N0 of GPS software receiver will change. Figure 3shows the curves that express the relation between the equivalent C/N0 and the interference to signal ratio (ISR), respectively, under conditions of continuous wave (CW) interference, broadband FM interference, narrowband FM interference, sweeping frequency CW interference, and uniform spectrum interference for the PRN3 satellite.

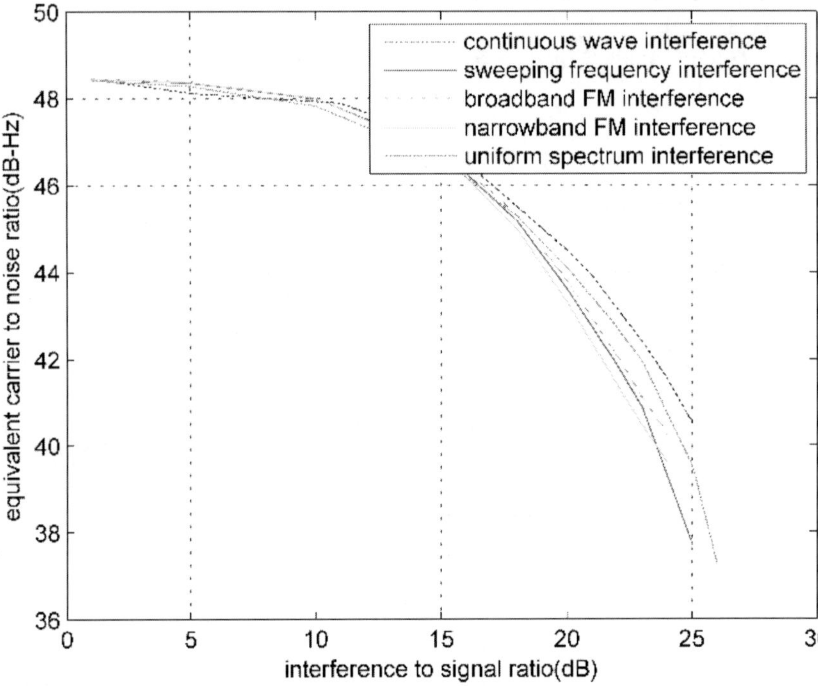

Figure 3. Influences of different RFIs on equivalent C/N0.

The experiment demonstrates that the influences on the equivalent C/N0 from different RFIs are different, which further proves that C/N0 can be used as the interference monitoring indicator.

5.1.2. Correlator Output

When noise and RFIs are superimposed on GPS digital intermediate frequency signal, with different RFIs and the increase of interference signal power, the output COP and COPσ of GPS software receiver will change.Figure 4 and Figure 5 show the PRN3 satellite curves of the relation between both COP and COPσ and the ISR, respectively, under conditions of CW interference, broadband FM interference, narrowband FM interference, sweeping frequency CW interference, and uniform spectrum interference.

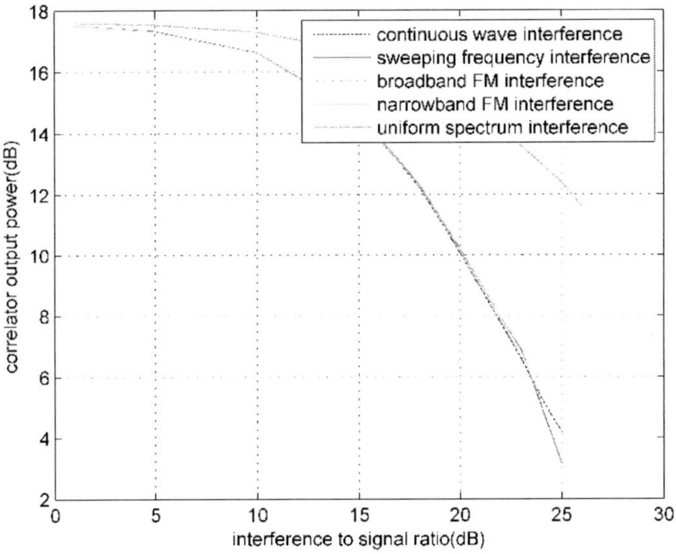

Figure 4. Influences of different RFIs on COP.

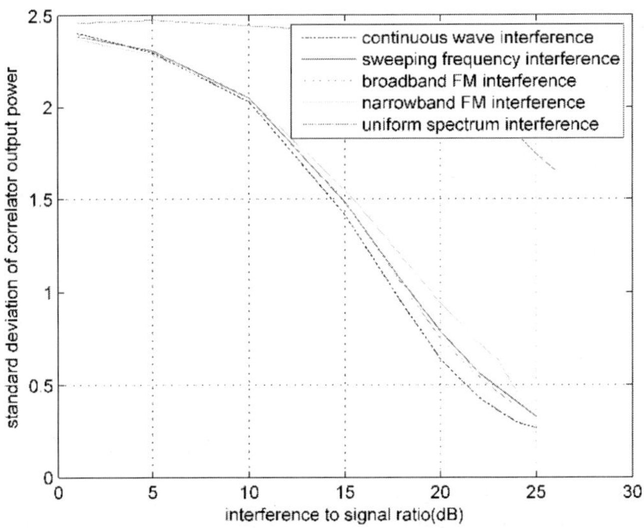

Figure 5. Influences of different RFIs on COPσ.

In order to better analyze the impact on COP from RFIs, Figure 6 depicts the relation curves of COPσ to COP mean ratio under conditions of different RFIs and interference strength.

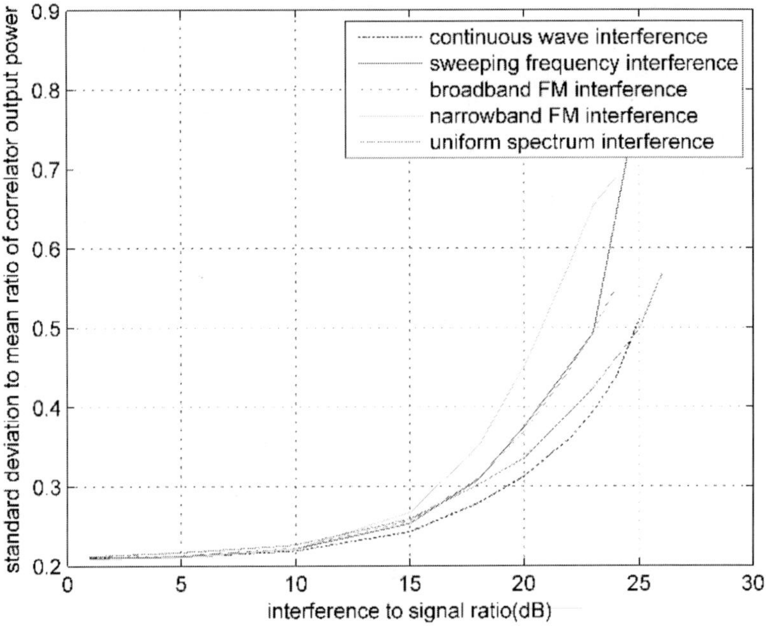

Figure 6. The variation curves of COPσ to COP mean ratio under different RFIs.

The experimental results show that, with the increase of interference strength, COP gradually decreases until the receiver loses lock. Therefore, the COP can be used as an indicator of interference monitoring. In the experiment, the COPσ measured after the receiver stably tracking does not increase with the rising of interference strength, but show a decreasing trend This is because of the gradual decreasing of COP. As is shown in Figure 6, with the increase of interference strength, the COP mean decreases faster than the COPσ. This indirectly illustrates that the COP jitter is aggravating.

5.2. Experiment of Interference Monitoring

To demonstrate the results discussed in the previous sections, three experiments were designed. The hardware setup for the experiments is shown in Figure 7. As can be seen the software receiver is used to capture the IF data to be analyzed and a SMU200A Vector Signal Generator (Rohde & Schwarz GmbH & Co. KG, Munich, Germany) is used to generate the broadband and narrowband interferences. The interferences are transmitted by a G5Ant-52PT1 transmitting antenna (Antcom corporation, Torrance, CA, USA) which are combined with the real GPS signal received by a HX-BS781A antenna (Harxon corporation, Shenzhen, China).

Figure 7. The hardware setup for the experiments.

5.2.1. Pre-Correlation

The power spectral density is utilized as the pre-correlation indicator. With the broadband and narrowband interference monitoring tests, the monitoring effect of the SVM is verified.

The testing parameters are set as follows:

Standard SVM: penalty parameter C=10, kernel function parameter σ=1;

LSTWSVM: penalty parameter c1=c2=10, kernel function parameter σ=1.

The monitoring results are shown in Table 1 and Table 2, respectively.

Table 1. Broadband interference monitoring result.

Training Algorithm	Interference Strength (dB)	Training Time (ms)	Monitoring Time (μs)	Monitoring Precision (%)
Standard SVM	$I/S < 5$	60	15	100
	$5 \leqslant I/S < 15$			100
	$15 \leqslant I/S < 25$			100
	$I/S \geqslant 25$			100
LS-TWSVM	$I/S < 5$	0.75	15	100
	$5 \leqslant I/S < 15$			100
	$15 \leqslant I/S < 25$			100
	$I/S \geqslant 25$			100

Table 2. Narrowband interference monitoring result.

Training Algorithm	Interference Strength (dB)	Training Time (ms)	Monitoring Time (μs)	Monitoring Precision (%)
Standard SVM	$I/S < 5$	5.3	15	100
	$5 \leqslant I/S < 15$			100
	$15 \leqslant I/S < 25$			100
	$I/S \geqslant 25$			100
LS-TWSVM	$I/S < 5$	0.71	15	100
	$5 \leqslant I/S < 15$			100
	$15 \leqslant I/S < 25$			100
	$I/S \geqslant 25$			100

Table 1 and Table 2 show the comparison of classification accuracy and speed between LSTWSVM and the standard SVM for the broadband and narrowband interference monitoring. The results indicate that both the LSTWSVM and the standard SVM are of high classification accuracy for the broadband and narrowband interferences, but the training speed of LSTWSVM is much faster than the standard SVM. Taking the broadband interference as an example, the training speed of LSTWSVM is 80 times faster than the standard SVM. The monitoring time of the two algorithms are the same (15 μs for both), which satisfies the requirement of the real-time interference monitoring for GNSS signals.

5.2.2. Post-Correlation

For the post-correlation interference monitoring, GPS signals with the broadband and narrowband interferences whose ISRs are 16 dB~24 dB are analyzed. The parameter settings are the same as pre-correlation, and the experimental results are shown in Table 3.

Table 3. Broadband and narrowband interference monitoring results.

Training Algorithm	Interference Strength (dB)	Training Time (ms)	Monitoring Time (μs)	Monitoring Precision (%)
Standard SVM	Broadband I/S : 16–24	54	15	100
	Narrowband I/S : 16–24			100
LS-TWSVM	Broadband I/S : 16–24	0.99	15	100
	Narrowband I/S : 16–24			100

Table 3 shows the comparison of classification accuracy and speed between LSTWSVM and the standard SVM under the broadband and narrowband interferences whose ISRs are both 16–24 dB, and the results indicate that both the LSTWSVM and the standard SVM are of high classification accuracy, but the training speed of LSTWSVM is about 55 times faster than standard SVM. The monitoring time of the two algorithms are the same (both equal to 15 μs), which satisfies the requirement of the real-time interference monitoring for GNSS signals.

6. CONCLUSIONS

Applying SVM in the interference monitoring of GNSS signals satisfies the requirement of objectivity and accuracy. However, the training speed of standard SVM cannot satisfy the requirement of being in real-time for interference monitoring. TWSVM, which is faster than the standard SVM in training speed, is introduced in this paper. The TWSVM model is solved by a modified algorithm, LSTWSVM. LSTWSVM is much faster than the standard SVM, and it can be applied in the field of interference monitoring of GNSS signals. The experimental results indicate that the training speed of TWSVM is on millisecond (ms) level and the classification speed of TWSVM is on microsecond (μs) level, which can be used in practice.

ACKNOWLEDGMENTS

This research work is supported by the National Natural Science Foundation of China (No. 61202078).

AUTHOR CONTRIBUTIONS

Wutao Li contributed to writing much of this paper and performed the whole experiment in Section 5.2. Zhigang Huang and Honglei Qin provided valuable feedback, advice, and mentoring during the manuscript modification. Rongling Lang provided some suggestions to the design of the experiments and especially the method on the choice of the penalty parameters in Section 3.1 and Section 5.2. Kai Zhou contributed to the data analyses in Section 5.1 and edited the manuscript. Yongbin Cao contributed to several sections of this paper and also gathered sources from the literature.

REFERENCES

1. Pinker, A.; Smith, C. Vulnerability of the GPS Signal to Jamming. *GPS Solut.* **1999**, *36*, 7535–7543.
2. Balaei, A.T.; Dempster, A.G. A Statistical Inference Technique for GPS Interference Detection. *IEEE Trans. Aerosp. Electron. Syst.* **2009**, *45*, 1499–1511.
3. O'Driscoll, C.; Murphy, C.C. A simple technique for the detection of multiple access interference in the parallel acquisition of weak GPS signals. In Proceedings of the 20th International Technical Meeting of the Satellite Division of the Institute of Navigation ION GNSS 2007, Fort Worth, TX, USA, 25–28 September 2007; pp. 282–291.

4. Capozza, P.T.; Holland, B.J.; Hopkinson, T.M.; Li, C.; Moulin, D.; Pacheco, P.; Rifkin, R. Measured effects of a narrowband interference suppressor on GPS receivers. In Proceedings of the 55th Annual Meeting of the Institute of Navigation, Cambridge, MA, USA, 27–30 June 1999; pp. 645–651.

5. Balaei, A.T.; Motella, B.; Dempster, A.G. GPS interference detected in Sydney-Australia. In Proceedings of the International Global Navigation Satellite Systems Society Symposium IGNSS 2007, Sydney, Australia, 4–6 December 2007.

6. Thompson, R.J.R.; Wu, J.H.; Balaei, A.T.; Dempster, A.G. Detection of RF interference to GPS using day-to-day C/No differences. In Proceedings of the International Symposium on GPS/GNSS, Taipei, Taiwan, 26–28 October 2010.

7. Ouzeau, C.; Macabiau, C.; Roturier, B.; Mabilleau, M. Performance Assessment of Multi correlators Interference Detection and Repair Algorithms for Civil Aviation. In Proceedings of the European Navigation Conference of EUGIN ENC-GNSS, Toulouse, France, 23–25 April 2008.

8. Faurie, F.; Giremus, A. Bayesian detection of interference in satellite navigation systems. In Proceedings of the IEEE International Conference on Acoustics, Speech, and Signal Processing ICASSP 2011, Prague, Czech Republic, 22–27 May 2011; pp. 4348–4351.

9. Wilson, D.; Ganguly, S. Flexible GPS receiver for jammer detection, characterization and mitigation using a 3D CRPA. In Proceedings of the 19th International Technical Meeting of the Satellite Division of the Institute of Navigation ION GNSS 2006, Fort Worth, TX, USA, 26–29 September 2006; pp. 189–200.

10. Baesens, B.; van Gestel, T.; Viaene, S.; Stepanova, M.; Suykens, J.; Vanthienen, J. Benchmarking state-of-the-art classification algorithms for credit scoring. *J. Oper. Res. Soc.* **2003**, *54*, 627–635.

11. Deng, N.Y.; Tian, Y.J. *A New Method in Data Mining-Support Vector Machines*; Science Press: Beijing, China, 2003.

12. Platt, J.C. Fast training of support vector machines using sequential minimal optimization. In *Advances in Kernel Methods-Support Vector Learning*; Scholkopf, B., Burgers, C.J.C., Smola, A.J., Eds.; MIT Press: Cambridge, MA, USA, 1999; pp. 185–208.

13. Lang, R.L.; Deng, X.L.; Gao, F. The Heuristic Algorithms for Selecting the Parameters of Support Vector Machine for Classification. *Chin. J. Electron.* **2012**, *21*, 485–488.

14. Shao, Y.H.; Zhang, C.H.; Wang, X.B.; Deng, N.Y. Improvements on Twin Support Vector Machines. *IEEE. Trans. Neural Netw.* **2011**, *22*, 962–968.

15. Tian, Y.J.; Ju, X.C.; Qi, Z.Q.; Shi, Y. Improved twin support vector machine. *Sci. China Math.* **2014**, *57*, 417–432.

16. Kumar, M.A.; Gopal, M. Least squares twin support vector machines for pattern classification. *Expert Syst. Appl.* **2009**, *36*, 7535–7543.

3

CHAPTER

DEFINITION OF AN ENHANCED MAP-MATCHING ALGORITHM FOR URBAN ENVIRONMENTS WITH POOR GNSS SIGNAL QUALITY

Felipe Jiménez, Sergio Monzón and Jose Eugenio Naranjo

University Institute for Automobile Research (INSIA), Ctra. Valencia, Km. 7, Madrid 28031, Spain,

ABSTRACT

Vehicle positioning is a key factor for numerous information and assistance applications that are included in vehicles and for which satellite positioning is mainly used. However, this positioning process can result in errors and lead to measurement uncertainties. These errors come mainly from two sources: errors and simplifications of digital maps and errors in locating the vehicle. From that inaccurate data, the task of assigning the vehicle's location to a link on the digital map at every instant is carried out by map-matching algorithms. These algorithms have been developed to fulfil that need and attempt to amend these errors to offer the user a suitable positioning. In this research; an algorithm is developed that attempts to solve the errors in positioning when the Global Navigation Satellite System (GNSS) signal reception is frequently lost. The algorithm has been tested with satisfactory results in a complex urban environment of narrow streets and tall buildings where errors and signal reception losses of the GPS receiver are frequent.

Keywords: satellite positioning systems; Global Navigation Satellite System (GNSS); navigation; information fusion; map matching.

1. INTRODUCTION

Vehicle positioning is a key element in the performance of numerous driver assistance and safety systems [1]. Depending on the particular demands, higher or lower levels of accuracy are required when positioning the vehicle [2]. Thus, informative applications for navigation systems need not be especially accurate while other systems such as those focused on the improvement of traffic safety might need submetric precisions [3]. Generally speaking, the location of a vehicle is mainly accomplished by satellite positioning systems (e.g., GNSS). With the current satellite positioning technology that can be found in all vehicle positioning systems, that submetric precision is not usually available and there are many situations where the environment causes the GNSS signal to be frequently lost or damaged (e.g., driving among tall buildings or in areas with dense tree cover). Even in less demanding applications, such as navigation systems, there are situations where positioning errors are made when locating the vehicle into a link. These errors come mainly from two sources: errors and simplifications of digital maps and errors in locating the vehicle. From that inaccurate data, the task of assigning the location of the vehicle to a link on the digital map at every instant is carried out by map-matching algorithms whose performance depends on many factors. As of today, locating a vehicle in an urban environment with high street density cannot be achieved at every instant, due to GNSS signal reception losses.

However, there are several solutions for improving the accuracy of satellite positioning. Differential correction techniques could improve GPS accuracy from 10 to 15 m of error to 1 cm of error. If this differential correction is obtained using only GPS pseudo-range code, it is named Differential GPS (DGPS) and can achieve accuracies of around 1 to 5 m. On the contrary, if it is obtained using the GPS carrier phase information, it is named Real Time Kinematics (RTK). DGPS and its accuracy is around 1 to 10 cm. However, these solutions are not limitless. In [3], the authors pointed out that the previous solutions accumulate a high amount of time without GPS coverage and the high precision levels offered by type 4 or type 5 positioning are available less than 37% of the time in some situations. On the other hand, the solution of using our own base stations for the differential correction has a very limited range and it is only acceptable for obtaining high levels of accuracy near the base station, so it is not an acceptable solution for real traffic conditions. Moreover, the equipment cost must be taken into account, which in addition to the connections required to obtain the differential corrections, makes them an invalid solution for their global implementation at present. In all previous situations the signal reception losses are still occurring and so it is necessary to turn to methods that enable positioning even in those areas. To do so, inertial systems are usually employed, though these systems have a cumulative error [3] that makes their reliability decrease as the travelling distance increases [2]. In particular, [4] analyse the effect of using low-performance and low-cost sensors. In [5], the authors also show that the GPS coverage in central Hong Kong can be less than 20%, which makes it essential to use other systems to keep the user informed of their route.

In dense urban areas it is still difficult to obtain good positioning using a single technology. This problem has led to the introduction of combining multiple positioning techniques. Intelligent Urban Positioning (IUP) is based on combining positioning algorithms augmented with three dimensional mapping techniques for distinguishing between non-line-of-sight (NLOS) and line-of-sight (LOS) signals and multi-constellation GNSS, using signals from all visible satellites. This can be used to predict the blockage and reflection of signals [6]. For example, a method using a 3D city model to improve cross-street GNSS positioning accuracy is presented in [7].

Additionally, digital maps have improved their accuracy, mainly when their construction methods have changed from air pictures [8,9] to instrumented vehicles [10,11]. The use of instrumented vehicles usually includes satellite positioning systems and inertial systems and both measurements are combined [2]. Although both sources are combined, their own particular errors mean the final map is not perfectly accurate. Furthermore, it must be taken into account that optimized storage of spatial road network data involves simplifications that can be larger or smaller according to the kind of approximation employed [11].

Finally, as stated in [12], "the process of continually estimating a user's position on a road segment is known as map matching". In this process, it has to be taken into account that both information sources (GNSS positioning and digital map data) have associated errors, and therefore the algorithm attempts to found the location of the vehicle on the link that most approaches its real position.

Efficient map-matching algorithms have many applications apart from the ones focused on the driver assistance systems that have been cited above. For example, for travel behaviour, it is required to unambiguously identify the correct road links followed by the traveller and all these identified links should form a meaningful travel route [13,14]. Depending on the final application, the requirements imposed on the map-matching algorithm could be diverse. In this sense, safety applications require high standards of accuracy and reliability that are not completely essential for personal navigation [15]. The previous systems require real time processing but traveller displacement studies allow offline processing.

Several map-matching algorithms have been developed up to the present [12,16]. The geometric algorithms are the simpler ones but they do not perform well at junctions, roundabouts and parallel roads [17,18,19,20]. There are three different approaches to geometric algorithms, *i.e.*, point-to-point matching, point-to-curve matching and curve-to-curve matching. However, their simplicity means they do not usually perform well in complex situations. More complex methods are used in topological algorithms, which use the relationship between entities of the digital map such as adjacency, connectivity and containment [21,22,23,24] to select the correct link. Hence, they take into account the previous information of the route to locate the vehicle into a link and allow the algorithm to keep the continuity of the trajectory. Probabilistic algorithms generate a confidence region around a position fix obtained from a GPS receiver, and are used mostly in the decisions concerning junctions [25,26]. Finally, advanced algorithms use Kalman Filters (KF), Extended Kalman Filters (EKF), Fuzzy Logic models (FL), Dempster-Shafer's mathematical theory of evidence, particle filters or Bayesian inference and outperform the other algorithms in terms of correct link identification and accuracy. Nevertheless, they require more input data and could be relatively slow and difficult to implement in real-time applications [27,28,29,30].

Due to the correlation between the different kinds of algorithms enumerated above, [29] proposed a distinction between simple method algorithms, weighted algorithms and advanced algorithms. The former include geometric and topological algorithms, such as probabilistic algorithms, as long as they do not use historic data of the route. The latter try to improve the map-matching results by applying weights when choosing the correct link in which the vehicle is located. These algorithms reach percentages of correct link identification of between 85% and 98.5%. Nonetheless, their performance is not enough to support some ITS applications. These kinds of algorithms are the ones developed by [21,23,31,32].

Though there are others, some algorithms that offer good performances are mentioned in the following lines. According to [12], the algorithm developed in [30] is one of the most accurate according to the correct number of segment identifications and the accuracy of the location of the vehicle in each link. This algorithm, which uses an integrated GPS/DR sensor identifies 98%–99% of links correctly. Another algorithm, developed in [31] reaches a percentage of around 96% of correct link identification. The complex algorithm developed in [32] ranges from 97% to 99.8%. This algorithm uses inputs generated by an integrated GNSS/DR and height data obtained from a Digital Elevation Model (DEM).

The difference between real-time algorithms and post-processing algorithms should be pointed out. The former, such as those focused on in-vehicle navigation systems should be simple and not computationally intense. The latter use all the position fixes generated by a GNSS along a route as an input, and compare them with the digital map data to produce the whole route travelled as an output. In this last scenario, rapid computer processing is not essential, so the algorithms can be more complex and, therefore, the results could outperform those executed in real-time.

Finally, due to the errors in GNSS position and digital maps, there can be a high degree of uncertainty associated with the map-matched locations and a quality indicator representing the level of confidence in map-matched locations could be necessary. An empirical method to derive the integrity of a map-matched location is proposed in [33].

In this research, a real-time map-matching algorithm is put forward. It is focused on environments that influence negatively on the GNSS signal reception (reception losses or accuracy degradation) and requires little information from the digital map. It can also work without using additional information from inertial systems, which allows it to be implemented in after-market navigators. The algorithm is tested in an urban environment where the GNSS coverage is poor and there are frequent signal reception losses. Likewise, the use of the speed information as a new input is tested, even without needing high accuracy in its measurement.

2. MAP MATCHING ALGORITHM PROPOSAL

2.1. Specifications

The map-matching algorithm developed in this research tries to simultaneously improve some of the limitations of other algorithms. Specifically, the following characteristics have been defined for it:

- The algorithm should be implemented in real time and with limited computational means.
- The algorithm is based on satellite positioning and digital map data, without using inertial sensors because these will limit its implementation in after-market devices.
- Optionally, it will include the possibility to use speed information, if it is available, to know when the vehicle is moving and when not. Using inertial sensors for speed and yaw angle could improve the final results but could increase the hardware cost if high performance sensors are integrated [4]. Furthermore, the aim is to develop an algorithm that could be implemented in any navigation device, and not all of them contain these sensors, even low performance ones.
- The information contained in the digital map has to be as simple as possible. The necessary information includes *x-y* coordinates, the legal turns and the segments priority.
- After making a change of segment, the algorithm should check for some time if it has been a good decision or if the vehicle needs to be relocated to another link.
- The algorithm should recover the positioning of the vehicle quickly after long periods of time without GPS coverage.
- In the event of an error in the positioning of the vehicle in a link, the algorithm should be able to detect the erroneous decision and correct it in a few seconds taking into account the previous correct location.
- The algorithm should be capable of working even with degraded GNSS signals.

2.2. Description

The designed algorithm uses topological information from the digital map, as well as historical information of the route. Basically, the map is divided into segments (defined between crossings), formed by links or sub-segments (for all two consecutive points on the map a link is created). The algorithm could be divided into four main steps:

1. Candidate segments and links definition
2. Preliminary segment selection:Topological and historical links parameters and final weight calculation of the segments
3. Definitive segment selection: (1) Correction of the belonging parameter impact; (2) Correction of inappropriate segment change
4. Vehicle positioning on the digital map

2.2.1. Step 1: Candidate Segments and Links Definition

For each GNSS positioning, the segments included in a confidence region (circumference of radius R) are identified as candidates. Each link contained in a candidate segment within the confidence region becomes a candidate link of that segment.

2.2.2. Step 2: Preliminary Segment Selection

In order to select the segment in which the vehicle is located, some parameters are calculated according to the physical characteristics and the relationship with the GNSS-generated point. Taking into account previous approaches found in the literature review of geometric, topological and probabilistic algorithms [17,18,19,20,21,22,23,24,25,26] and tests that provide the relative relevance of each parameter on the final positioning results, the following have been chosen:

1. **Distance parameters:**

- Point-to-point average distance parameter: It uses the average distance between the two extreme points of the candidate link and the location given by the GNSS receiver. This value is useful to improve situations where the perpendicular distance is small but the link is not actually near the GNSS location.

- Perpendicular distance parameter: It uses the perpendicular distance between the GNSS positioning and the line that contains the candidate link.

2. **Parameters depending on the movement history:**

- Direction parameter: It compares the difference between the GNSS trajectory and the link trajectory by calculating the angle between them. To estimate the trajectory of the vehicle the last GPS positions are taken into account, instead of just using the last two. Should the GNSS reception losses be longer than 1 s, this parameter decreases its relevance linearly because its relevance is lower as the time without the GNSS signal increases.

- Segment belonging parameter: It assigns values to the candidate links according to their priority in a hypothetical junction (stored in the digital map information) and penalizes those links that are not physically accessible or those that involve illegal turns. At a junction, those links that involve going straight (keeping in the same street that is, not doing any turns) are considered priority links and get a higher value than those legal turns not considered priority links. The illegal turns receive a negative value to penalize them and avoid their being taken as the vehicle location by the algorithm. However, if the user makes an illegal turn, the algorithm will locate the vehicle correctly after a few seconds. Finally, the links that belong to the segment where the vehicle was positioned before receive the highest value. Like the Direction parameter, should the GNSS reception losses be longer than 1 s, this parameter decreases its relevance linearly because its relevance is lower as the time without the GNSS signal increases.

3. **Additional parameter for segments:**

- Perpendicular parameter: An additional value is assigned to a candidate segment in case the perpendicular projection of the GNSS point on one of the lines defined by each candidate link is among the two points of the digital map that define the link.

The sum of the first four parameters referred to links gives the final weight for each candidate link, and the average value of all links contained in a candidate segment, adding the perpendicular parameter, gives the final segment weight. Table 1 shows a summary of all the parameters and the final equation for the final segment weight. Table

2 includes the nomenclature used in the equations. The algorithm obtains a preliminary decision in which the vehicle is located in the segment with the highest weight.

Table 1. Parameters included in the map-matching algorithm.

Parameters		
Point-to-point distance	$W_{d2} = K_{d}\cdot\left[1 - \frac{\bar{d}(GNSS, link\ k)}{R}\right]$	(1)
Perpendicular distance	$W_{d} = K_{d2}\cdot\left[1 - \frac{d(GNSS, link\ k)}{R}\right]$	(2)
Direction	$W_a = K_a\cdot\lvert\cos\alpha_k\rvert$	(3)
Belonging	$W_b = K_b\cdot\beta_k$	(4)
Perpendicularity	$W_p = \begin{cases} K_p\ if\ the\ condition\ is\ satisfied \\ 0\ if\ the\ condition\ is\ not\ satisfied \end{cases}$	(5)
Segment s Final Weight		
$W_{T,s} = \frac{\sum_{k=1}^{n_s}(W_{a,k} + W_{b,k} + W_{d,k} + W_{d2,k})}{n_s} + W_{p,k}$		(6)

Table 2. Nomenclature.

$d(GNSS,\ link\ k)$	Distance between the GNSS point and link k
α_k	Angle between the GNSS trajectory and the link k trajectory
β_k	Denotes link k priority
n_s	Number of candidate links contained in segment s
$K_d,\ K_{d2}, K_a,\ K_b,\ K_p$	Parameter coefficients $K_a,\ K_b = \begin{cases} NV\ if\ t_l \leq 1s \\ (NV/9)(t_l - 10)\ if\ 1s \leq t_l \leq 10s \\ 0\ if\ t_l \geq 10s \end{cases}$ $NV = nominal\ value$ $t_l = GNSS\ signal\ reception\ loss\ time$

2.2.3. Step 3: Definitive Segment Selection:

In order to corroborate the current segment selection, some scenarios of possible previous incorrect segment selections are considered and corrections can be performed in a short period of time.

The fact that the belonging parameter has a high influence when calculating the average value of the segment can cause the algorithm to take a little more time than necessary to detect when the vehicle has turned at a junction, so the segment where the vehicle is located has changed. To solve this problem, the algorithm also calculates the weights of all candidate segments without considering the belonging parameter:

$$W'_{T,s} = \frac{\sum_{k=1}^{n_s} \left(W_{a,k} + W_{d,k} + W_{d2,k}\right)}{n_s} + W_{p,k} \tag{7}$$

where $W'_{T,s}$ is the weight of the segment s w

ithout considering the belonging parameter.

This way, if this new total weight is higher for a different segment s' than for the one that has been chosen to locate the vehicle, the algorithm starts to store an uncertainty variable that adds to the first calculated weight of the segment (which considers the belonging parameter) to check if now the first weight plus the uncertainty variable are higher than the original weight assigned to the chosen segment. Then, the algorithm applies the following decisions:

If:

$$W'_{T,s'} > W'_{T,s} \tag{8}$$

if:

$$W_{T,s'} + U_{s'} > W_{T,s} \Rightarrow Change\ of\ segment\ from\ s\ to\ s' \tag{9}$$

where $U_{s'}$ is the uncertainty variable of segment s'.

else:

$$U_{s'} = U_{s'} + \Delta U_{s'}\ and\ there\ is\ no\ change\ of\ segment \tag{10}$$

else there is no change of segment and the vehicle is located in segment s.

Thanks to the uncertainty, the change of segment will be done quickly. Moreover, to let the program restore any mismatching in a quick way, in the event it has made an incorrect change of segment, when a change of segment is made, and if Equation (8) is satisfied, the algorithm calculates, for each point generated by the GPS, the total weights of the candidate segments not only considering the new belonging parameter but also the belonging parameter that the candidate segments had before the change was made. Therefore, considering the following criteria, the algorithm is capable of restoring from its positioning errors quickly. If Equation (8) is satisfied the algorithm applies the following calculations and decisions:

$$W''_{T,s} = \frac{\sum_{k=1}^{n_{links}} \left(W_{a,k} + W_{b\prime,k} + W_{d,k} + W_{d2,k}\right)}{n_{links}} + W_{p,k} \quad (11)$$

where $W''_{T,s}$ is the weight of the segment s considering the belonging parameter that this segment had before the change of segment was made (symbolized by $W_{b',k}$):

$$if\ W''_{T,s\prime} > W''_{T,s} \Rightarrow Change\ of\ segment\ from\ s\ to\ s\prime \quad (12)$$

else there is no change of segment and the vehicle is located in segment ***s***.

Figure 1 shows a diagram that illustrates the decision algorithm that corroborates or changes the preliminary selection of segment *s*.

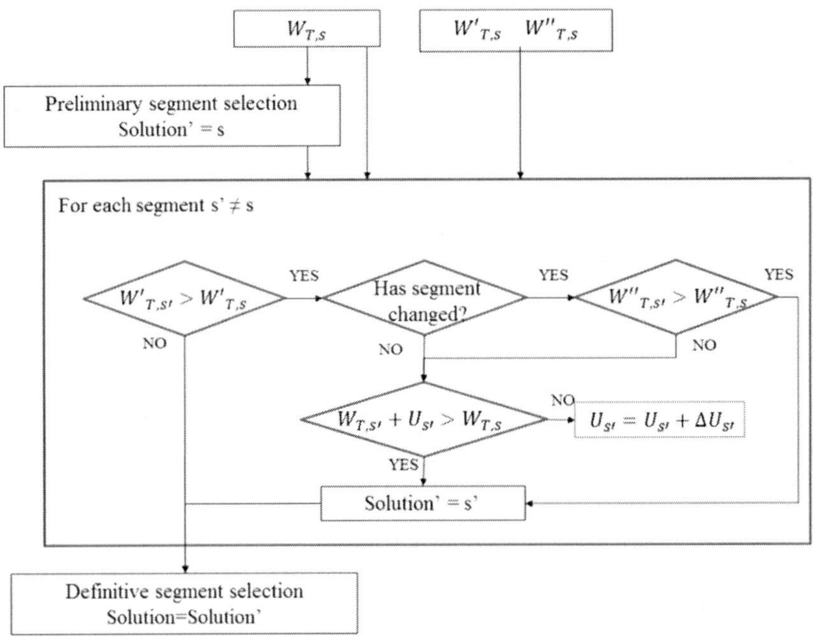

Figure 1. Preliminary segment verification algorithm.

2.2.4. Step 4: Vehicle Positioning on the Digital Map

Finally, once the preferred segment is selected, the positioning of the vehicle is carried out with the following criteria:

- If any of the candidate links of the chosen segment contain the perpendicular projection of the GNSS positioning between their two extremes, then this point is used to locate the vehicle.
- If the condition above does not apply for any link, the nearest point of the digital map to the GNSS-generated point that belongs to the selected segment is assigned to locate the vehicle.

3. PRACTICAL APPLICATION

3.1. Urban Scenario and Digital Map

To test the algorithm's performance, two areas of central Madrid (Spain) were selected where the GNSS coverage is poor, with frequent signal reception losses and signal damage because of the constant change of satellites detected caused by the narrow streets with tall buildings and dense tree cover (Figure 2).

Figure 2. Test areas. (**a**) Route 1; (**b**) Route 2.

The digital map was generated using an instrumented vehicle with the following equipment:

1. Inertial measurement system: (1) Correvit L-CE- non-contact speed sensor to measure speed and the distance travelled; (2)

RMS FES 33 gyroscopic platform to provide measurement of the angles drawn about three axes.

2. Astech G12 GPS autonomous receiver with an update frequency of 10 Hz.

The GPS position fixes of the analysed zones have high uncertainties. Specifically, on route 1, the GPS coverage was active 46.8% of the time, and on route 2 it reached a 71.1% value. Nevertheless, if the wrong positioning received by the GPS receiver, that is, those positions that are too far from the actual location to be considered valid, are eliminated, the percentages of relatively good GPS position fixes decrease to 33.5% and 56.0%, respectively. In both cases, they are low values consistent with other studies [5]. For this reason, the digital map was generated as stated in [2,11]. Nonetheless, in order to reduce the cumulative error of this process, and because the results obtained by the inertial sensor need to be turned and translated [2], this turn is done by steps to minimise the deviations between the generated trajectory and the positioning points obtained by the satellite positioning navigator.

In Figure 3, it can be seen that, for the adjustment that has been made, when the route passes more than once along a certain street, the different inertial sensor coordinates overlap one another quite exactly, and there are no deviations such as those where the different passes show different orientations, as usually occurs because of the cumulative error of the inertial systems [4].

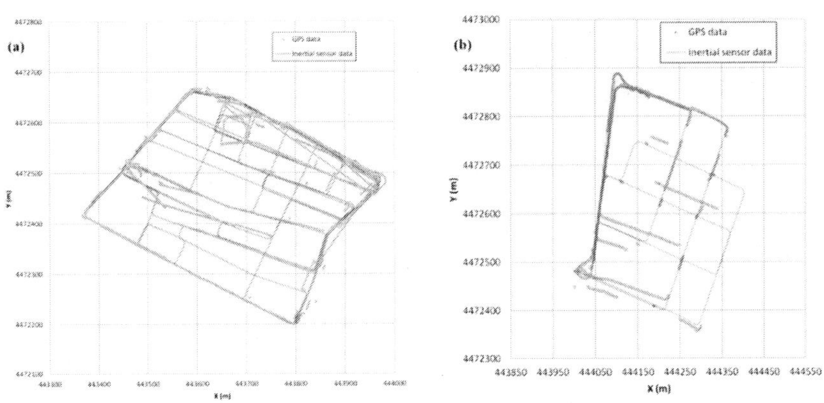

Figure 3. Data adjustment using inertial measurement system and GPS receiver. (**a**) Route 1; (**b**) Route 2.

Finally, Figure 4 shows the points of the digital map stored with which the map-matching algorithm will be evaluated. In order to test the robustness of the algorithm, intersections in the digital map include some missing data so that poor information in these critical sections is consciously provided. On the other street stretches, points of the digital map are separated by 2 m.

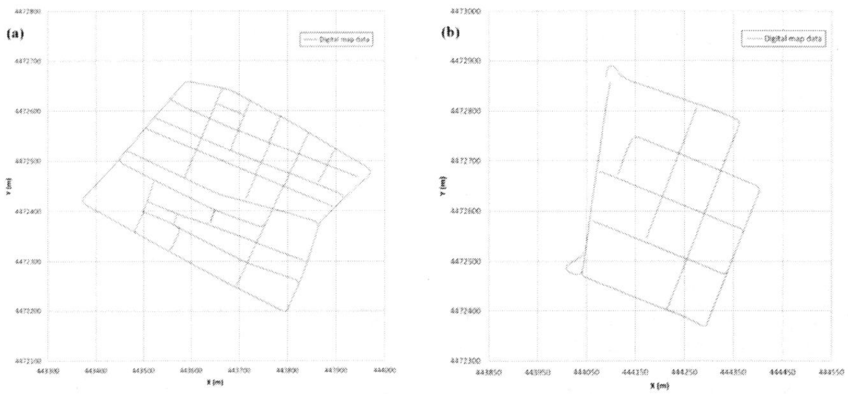

(a) (b)

Figure 4. Digital maps for the test areas. (**a**) Route 1; (**b**) Route 2.

3.2. Results of the Map Matching Algorithm

In order to analyse the performance of the developed algorithm, some tests were carried out in the previously mentioned areas, going along a different route from the one used to create the digital map, to avoid any correlation between them. Besides, the tests were carried out at different times and on different dates from those used to generate the map. The equipment used in the test was the same Astech G12 GPS receiver and additionally the speed of the vehicle was obtained from the vehicle internal communication bus, because its measurement did not need to be too accurate as it was only used to distinguish when the vehicle was moving and when not. The confidence region for this GPS receiver was fixed in a circumference whose radius is 12 m. This value was obtained after completing tests with the receiver in different static locations over a long period of time and analysing the dispersion of the positioning points. The static locations included obstacle-free areas and areas with some high obstacles in the nearby and densely covered areas. The defined circumference comprises more that 75% valid points. Figure 5 shows the results of one of the examples during a test of the position points obtained from the GPS receiver, along with the digital map points and the positioning of the vehicle on it using the map-matching algorithm.

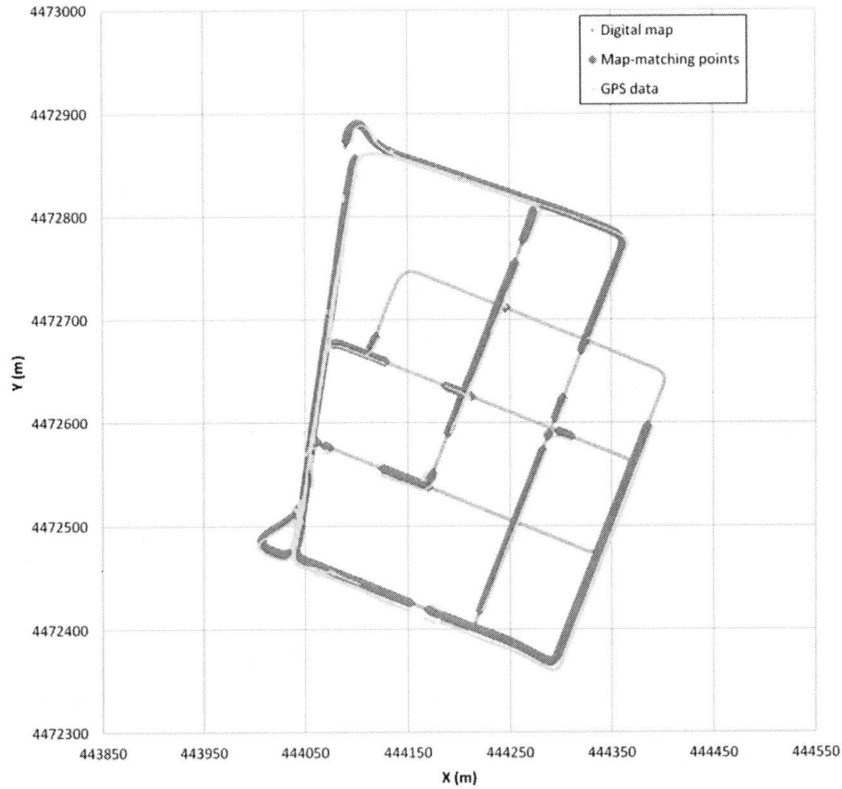

Figure 5. Example of the map-matching algorithm performance.

Link identification is considered to be correct when the algorithm assigns a certain GPS positioning to its correct segment on the map. With this consideration it has been observed that the algorithm correct link identification percentage ranges from 98.1% to 99.3%, the higher value being achieved when the speed information is taken as an input. These results are quite satisfactory and are among the ones obtained by the map-matching algorithms developed in [30,32], among others. Unlike these other algorithms, no information is required from inertial sensors (the speed is only used to detect when the vehicle is not moving) and it is quick enough to be executed in real-time, and also, it is possible to have high control over its performance by changing the parameters and weights defined along the algorithm. Additionally, the causes of the different errors have been analysed.

Table 2 shows the results for tests on the two test areas and the kind of errors detected. The causes of the errors are quite specific and their frequency very low. Besides, the time needed to restore the correct positioning is quite short. The main errors that appear can be divided into three different types:

- Type I: The first error (Figure 6) occurs when a junction is crossed and a legal turn is made instead of going straight. Since the priority link at this scenario is, in general, the one that involves going straight (hence, getting a higher belonging value) if the positioning generated by the GPS is not quite accurate, it could occur that for a few GPS points the algorithm locates the vehicle in an incorrect link (the one that goes straight). However, when the direction of the GPS points is parallel to the segment where the vehicle actually is, and when the vehicle moves away from the junction, the algorithm changes and locates the vehicle in the correct link. This error is quite unusual and the algorithm restores the correct positioning in a short period of time.
- Type II: The second error (Figure 7) occurs when the GPS coverage is lost for a period of time and gets back near a junction and the algorithm decides incorrectly which segment is the good one after the vehicle turn. In this case, the map-matching process is performed as if it were the initial point. In some cases, this causes a mismatching, although after a few seconds the algorithm corrects its mistake locating the vehicle in the correct link. After losing the GPS signal, if the vehicle is in another segment when the signal is recovered and the belonging value does not decrease its weight, this could make the algorithm mismatch the position of the vehicle. After GPS positioning is recovered, the algorithm locates the vehicle in the incorrect segment because it is a priority link (taking into account the segment where the vehicle was located before) and after a few points it changes to the other segment, which is obviously where the car is. This error is quite unusual and the algorithm restores the correct positioning in a short period of time.

Table 3. Map-matching algorithm results.

	Test Area 1	Test Area 2	
Test duration (s)	1060.0	1050.1	
GPS signal availability (%)	61.2%	61.1%	
–	# error appears (without speed information / with speed information)	# error appears (without speed information / with speed information)	longest duration (s)
Error type I	4 / 4	2 / 2	2.1 / 2.1
Error type II	3 / 3	2 / 2	1.6 / 1.6
Error type III	7 / 2	3 / 2	11.8 / 3.0

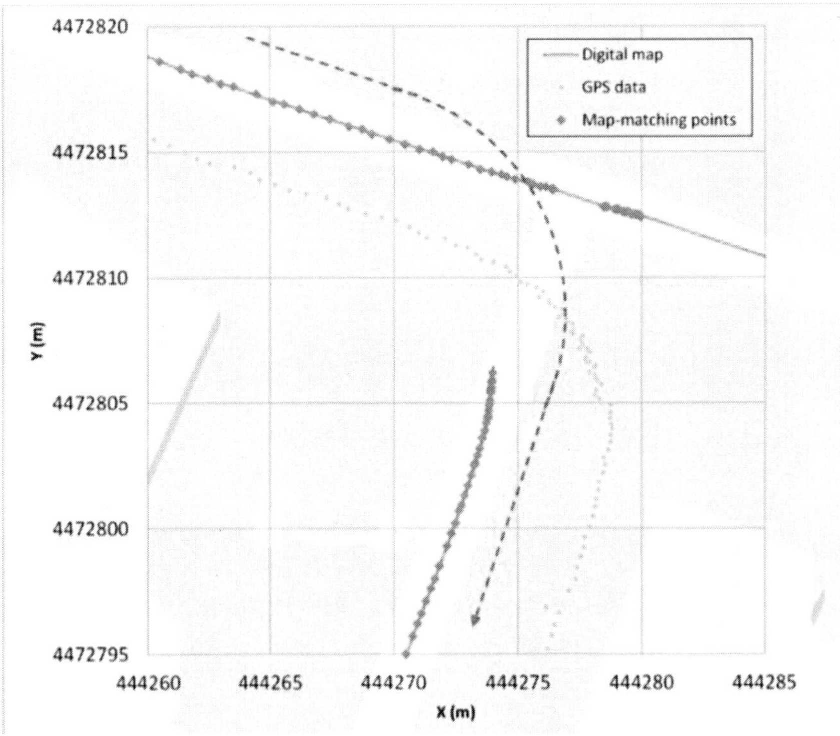

Figure 6. Type I error example.

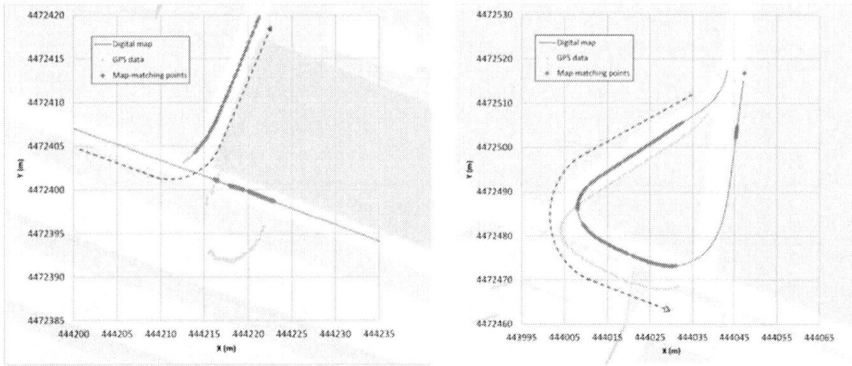

(a) (b)

Figure 7. Type II error examples. (**a**) T-junction; (**b**) V-junction.

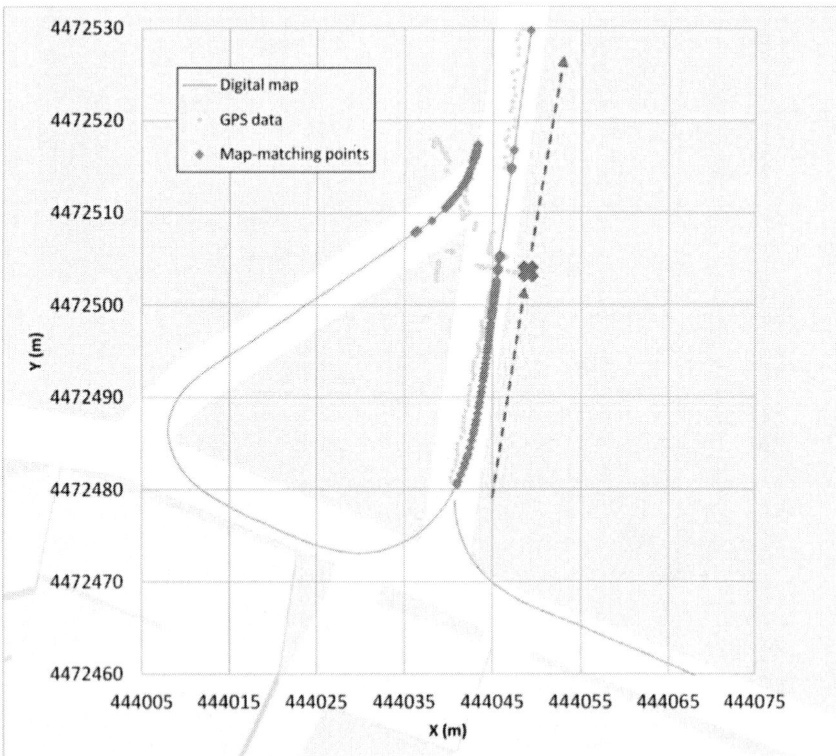

Figure 8. Type III error example.

- Type III: The third error (Figure 8) occurs when the vehicle is not moving. In these cases the GPS position fixes could appear scattered and if the vehicle is near a junction, they could get closer to another link than the one in which the vehicle is stopped, eventually changing its matching position. Also, if the stop takes a long time, when the vehicle starts to move again the algorithm is not yet considering the possibility of relocating the vehicle (this only happens for a few seconds after a change of segment is made) so it will take more time to relocate the vehicle correctly than in the other errors. It is important to point out that this kind of error can be solved by taking into account the speed information, even if it is not too accurate. It should also be pointed out that the 11.8 s error that appears in Table 3 does not occur when the vehicle is moving but is solved when the vehicle starts moving, but it takes more time than in the other errors to relocate the vehicle (around 3 s).

To conclude, some characteristics regarding the performance of the algorithm could be pointed out:

- The algorithm does not mismatch the vehicle's position when the GPS inputs have large errors in positioning.
- After a change of segment when locating the vehicle, the algorithm considers the possibility of being mistaken and analyses other solutions for a few points after the change is made. Therefore, this allows it to relocate the vehicle if necessary.
- Knowing that the vehicle is stopped could significantly improve the results, even without knowing the speed of the vehicle with high accuracy. This allows the algorithm to eliminate the error that can be introduced by the spreading of the GPS signal, which in bad conditions of GPS coverage could be quite high.
- The perpendicular distance value penalizes those digital maps with a high density of points, that is, with little distance between them, as in the map used in this research, improving the results if this distance is higher, which is more usual. However, the algorithm proved to be robust even in these unpromising circumstances.

4. CONCLUSIONS

A map-matching algorithm that does not require inertial sensors and performs well even in areas with low GNSS coverage is put forward in this research. It has been observed that the errors that appear have a repetitive pattern and that these errors are corrected in a few seconds, which means the user does not notice them. Also, it can solve a common problem that navigator systems have, in which once the system has decided to locate the vehicle in a link, it takes time for it to relocate the vehicle in the event of a mismatch. After a change of segment, this new algorithm considers the possible alternatives that existed before the change, relocating the vehicle if necessary, when the results of the weights seem more coherent for the vehicle to be located in another link. This makes the correction process quicker than with other algorithms.

The algorithm presents limitations in complex scenarios with Y-junctions and frequent signal reception losses that sometimes make it mismatch the vehicle position. Nevertheless, the results obtained are very satisfactory and the corrections, in the event of there being an error, are carried out quickly, thus, avoiding the need to use inertial sensors.

The map-matching algorithm is also valid for commercial digital maps. Considering the analysis of the most frequent errors detected, their main cause is the degradation of the GNSS signal while the digital map accuracy is responsible only for a reduced set of incorrect locations. Obviously, in the event that the map has worse accuracy, vehicle positioning could be negatively influenced but the structure of the algorithm could provide better results than other methods that work in real time.

Finally, it should be pointed out that in spite of being appropriate for informative functions, these algorithms are not useful for applications where safety is a critical element. In those kinds of applications more accuracy is needed, both in the digital map data and in positioning the vehicles into them. The map-matching algorithms, as conceived today, will not make sense in these scenarios because there will be more accurate maps and also more accurate satellite positioning, making locating the vehicle easier for the algorithm. Nonetheless, to achieve these values of accuracy in the automotive field is a different technological problem. Therefore, as of today, it is necessary to turn to these map-matching algorithms that try to absorb and minimize the limitations of positioning systems and digital maps.

ACKNOWLEDGMENTS

This work has received the support of the Spanish Ministerio de Economía y Competitividad (TRA2013-48314-C3-2-R) and the Region of Madrid Excellence Network SEGVAUTO-TRIES.

AUTHOR CONTRIBUTIONS

Felipe Jimenez is the project's main researcher and proposed the system. Sergio Monzón implemented the map matching system algorithms. José E. Naranjo was in charge of coordinating the field tests.

REFERENCES

1. Jiménez, F.; Naranjo, J.E. Nuevos requerimientos de precisión en el posicionamiento de vehículos para aplicaciones ADAS. *Dyna* **2009**, *84*, 245–250.
2. Jiménez, F. Improvements in road geometry measurement using inertial measurement systems in datalog vehicles. *Measurement* **2011**, *44*, 102–112.
3. Naranjo, J.E.; Jiménez, F.; Aparicio, F.; Zato, J. GPS and inertial systems for high precision positioning on motorways. *J. Navig.* **2009**, *62*, 351–363.
4. Jiménez, F.; Naranjo, J.E.; García, F.; Armingol, J.M. Can low-cost road vehicles positioning systems fulfil accuracy specifications of new ADAS applications? *J. Navig.* **2011**, *64*, 251–264.
5. Chao, J.; Chen, Y.Q.; Chen, W.; Ding, X.; Li, Z.; Wang, N.; Yu, M. An experimental investigation into the performance of GPS based vehicle positioning in very dense urban areas. *J. Geospatial Eng.* **2001**, *3*, 59–66.
6. Groves, P.D.; Jiang, Z.; Wang, L.; Ziebart, M.K. Intelligent Urban Positioning using Multi Constellation GNSS with 3D Mapping and NLOS Signal Detection. In Proceedings of the 25th International Technical Meeting of the Satellite Division of The Institute of Navigation, Nashville, TN, USA, 17–21 September 2012.

7. Groves, P.D.; Jiang, Z.; Wang, L.; Ziebart, M.K. Intelligent urban positioning, shadow matching and non-line-of-sight signal detection. In Proceedings of the 2012 6th ESA Workshop on Satellite Navigation Technologies and European Workshop on GNSS Signals and Signal Processing (NAVITEC), Noordwijk, The Netherlands, 5–7 December 2012.

8. Bendafi, H.; Hummelsheim, K.; Sabel, H.; Van de Ven, S. *Classification of Data Capturing/Production Techniques. NextMap Project Deliverable D 3.1*; NextMap Consortium: Brussels, Belgium, 2000.

9. Miles, J.C.; Chen, K. *ITS Handbook*, 2nd ed.; PIARC: Paris, France, 2004.

10. Yerpez, J.; Ferrandez, F. *Road Characteristics and Safety. Identification of the Part Played by Road Factors in Accident Generation*; INRETS: Arcueil, France, 1986.

11. Jiménez, F.; Aparicio, F.; Estrada, G. Measurement uncertainty determination and curve fitting algorithms for development of accurate digital maps for Advanced Driver Assistance Systems. *Transport. Res. C Emerg.* **2009**, *17*, 225–239.

12. Hashemi, M.; Karimi, H.A. A critical review of real-time map-matching algorithms: Current issues and future directions. *Comput. Environ. Urban Syst.* **2014**, *48*, 153–165.

13. Blazquez, C.A.; Miranda, P.A. A Real Time Topological Map Matching Methodology for GPS/GIS-Based Travel Behavior Studies. In *Mobile Technologies for Activity-Travel Data Collection and Analysis*; Rasouli, S., Timmermans, H., Eds.; Information Science Reference: Hershey, PA, USA, 2014; pp. 152–170.

14. Wolf, J.; Bachman, W.; Oliveira, M.S.; Auld, J.; Mohammadian, A.; Vovsha, P. *Applying GPS Data to Understand Travel Behavior. Transportation Research Board*; National Cooperative Highway Research Program, Report 775: Washington, WA, USA, 2014.

15. Syed, S.; Cannon, M.E. Fuzzy logic based-map matching algorithms for vehicle navigation system in urban canyons. In Proceedings of the ION National Technical Meeting, San Diego, CA, USA, 26–28 January 2004.

16. Quddus, M.A.; Ochieng, W.Y.; Noland, R.B. Current map-matching algorithms for transport applications: State-of-the art and future research directions. *Transport. Res. C Emerg.* **2007**, *15*, 312–328.

17. Kim, J.S.; Lee, J.H.; Kang, T.H.; Lee, W.Y.; Kim, Y.G. Node based map-matching algorithm for car navigation system. In Proceedings of the 29th ISATA Symposium, Florence, Italy, 3–6 June 1996.

18. Bernstein, D.; Kornhauser, A. *An Introduction to Map-Matching For Personal Navigation Assistants*; New Jersey TIDE Center: Newark, NJ, USA, 1996.

19. White, C.E.; Bernstein, D.; Kornhauser, A.L. Some map-matching algorithms for personal navigation assistants. *Transport. Res. C Emerg.* **2000**, *8*, 91–108.

20. Phuyal, B. Method and use of aggregated dead reckoning sensor and GPS data for map-matching. In Proceedings of the Institute of Navigation (ION) Annual Conference, Portland, OR, USA, 20–27 September 2002.

21. Greenfeld, J.S. Matching GPS observations to locations on a digital map. In Proceedings of the 81st Annual Meeting of the Transportation Research Board, Washington, WA, USA, 13–17 January 2002.

22. Chen, W.; Yu, M.; Li, Z.; Chen, Y.Q. Integrated vehicle navigation system for urban applications. In Proceedings of the 7th International Conference on Global Navigation Satellite Systems (GNSS), Graz, Austria, 22–25 April 2003.

23. Quddus, M.A.; Ochieng, W.Y.; Zhao, L.; Noland, R.B. A general map-matching algorithm for transport telematics applications. *GPS Solut.* **2003**, *7*, 157–167.

24. Blazquez, C.A.; Vonderohe, A.P. Simple map-matching algorithm applied to intelligent winter maintenance vehicle data. *Transport. Res. Rec.* **2005**, *1935*, 68–76.

25. Honey, S.K.; Zavoli, W.B.; Milnes, K.A.; Phillips, A.C.; White, M.S.; Loughmiller, G.E. Vehicle Navigational System and Method. U.S. Patent No. 4,796,191, 3 January 1989.

26. Ochieng, W.Y.; Quddus, M.A.; Noland, R.B. Map-matching in complex urban road networks. *Rev. Bras. Cartogr.* **2004**, *55*, 1–18.

27. Zhao, L.; Ochieng, W.Y.; Quddus, M.A.; Noland, R.B. An extended Kalman filter algorithm for integrating GPS and low-cost dead reckoning system data for vehicle performance and emissions monitoring. *J. Navig.* **2003**, *56*, 257–275.

28. Kim, W.; Jee, G.; Lee, J. Efficient use of digital road map in various positioning for ITS. In Proceedings of the IEEE Symposium on Position Location and Navigation, San Diego, CA, USA, 13–16 March 2000.

29. Gustafsson, F.; Gunnarsson, F.; Bergman, N.; Forssell, U.; Jansson, J.; Karlsson, R.; Nordlund, P. Particle filters for positioning, navigation, and tracking. *IEEE Trans. Signal Process.* **2002**, *50*, 425–435.

30. Quddus, M.A.; Noland, R.B.; Ochieng, W.Y. A high accuracy fuzzy logic-based map-matching algorithm for road transport. *J. Intell. Transport. Syst.* **2006**, *10*, 103–115.

31. Velaga, N.R.; Quddus, M.A.; Bristow, A.L. Developing an enhanced weight-based topological map-matching algorithm for intelligent transport systems. *Transport. Res. C Emerg.* **2009**, *17*, 672–683.

32. Li, L.; Quddus, M.; Zhao, L. High accuracy tightly-coupled integrity monitoring algorithm for map-matching. *Transport. Res. C Emerg.* **2013**, *36*, 13–26.

33. Quddus, M.A.; Ochieng, W.Y.; Noland, R.B. Integrity of map-matching algorithms. *Transport. Res. C Emerg.* **2006**, *14*, 283–302.

4

CHAPTER

FIRST POLARIMETRIC GNSS-R MEASUREMENTS FROM A STRATOSPHERIC FLIGHT OVER BOREAL FORESTS

Hugo Carreno-Luengo*, Adriá Amèzaga, David Vidal, Roger Olivé, Juan Fran Munoz and Adriano Camps

Barcelona Tech Remote Sensing Laboratory and IEEC/UPC, Universitat Politècnica de Catalunya, UPC Campus Nord, D3, Barcelona 08034, Spain

ABSTRACT

The first-ever dual-frequency multi-constellation Global Navigation Satellite Systems Reflectometry (GNSS-R) polarimetric measurements over boreal forests and lakes from the stratosphere are presented. Data were collected during the Swedish National Space Board (SNSB)/European Space Agency (ESA) sponsored Balloon Experiments for University Students (BEXUS) 19 stratospheric balloon experiment using the P(Y) and C/A ReflectOmeter (PYCARO) instrument operated in closed-loop mode. Maps of the polarimetric ratio for L1 and L2 Global Positioning System (GPS) and GLObal Navigation Satellite System (GLONASS), and for E1 Galileo signals are derived from the float phase at 27,000 m height, and the specular points are geolocalized on the Earth's surface. Polarimetric ratio (Γ_{rl}/Γ_{rr}) maps over boreal forests are shown to be in the range 2–16 dB for the different GNSS codes. This result suggests that the scattering is taking place not only over the soil, but over the different forests elements as well. Additionally to the interpretation of the experimental results a theoretical investigation of the different contributions to the total reflectivity over boreal forests is performed using a bistatic scattering model. The simulated cross- (reflected Left Hand Circular Polarization LHCP) and co-polar (reflected Right Hand Circular Polarization RHCP) reflectivities are evaluated for the soil, the canopy, and the canopy–soil interactions for three different biomass densities: 725 trees/ha, 150 trees/ha and 72 trees/ha. For elevation angles larger than the Brewster angle, it is found that the cross-polar signal is dominant when just single reflections over the forests are evaluated, while in the case of multiple reflections the co-polar signal becomes the largest one.

Keywords: GNSS-R; stratospheric balloon; boreal forests; lakes

1. INTRODUCTION

Scientific applications of GNSS-R for Earth Observation include mesoscale ocean altimetry [1], wind speed measurements [2], ice altimetry [3], and soil moisture and vegetation determination [4]. There are different data acquisition techniques [5]: Conventional GNSS-R (cGNSS-R) [6], interferometric GNSS-R (iGNSS-R) [7], and reconstructed-code GNSS-R (rGNSS-R) [8]. At present, data collected by the Surrey Satellite Technology Ltd SSTL's TechDemoSat-1 mission [9] are being processing by the community. Scattering over ocean is found to be highly diffuse, while over land and cryosphere surface targets the reflection becomes specular [10]. These studies consolidate the pioneer results of the United Kingdom Disaster Monitoring Constellation (UK-DMC) and push forward the popularity of GNSS-R in the Remote Sensing community. In the near future several new space-borne missions will include GNSS-R payloads: National Aeronautics and Space Administration (NASA) Cyclone Global Navigation Satellite System (GyGNSS) mission [11], ESA's GNSS rEflectometry, Radio Occultation and Scatterometry experiment on-board the International Space Station (GEROS-ISS) [12], ESA's Passive Reflectometry and Interferometry System In-Orbit Demonstrator (PARIS-IoD) [13], and UPC's ^{3}Cat-2 6U CubeSat [14].

The use of GNSS-R polarimetry over land was first proposed in 2000 for soil moisture monitoring [15]. Later, theoretical simulations were carried out to evaluate the performance of GNSS-R polarimetric measurements for biomass monitoring [16]. Coherent and incoherent scattering were considered in the simulations. In particular, the coherent electromagnetic field was modeled as the reflection of the GNSS signals over the soil, attenuated by the vegetation above it. In this work [16], it was stated that the coherent component of the co-polar reflected signal is 30 dB lower than the cross-polar one. The incoherent component is dominant for co-polar signatures and for a biomass density larger than 50 ton/ha, while the coherent component is the highest for cross-polar signals up to 200 t/ha. Later, an experimental study [17] showed that the co-polar coherent reflectivity is roughly constant for biomass densities from 100 t/ha to 350 t/ha and for an elevation angle in the range θ_e = (50°, 80°). On the other side, the cross-polar component is shown to be reduced to approximately 5 dB. Recently, a different approach has proposed that the forward scattering coefficient is governed by the scattering properties of the vegetation elements and the soil surface, as well as by the interaction between the canopy and the soil, and the soil with the trunks [18]. In this study, the total cross- and the co-polar scattering coefficients are shown to be, respectively, approximate −8 to 8 dB, and approximate −2 to −15 dB for elevation angles in the range θ_e = (50°, 80°).

This work presents the first-ever measurements of polarimetric GNSS-R signatures using data from a stratospheric balloon experiment. Section 2 describes the experimental set-up. Section 3 presents the experimental results obtained with the PYCARO instrument and their interpretation. Section 4 provides final discussions, and conclusions are included in Section 5. Additionally, theoretical simulations of the cross- and co-polar reflectivities are presented in the appendix. The EMISVEG simulator [19,20] uses a bistatic scattering model to evaluate separately the forward scattered field over the soil, and the different forest's trees elements.

2. EXPERIMENTAL SET-UP AND SCENARIO

The ESA BEXUS 19 stratospheric flight (Figure 1) took place on 8 October 2014, at 6:00 p.m. (GPS Time), the experiment duration was 4 hours, the maximum flight height was ~27,000 m, and the trajectory was a single track from Esrange Space Center (latitude 67°53′N, longitude 21°04′E) to Finland (latitude 68°04′N, longitude 25°81′E) over lakes and boreal forest, which is characterized by coniferous forests consisting mostly of pines, spruces, and larches with needle-shaped leaves. Note that biomass of boreal forests is high, but the plant species diversity is less than that in tropical wet forests. The structure of these forests is quite homogeneous which allows the use of allometric equations to characterize the biomass. An updated version of the PYCARO reflectometer [8] including GLONASS and Galileo correlation channels was used. Additionally, both dual-band (L1, L2) and dual-polarization (RHCP, LHCP) zenith-looking antenna patch to collect the direct GNSS signals and nadir-looking antenna array to collect the Earth-reflected signals (Figure 1) were installed in the gondola. The nadir-looking antenna was composed of 6 elementary antenna patches. The total gain of the array was 12.9 dB at L1-LHCP, 13.3 dB at L1-RHCP, 11.6 dB at L2-LHCP and 11.6 dB at L2-RHCP. The On-Board Data Handling (OBDH) subsystem was composed of a Commercial Off The Shelf (COTS) microcontroller for housekeeping and scientific data management, communications with the ground station, and data storage in a micro-Secure Digital (SD) card. The collected data were registered on-board, and simultaneously they were sent to the ground station using the E-Link system [21]. Additionally, an active thermal control was used because the outdoor temperature was so cold down to −70 °C.

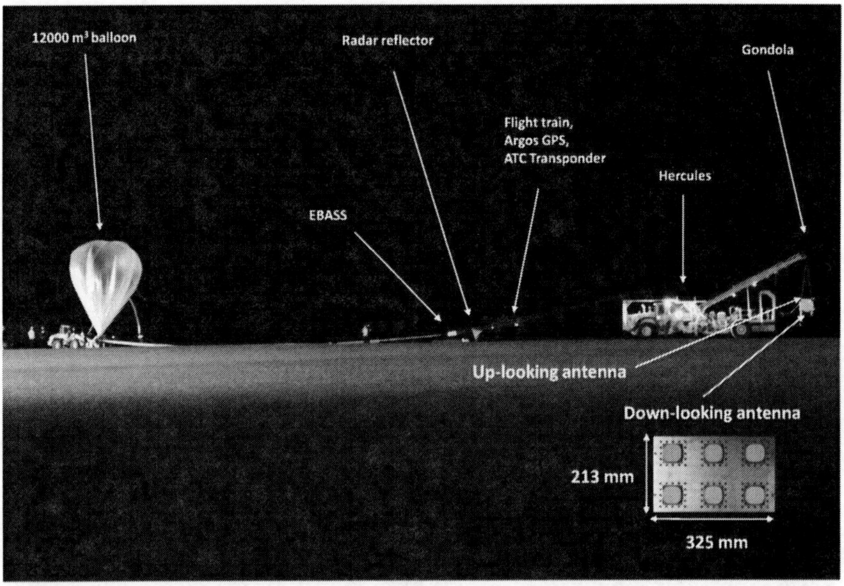

Figure 1. Configuration of the BEXUS 19 stratospheric balloon at Esrange Space Center: 12,000 m^3 balloon, valve, cutter, parachute, Esrange Balloon Service System (EBASS), flight train, Argos GPS and Air Traffic Control (ATC) Transponder (AGT), strobe light, radar reflector, and gondola. Total length of the system is up to 75 m [21]. Photo credits: European Space Agency (ESA).

3. RESULTS OF A STRATOSPHERIC BALLOON EXPERIMENT OVER BOREAL FORESTS

The PYCARO reflectometer was operated in closed-loop mode during this experiment, with delay and phase tracking loops activated. It uses cGNSS-R technique to process the open-access codes, and the rGNSS-R one for the encrypted codes. The polarimetric study is performed using two different observables: the polarimetric ratio [15] and the polarimetric phase [22] using the measurements provided by the phase tracking loop. The polarization ratio is more sensitive to soil dielectric properties and can cancels roughness effects although it does not do so perfectly for arbitrary scattering media. It is defined as the ratio of the cross-over the co-polar reflectivities Γrl/Γrr, and it is estimated as the ratio of the peak of the reflected signal power waveform at LHCP $||\langle Y^{l}_{reflected,Peak}\rangle||^2$ over the peak of the reflected signal power waveform at RHCP $||\langle Y^{r}_{reflected,Peak}\rangle||^2$, after proper compensation of the noise power floor and the antenna radiation pattern:

$$\frac{\Gamma_{rl}}{\Gamma_{rr}} = \frac{\left|\left\langle Y^{l}_{reflected,Peak}\right\rangle\right|^2}{\left|\left\langle Y^{r}_{reflected,Peak}\right\rangle\right|^2} \quad (1)$$

where l and r denote Left and Right Hand Circular Polarization, respectively. Note that the noise power floor of the reflected waveforms $\langle|Y_{noise}|\rangle^2$has to be subtracted to the reflected power waveform $\langle|Y|^2\rangle$ to obtain the power of the signal itself $\langle|Y_{signal}|^2\rangle$. Additionally, the different gain of the antenna array at LHCP (12.9 dB at L1-LHCP and 11.6 at L2-LHCP) and RHCP (13.3 dB at L1-RHCP and 11.6 dB at L2-RHCP) have to be compensated by subtracting the antenna gain to the reflected signal power. Once the effect of the noise and the antenna have been compensated, the incoherent power component is omitted in the reflectivity, as obtained using Equation (1), by subtracting to each incoherently averaged waveform's peak $\langle|Y_{reflected,Peak}|^2\rangle$ the amplitude variance of the complex waveforms' peaks $\sigma^2||Y_{reflected,Peak}||$ [23]:

$$\left|\left\langle Y_{reflected,Peak}\right\rangle\right|^2 = \left\langle\left|Y_{reflected,Peak}\right|^2\right\rangle - \sigma^2_{|Y_{reflected,Peak}|} \quad (2)$$

Thus, reflectivity values are associated to the first Fresnel zone. The semi-major axis of the first Fresnel zone during the float phase of the flight and for an elevation angle $\theta_e = 70°$ is: GPS L1 Coarse/Acquisition C/A (74 m), GPS L2 Precise/Secure P(Y) (83 m), GPS L1 P(Y) (74 m), GPS L2 Civilian C (83 m), GLONASS L1 C/A (74 m), GLONASS L2 C/A (83 m), GLONASS L2 Precise P (83 m) and Galileo E1 BC (74 m). Additionally, the difference of the unwrapped phases of the complex waveforms peak at LHCP Ψln and RHCP Ψrn can be used. This phase has two terms; one induced during the scattering process, which is roughly constant at high elevation angles (e.g., $\theta_e \geq 60°$), and another one due to the propagation. As compared to the polarimetric ratio the main advantage is that the phase difference between the RHCP and LHCP signals can be modeled independently of the elevation angle [22].

After compensation of the first term (using as a first approximation a flat soil model (e.g., Equations (18) and (19) in [24]) the phase difference $\delta\Psi n$ between the peak amplitude of the LHCP and the RHCP reflected complex waveforms at time t_n can be used to infer the geometric delay difference $\rho geo,n$ as:

$$\rho_{geo,n} = \frac{\lambda \delta \Psi_n}{2\pi} \quad (3)$$

where λ is the signal wavelength. Height differences h_n of the center of phase of the scatterers at LHCP and RHCP are related to the geometric delay difference $\rho geo,n$ as:

$$h_n = \frac{\rho_{geo,n}}{2\sin\theta_e} \quad (4)$$

It is found that the mean of the polarimetric phase corresponding to GPS, GLONASS and Galileo signals over boreal forests is in the range from approximate −1.4 m to −9.6 m, which suggests that the phase center of the reflected signals at LHCP is higher than the one at RHCP, that is the scattering process takes place over the canopy and the soil [18]. The trajectory of the balloon was provided by Swedish Space Corporation (SSC) using a GPS receiver on-board, and small platform height variations were compensated for. Note that the vertical speed of the gondola during the float phase was smaller than 1 m/s, which prevented phase jumps. Single reflections (from RHCP to LHCP) are mainly due to interactions with the upper scatterers on the forests. On the other side, signals collected at RHCP involve multiple scattering soil–leaves and soil–branches (first from RHCP to LHCP and then from LHCP to RHCP). This is the reason that explains that the polarimetric phase has negative values (Equation (4)).

Figure 2 and Figure 3 show the first-ever maps of the polarimetric ratio using dual-band multi-constellation signals over boreal forests and lakes. The polarimetric ratio was provided over lakes and boreal forests to give a more complete information and description of the polarimetric properties in the GNSS-R case for different types of scattering media. In Figure 2 there are two different color scales to show a general overview of the polarimetric ratio and also the sensitivity of the technique over boreal forests as it can be seen in the embedded scales. For this study, the elevation angle of the selected satellites was $\theta_e = 70°$ for GPS and GLONASS, and $\theta_e = 60°$ for Galileo. This selection was made because for lower elevation angles the performance of the technique was degraded due to the high directivity of the down-looking antenna array. The maps correspond to different ground tracks. The polarimetric ratio values at the specular reflection points were geolocated and represented over the Earth's surface using Google Maps for simpler interpretation. The ephemerides as provided by an on-board positioning receiver were used to derive the orbit parameters of the GNSS satellites, while the PYCARO trajectory was measured by the on-board receiver. cGNSS-R was used for data acquisition of GPS L1 C/A (Figure 2a), GPS L2 C (Figure 2b), GLONASS L1 C/A (Figure 3a), GLONASS L2 C/A (Figure 3b), GLONASS L2 P (Figure 3c) and Galileo E1 BC (Figure 3d) signals, while rGNSS-R for GPS L1 P(Y) (Figure 2c) and GPS L2 P(Y) (Figure 2d). The mean polarimetric ratio (PR) for GPS L1 C/A signals is ~8 dB and ~4.2 dB over lakes and boreal forests, respectively (Table 1). Additionally, it is found that for the so-called data-less signal GPS L2 C the ratio is, respectively, ~12.7 dB and ~8.1 dB over lakes and boreal forests. The reason that explains the higher values of PR of GPS L2 C as compared to GPS L1 C/A signals is that the depolarization of the direct signal (Table 2) is higher for L1 C/A than for L2 C signals SNRGPS,L1C/A,l = 13 dB and SNRGPS,L2C,l = 3 dB). The Signal-to-Noise Ratio (SNR) values of the direct signals are higher for L1 than for L2, as can be appreciated in Table 2. This is empirical evidence showing that the degree of depolarization is lower for GPS L2 C signals.

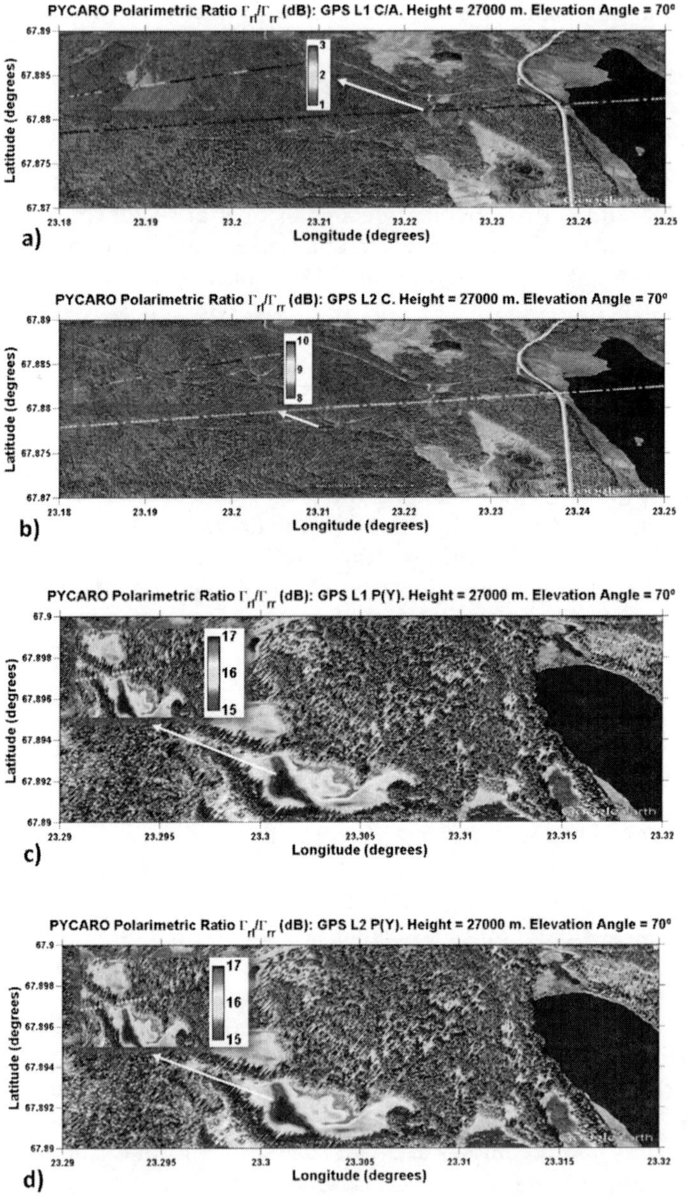

Figure 2. Measured polarimetric ratios for a flight height of 27,000 m and an elevation angle θ_e = 70° for (**a**) GPS L1 C/A, (**b**) GPS L2 C, (**c**) GPS L1 P (Y), (**d**) GPS L2 P (Y).

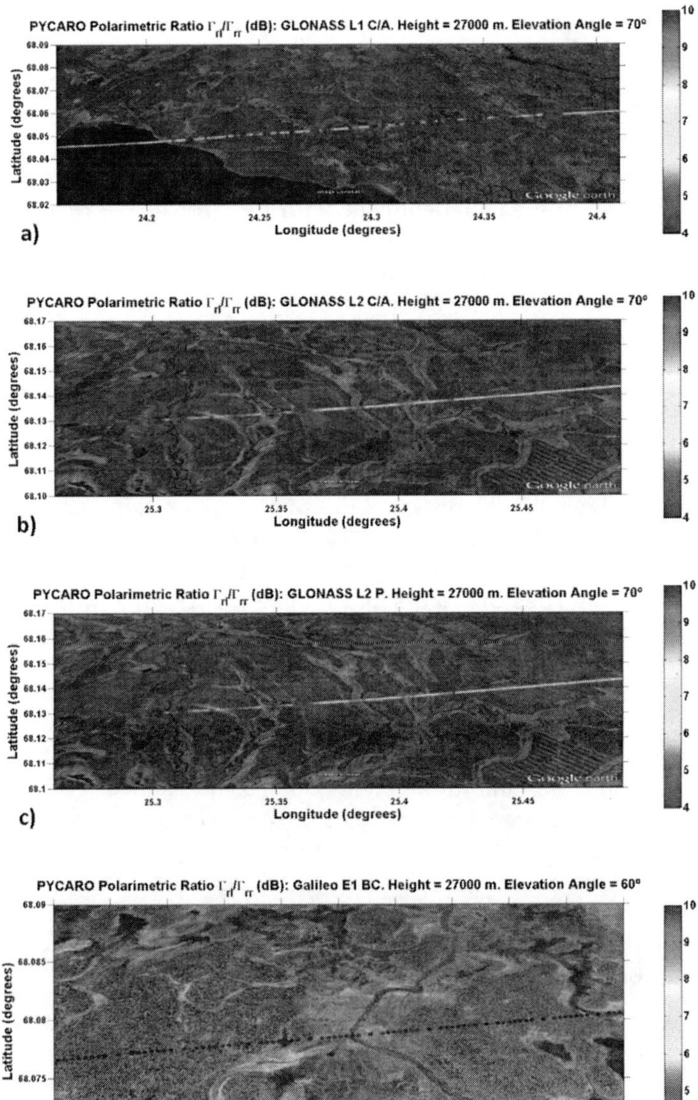

Figure 3. Measured polarimetric ratios for a flight height of 27,000 m and an elevation angle θ_e = 70° for (**a**) GLONASS L1 C/A, (**b**) GLONASS L2 C/A, (**c**) GLONASS L2 P; and for an elevation angle θ_e = 60° for (**d**) Galileo E1 BC.

The rGNSS-R is evaluated successfully for first time over forests, despite the large dispersion of the signal induced by the scattering media. The polarimetric ratio is larger as compared to the GPS L1 C/A signals, with a value ~20.4 dB and ~14.6 dB over lakes and boreal forests, respectively. It is worth pointing out some considerations about the squaring losses of the P (Y) code correlation technique implemented in PYCARO (rGNSS-R). They have a non-linear dependence with the SNR of the received signal [8]: The lower the SNR, the larger the squaring losses [25]. Therefore, the polarimetric ratio as derived from the P (Y) code is higher because of the SNR of the collected RHCP signals is lower than the LHCP ones. Additionally, the direct P (Y) signal depolarization is much lower as compared to the L1 C/A and the L2 C, which also contributes to a higher PR.

Maps of the mean polarimetric ratio for GLONASS L1 C/A (Figure 3a), L2 C/A (Figure 3b) and L2 P (Figure 3c) are also included. Values over lakes are found to be ~8.2 dB for GLONASS L1 C/A signals, in agreement with GPS L1 C/A signals (Table 1 and Table 2) because the corresponding direct signals have a similar level of depolarization, and the lakes scattering properties are the same independently of the track. On the other side, the effect of the different tracks (different forest structure) is manifested in the different values of the polarimetric ratio over forests because the signal depolarization due to the scattering, the propagation and the effect of multiple reflections. However, there were no signal acquisitions over lakes at RHCP for GLONASS L2 C/A and L2 P because the direct signal is found to be highly polarized (Table 2). Table 2 shows that there were no signal acquisitions of the direct GLONASS L2 C/A and L2 P signals at LHCP. This is an indication showing that these signals are highly polarized so that the SNR at LHCP is so low that could not be detected by the PYCARO instrument during the flight. On the other side, the following values are found over boreal forests: $PR_{GLONASS,L1C/A} \sim$ 6.7 dB, $PR_{GLONASS,L2C/A} \sim$ 6.3 dB and $PR_{GLONASS,L2P} \sim$ 6.3 dB.

In this work, authors use the available ground truth data to interpret the results, but unfortunately there are no ground truth data for each track. The PR for GLONASS L2 C/A and L2 P are found to be equal over the same track because the direct signals were no depolarized (Table 2). Finally, it is found that the polarimetric ratio for Galileo E1 BC over boreal forests for a different track and time of signal acquisition is $PR_{Galileo,E1BC} \sim$ 4.1 dB.

Table 1. Mean polarimetric ratio over forests and lakes for GPS (L1 C/A, L2 C, L1 P (Y) and L2 P (Y)), GLONASS (L1 C/A, L2 C/A and L2 P) and Galileo (E1 BC) signals during the float phase of BEXUS 19 flight.

GNSS Code	Elevation Angle	PR (dB) (Forests)	PR (dB) (Lakes)
GPS L1 C/A	$\theta_e \sim 70°$	4.2	8
GPS L2 P(Y)	$\theta_e \sim 70°$	14.6	20.4
GPS L1 P(Y)	$\theta_e \sim 70°$	14.6	20.4
GPS L2 C	$\theta_e \sim 70°$	8.1	12.7
GLONASS L1 C/A	$\theta_e \sim 70°$	6.7	8.2
GLONASS L2 C/A	$\theta_e \sim 70°$	6.3	-
GLONASS L2 P	$\theta_e \sim 70°$	6.3	-
Galileo E1 BC	$\theta_e \sim 60°$	4.1	-

Table 2. Signal-to-Noise Ratio at RHCP y LHCP of the direct GPS (L1 C/A, L2 C, L1 P (Y) and L2 P (Y)), GLONASS (L1 C/A, L2 C/A and L2 P) and Galileo (E1 BC) signals as function of the elevation angle during the float phase of BEXUS 19 flight.

GNSS Code	SNR_r (dB) $\theta_e \sim 70°$	SNR_l (dB) $\theta_e \sim 70°$	SNR_r (dB) $\theta_e \sim 60°$	SNR_l (dB) $\theta_e \sim 60°$	SNR_r (dB) $\theta_e \sim 50°$	SNR_l (dB) $\theta_e \sim 50°$	SNR_r (dB) $\theta_e \sim 40°$	SNR_l (dB) $\theta_e \sim 40°$
GPS L1 C/A	34	13	33	23	32	23	30	23
GPS L2 P(Y)	19	-	16	-	13	-	10	-
GPS L1 P(Y)	19	-	16	-	13	-	10	-
GPS L2 C	28	3	25	3	23	8	21	8
GLONASS L1 C/A	31	12	31	18	29	25	21	15
GLONASS L2 C/A	16	-	16	-	20	-	20	-
GLONASS L2 P	12	-	12	-	16	-	16	-
Galileo E1 BC	15	-	14	-	13	-	11	-

A common characteristic for all the GNSS signals is that the polarimetric ratio is lower over boreal forests (PRGPS,L1C/A ~ 4.2 dB, PRGPS,L2C ~ 8.1 dB, PRGPS,P(Y) ~ 14.6 dB, and PRGLONASS,L1C/A ~ 6.7 dB) than over lakes PRGPS,L1C/A ~ 8 dB, PRGPS,L2C ~ 12.7 dB, PRGPS,P(Y) ~ 20.4 dB, and PRGLONASS,L1C/A ~ 8.2 dB). In this scenario, in addition to depolarization effects due to scattering and propagation through the vegetation, multiple reflections involving canopy-soil and soil-canopy increase the amount of co-polar signals in the final scattered field reaching the receiver.

4. FINAL DISCUSSIONS

The BEXUS 19 stratospheric balloon experiment with an apogee of ~27,000 m has provided a unique opportunity to study for first time the scattering of GNSS signals over boreal forests. Multi-constellation (GPS, GLONASS and Galileo) reflected signals were collected by the PYCARO instrument at dual-band (L1, L2) and dual-polarization (RHCP, LHCP). These measurements provide the first experimental evidence towards the further evaluation of GNSS-R for polarimetric applications. First results show a promising performance of the PYCARO instrument as the payload for the ^{3}Cat-2 mission. The scientific evaluation of this dataset offers the opportunity to evaluate the feasibility of the GNSS-R to perform biomass monitoring, which is a key-factor for analyzing the carbon cycle. The main added value is the measurement of polarimetric signatures, which shows sensitivity over forests and lakes. Additionally, different data acquisition techniques have been used: cGNSS-R for the open-source codes and the novel rGNSS-R for the encrypted P(Y) GPS code. The scattering of the GNSS-R signals takes place over the soil and the canopy but also through multiple reflections involving canopy-soil and soil-branches interactions. This is an important issue that has to be considered to perform biomass monitoring, since the vegetation provides a scattered field additionally to the effect of the attenuation of the signals reflected over the soil. Future work should include a study of: (a) The potential advantages of the synergy between both data access techniques, and (b) scattering over different types of vegetated soils.

5. CONCLUSIONS

The polarimetric ratio and the mean polarimetric phase over boreal forests with a biomass density of ~2,500 trees/ha and for an elevation angle of $\theta_e = 70°$ for GPS and GLONASS and $\theta_e = 60°$ for Galileo vary in the ranges from approximate 2 to 16 dB and from approximate −1.4 to −9.6 m, respectively. This is due to the effect of different tracks, periods of signal acquisition, levels of depolarization of the direct signals and because of the squaring losses of the rGNSS-R. The polarimetric phase is found to be negative, which means that the center of phase of the reflected signals at LHCP is higher in the vertical profile of the forests as compared with RHCP signals. As the main conclusion, GNSS-R has been shown to have sensitivity to perform polarimetric measurements over lakes and boreal forests from a stratospheric balloon flight with an apogee of 27,000 m using dual-band multi-constellation signals.

Additionally, a theoretical investigation of the different contributions to the total reflectivity over boreal forests has been performed and it is included in the appendix. The evolution of the reflectivity as a function of the elevation angle over boreal forests using simulations of the cross- (reflected LHCP) and co-polar (reflected RHCP) scattered fields over soil and canopy, including canopy-soil interactions, for three different levels of biomass density (725 trees/ha, 150 trees/ha and 72 tress/ha) has been analyzed. The total reflectivity is dominated by that of the soil. The canopy (branches, leaves) reflectivity variations for larger density levels are found to have a dependence with the elevation angle, the polarization of the reflected signal and the scatterer type. Cross-polar canopy reflectivity increments are from approximate 10 to 20 dB for elevation angles in the range θ_e = (10°, 80°). On the other side, attenuation due to signal propagation through forests leads to lower reflectivity values over soil as lower is the elevation angle, independently of the polarization. However, the polarization is found to be an important parameter that determines the reflectivity levels of canopy-soil interactions. Increments of canopy-soil cross-polar reflectivity values have a similar trend than those corresponding to soil scattering. Nonetheless, for elevation angles larger than ~30°, the former scattering mechanism shows a higher increment of reflectivity as compared to scattering only over the soil. Additionally, it is found that leaves–soil co-polar reflectivity levels reduces from 10 dB to 0 dB for elevation angles in the range θ_e = (10°, 80°), while for branches–soil interactions is roughly constant around zero.

ACKNOWLEDGMENTS

The authors would like to thanks Mrs. Pons and Khoe from the administration of the Department of Signal Theory and Communications (TSC) of the Universitat Politècnica de Catalunya-Barcelona Tech (UPC) and the REXUS/BEXUS (ESA, DLR, SNSB, SSC, and ZARM) team for their collaboration, help and advice during all the project. Special thanks to Dra. Natacha Callens (ESA) and Mr. Alexander Kinnaird (SSC) for their help during the flight campaign, and to Mr. Sylvain Vey (ESA) for his advice during the thermal analysis.

This project has received funding from the European Union's Seventh Framework Programme for research, technological development and demonstration under grant agreement "European GNSS-R Environmental Monitoring" no FP7-607126-E-GEM (Online Available: http://www.e-gem.eu), and by the project "Aplicaciones avanzadas en radio ocultaciones y dispersometría utilizando señales GNSS y otras señales de oportunidad" of the Spanish Ministerio de Ciencia e Innovación (MICINN), Grant No AYA2011-29183-C02-01.

AUTHOR CONTRIBUTIONS

Hugo Carreno-Luengo and Adriano Camps conceived and designed the experiment; Hugo Carreno-Luengo performed the experiment; Hugo Carreno-Luengo analyzed the data; Hugo Carreno-Luengo performed the theoretical simulations; Hugo Carreno-Luengo wrote the paper.

APPENDIX

Part I: Simulations of the Reflectivity over Boreal Forests

Part I of the appendix provides a theoretical analysis of the reflectivity over boreal forests. The bistatic geometry of GNSS-R is represented in Figure A1a, and a sample 3-D structure of the tree type used in the simulations is represented in Figure A1b. Simulations of the reflectivity at L1-band over forests have been performed with a modified version of EMISVEG (appendix Part II) using signals at Right Hand Circular Polarization (RHCP) and Left Hand Circular Polarization (LHCP). In the considered scattering model, individual tree components located above a tilted dielectric rough plane are illuminated by the electromagnetic field and the scattered fields are computed, and then added coherently, as in [26]. Branches and tree trunks are modeled by stratified dielectric cylinders, arranged following a fractal geometry described by Lindenmayer systems [27]. The leaves are modeled by dielectric needles, and are added to the tree model. This scattering model preserves the phase of scattering fields from scatterers (leaves, branches, and trunks) in the simulation of this coherent signature from the forest. The effect of attenuation and phase change of the coherent wave propagating in the forest media is also taken into account using the Foldy's approximation [28].

The scattering matrix $\overline{\overline{S}}$ was transformed from linear to circular polarization. This matrix relates the components of the scattered field $\overline{E}^{s}$ and the incident field $\overline{E}^{s}$ for all the polarimetric combinations:

$$\begin{bmatrix} E_r^s \\ E_l^s \end{bmatrix} = \frac{e^{jkR_0}}{R_0} \begin{bmatrix} S_{rr} & S_{rl} \\ S_{lr} & S_{ll} \end{bmatrix} \begin{bmatrix} E_r^i \\ E_l^i \end{bmatrix} \tag{A1}$$

where R0 is the distance from the specular point to the receiver, which is a function of the scatterer. Subscripts r and l denote Right Hand Circular Polarization and Left Hand Circular Polarization, respectively.

These simulations provide reflectivity values for the soil, branches (tree trunks are also included into the simulations), leaves and multiple reflections involving leaves-soil and branches-soil interactions, as in [18]. EMISVEG was originally developed for full-polarimetric microwave emission studies at L-band, but it has been now adapted to compute the forward scattering. L-band (1.400–1.427 MHz) data from the SMOS REFLEX 2004 field experiment [29] were used to validate the model. In the frame of this work, it has been modified to evaluate the forward scattering coefficient over boreal forests at circular polarization using the Forward Scattering Alignment (FSA) convention.

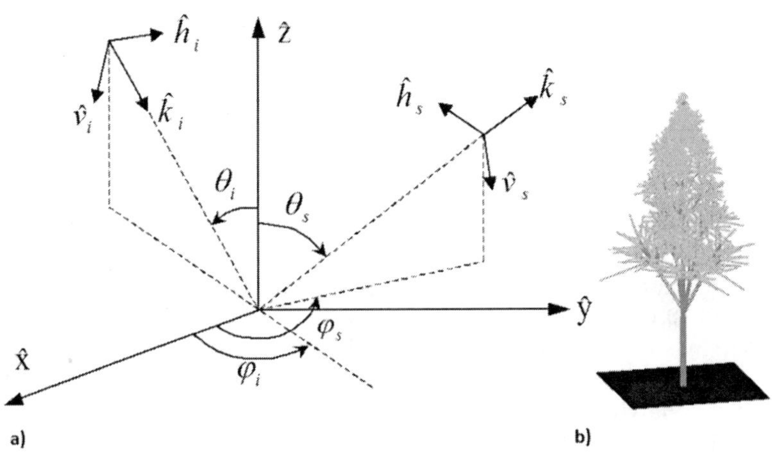

Figure A1. (**a**) Reference system following the Forward Scattering Alignment (FSA) criterion. (**b**) Sample 3-D structure of the tree-type used along the simulations performed to estimate the reflectivity.

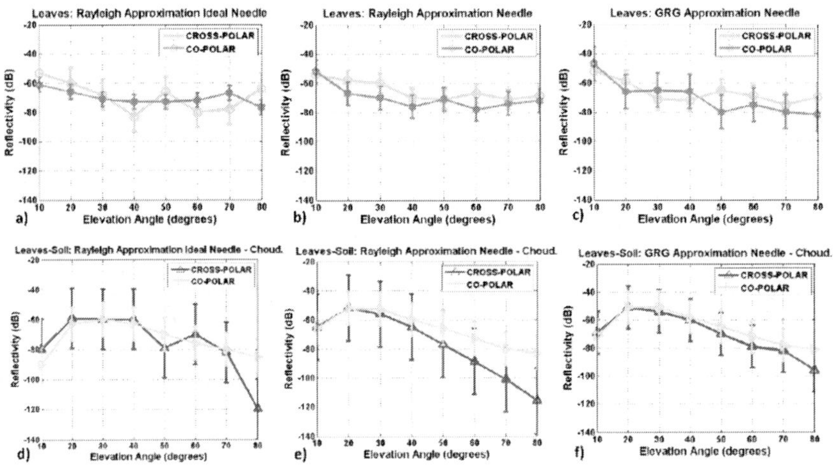

Figure A2. Simulated reflectivity for a biomass density of 725 tress/ha over leaves: (**a**) Rayleigh approximation ideal needle, (**b**) Rayleigh approximation needle, (**c**) Generalized Rayleigh Gans approximation needle; and leaves-soil interactions: (**d**) Rayleigh approximation ideal needle-Choudhury, (**e**) Rayleigh approximation needle-Choudhury, and (**f**) Generalized Rayleigh Gans approximation needle-Choudhury.

The vegetation height was set to be ~20 m (Figure A1b) based on the available ground truth data [30,31], and the scattering area was set to be equal to the first Fresnel zone AFresnel. The attenuation coefficient over canopy at L band is ~1.5 dB/m. A much lower tree density for a biomass density of 100 t/ha (725 trees/ha instead ~2,700 trees/ha as in [32] for a tree height ~20 m) was required to make the simulations feasible. The electromagnetic models selected for the expression of the bistatic scattering matrices were the Semi-Exact solution for the branches S=branches [33], the Generalized Rayleigh-Gans (GRG) approximation for a needle in the case of leaves S=needle [34], and the Choudhury method for the soil S=soil [35]. The Semi-Exact solution was selected because the radii of curvature of the branches of Northern Sweden forests is not an order of magnitude larger than the signals wavelength used in the simulations λL1 = 19 cm). The Physical Optics approximation for higher frequencies provides a fast algorithm when the radius of tree branches is large compared to the wavelength [33]. The GRG approximation for a needle was selected to evaluate the scattering on leaves because: (a) Needles are the best geometrical approximation for leaves in coniferous vegetation, and (b) GRG is valid for a scatterer with at least one of its dimensions small as compared to the signals wavelength [34]. Nonetheless, several simulations were performed to evaluate the differences with other methods: Rayleigh approximation for an ideal needle (Figure A2a), Rayleigh approximation for a needle (Figure A2b), and the GRG approximation for a needle (Figure A2c). Additionally, the leaves–soil scattering mechanism was evaluated using the same scattering methods for the leaves, and the Choudhury method [35] to account for the electromagnetic interaction with the soil (Figures A2d–f). These simulations show that there are several differences: (a) A larger dynamic range of the co-polar reflectivity over leaves using the GRG method (~30 dB) as compared to that obtained with the Rayleigh approximation for an ideal needle (~20 dB); and (b) both the cross- and the co-polar reflectivities for leaves–soil interactions.

The cross-polar values are under-estimated as compared to the GRG-Choudhury method (from ~10 dB at low elevation angles $\theta_e \sim 10°$, to ~20 dB for high elevation angles, $\theta_e \sim 80°$). On the other side, the co-polar reflectivity levels are under-estimated ~20 dB using the Rayleigh approximation for an ideal needle for low elevation angles, $\theta_e \sim 10°$.

After these previous considerations, the cross- and co-polar reflectivity coefficients are analyzed as a function of the elevation angle in the range $\theta_e = (10°, 80°)$ for three different biomass densities 725 trees/ha (Figure A3a, single reflections; and Figure A3d, multiple reflections), 150 trees/ha (Figure A3b, single reflections; and Figure A3e, multiple reflections) and 72 trees/ha (Figure A3c, single reflections; and Figure A3f, multiple reflections). The co-polar component over soil-surface is dominant for low elevation angles up to $\theta_e \sim 30°$, while the cross-polar is the highest component for larger elevation angles. Actually, this corresponds to the Brewster angle, which is a property of the reflector type and indicates the change of polarity of the vertical component of incident electromagnetic field after being reflected [36]. A common characteristic of the full ranges of biomass densities and elevation angles under study is the order of reflectivity levels. First of all, the soil surface (on average ~10 dB more than other contributions), which is followed by reflectivity levels over branches and leaves.

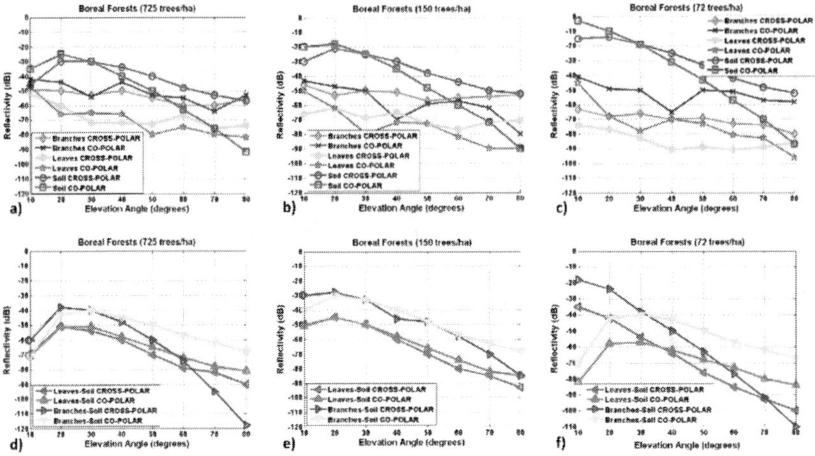

Figure A3. Cross- and co-polar reflectivity simulations over forests: (**a**–**c**) direct scattering (canopy, soil) and (**d**–**f**) multiple scattering (canopy-soil interactions), for a biomass density of (**a**,**d**) 725 trees/ha, (**b**,**e**) 150 trees/ha, and (**c**,**f**) 72 trees/ha.

The study of multiple reflections involving both leaves-soil and soil-leaves, and both branches-soil and soil-branches shows that the cross-polar component is the highest for low elevation angles up to $\theta_e \sim 30°$, while the co-polar one is the highest component for higher elevation angles. This inversed-behavior as compared with single reflections is due the double polarization changes induced by, first from RHCP to LHCP, and then from LHCP to RHCP.

Part II: Theoretical Electromagnetic Model

The presented model has to be validated and it represents a theoretical investigation of the different contributions to total reflectivity. The model should satisfy the following premises to properly simulate the properties of GNSS reflectometry during the experiment using the PYCARO instrument: (a) Realistic 3-D spatial structure of a forest, in which every scatterer has its deterministic location; (b) accurate generation of tree structures based on physical parameters; (c) account for the coherent effects that may exist in different scatterers; (d) account for the scattering contribution from the scatterers in the forest canopy, and also for the direct scattering of the rough ground surface; (e) the scattered fields of adjacent trees in a forest stand are assumed to be uncorrelated; (f) the effect of attenuation and phase change of the coherent wave propagating in the random media; and (g) only the scattering in the forward direction has to be accounted for.

The total scattered field within the first Fresnel zone can be written as a function of the position of the nominal specular point ρ¯ as follows:

$$\bar{E}^{s}(\bar{\rho}(t)) = \bar{E}^{s}_{soil}(\bar{\rho}(t)) + \bar{E}^{s,leaves}_{forest}(\bar{\rho}(t)) + \bar{E}^{s,leaves\_soil}_{forest}(\bar{\rho}(t)) + \bar{E}^{s,soil\_leaves}_{forest}(\bar{\rho}(t)) + \bar{E}^{s,branches}_{forest}(\bar{\rho}(t)) + \bar{E}^{s,branches\_soil}_{forest}(\bar{\rho}(t)) + \bar{E}^{s,soil\_branches}_{forest}(\bar{\rho}(t))$$

(A2)

$\bar{E}^{s}_{soil}$, $\bar{E}^{s,leaves}_{forest}$, $\bar{E}^{s,leaves\_soil}_{forest}$, $\bar{E}^{s,soil\_leaves}_{forest}$, $\bar{E}^{s,branches}_{forest}$, $\bar{E}^{s,branches\_soil}_{forest}$, and $\bar{E}^{s,soil\_branches}_{forest}$

are the total forward scattered fields by soil surface, leaves, multiple interactions involving both soil and leaves, branches, and multiple interactions involving branches and soil, respectively. Since the specular reflection point changes with time, the values of the electric fields in Equation (A2) are also a function of time.The Choudhury method [35] is used to simulate the direct scattering of the GPS signals E¯¯¯ssoil over the ground surface. To the first order of approximation, the scattering from every type of forest element is approximated by the superposition of the scattered field from each scatterer within the tree structure. Hence, neglecting the effect of multiple scattering among the scatterers, the scattered field from a single tree for every type of forest element can be evaluated by [26]:

$$\bar{E}^{s,element} = \frac{e^{jkr}}{r}\sum_{n=1}^{N} e^{j\phi_n}\,\bar{\bar{S}}_n^{interaction}\bar{E}_0^{i,element} \quad \text{(A3)}$$

where $j = \sqrt{-1}$ is the imaginary unit, N is the total number of the scatterers within the tree structure, S=interactionn is the scattering matrix in circular polarization of the nth scatterer above the soil corresponding to a forest element, ϕn is a phase compensation term accounting for the shift of the phase reference from the local coordinate system of the nth scatterer to the global coordinate phase reference, and $\bar{E}_0^{i,element}$ is the amplitude of the incident electromagnetic wave. Denoting the position of the nth scatterer in the global coordinate system by r¯n, ϕn is given by [26]:

$$\phi_n = (\bar{k}_i - \bar{k}_s)\bar{r}_n \tag{A4}$$

where $\bar{k}_i$ and $\bar{k}_s$ are, respectively, the unit vectors representing the propagation direction of the incident and the scattered fields.

The forest element scattered field $\bar{E}_{forest}^{s,element}$ is modeled taking into account the effect of attenuation and phase change of the coherent wave $\bar{E}^{s,element}$ propagating in the random media (boreal forest). Based on Foldy's approximation [28], the variation of the mean field $\bar{E}_{forest}^{s,element}$ with respect to the distance s along the direction $\bar{k}$ is generally governed by:

$$\frac{d\bar{E}_{forest}^{s,element}}{ds} = jK\bar{E}_{forest}^{s,element} \tag{A5}$$

where K is the effective propagation constant. Using eigen-analysis, the differential Equation (A5) is solved and the solution is given by:

$$\bar{E}_{forest}^{s,element} = e^{jk_o s}\bar{\bar{T}}\left(s, \bar{k}\right)\bar{E}_{forest}^{s,element,0} \tag{A6}$$

where $\bar{\bar{T}}$ is the transmissivity matrix accounting for the extinction due to scattering and absorption, and $\bar{E}_{forest}^{s,element,0}$ is the forest scattered field at s=0.

To account for this last term and for the bistatic nature of the GNSS reflectometry, the expressions for the scattering matrix as proposed in [26] in the forward-scatter direction of each different forest element interactions S=interactionn are modeled as follows:

$$\bar{\bar{S}}_n^{element} = \bar{\bar{T}}_n^i \bar{\bar{S}}_n^0(\bar{k}_s, \bar{k}_i)\bar{\bar{T}}_n^i \tag{A7}$$

$$\bar{\bar{S}}_n^{element,soil} = e^{j\tau_n}\bar{\bar{T}}^t\bar{\bar{R}}(\bar{k}_s, \bar{k}_{gs})\bar{\bar{T}}_n^r\bar{\bar{S}}_n^0(\bar{k}_{gs}, \bar{k}_i)\bar{\bar{T}}_n^i \tag{A8}$$

$$\bar{\bar{S}}_n^{soil,element} = e^{j\tau_n}\bar{\bar{T}}_n^i\bar{\bar{S}}_n^0(\bar{k}_s, \bar{k}_{gi})\ \bar{\bar{T}}_n^r\bar{\bar{R}}(\bar{k}_{gi}, \bar{k}_i)\bar{\bar{T}}^t \tag{A9}$$

where $\overline{\overline{S}}_n^{element}$, $\overline{\overline{S}}_n^{element,soil}$, and $\overline{\overline{S}}_n^{soil,element}$ are the direct scattering, the element-soil scattering and the soil-element scattering matrices, respectively. $\overline{\overline{T}}_n^{i}$, $\overline{\overline{T}}_n^{r}$, and $\overline{\overline{T}}^{t}$ are the transmissivity matrices, respectively, for the direct, reflected, and total travelling path, and R= is the reflection matrix of the rough tilted ground plane whose elements are derived as per Wang and Choudhury [37], and:

$$\bar{k}_{gi} = \bar{k}_i - 2\bar{n}_g(\bar{n}_g\bar{k}_i) \quad \text{(A10)}$$

$$\bar{k}_{gs} = \bar{k}_s - 2\bar{n}_g(\bar{n}_g\bar{k}_s) \quad \text{(A11)}$$

$$\tau_i = -2k_0(\bar{r}_n\bar{n}_g)(\vec{n}_g\bar{k}_i) \quad \text{(A12)}$$

$$\tau_s = 2k_0(\bar{r}_n\bar{n}_g)(\bar{n}_g\bar{k}_s) \quad \text{(A13)}$$

where the phase terms τ_i and τ_s account for the extra path lengths of the image excitation and the image scattered waves, respectively, and $\bar{k}_i$, $\bar{k}_s$, $\bar{k}_{gi}$, $\bar{k}_{gs}$ and $\bar{n}_g$ are defined in Figure A4.

The electromagnetic models selected for the expression of the bistatic scattering matrices $\overline{\overline{S}}_n^0$ are the Semiexact solution for the branches [33], and the Generalized Rayleigh-Gans approximation for a needle in the case of leaves [34].

The incident L1 GPS signals after the scattering over the ground surface ((a) in white in Figure A4a) are collected by the receiver. The direct scattering over the canopy attenuates and scatters forward the incident GPS signals without reaching the trunks. The incident GPS wavefront is first scattered by the canopy ((b) in white inFigure A4b). There is a component in the specular direction, but also, a fraction of it penetrates the trunks, reaches the ground ((c) in white in Figure A4c) and is reflected on the ground, and travels up through the upper layers to the receiver ((d) in white in Figure A4c). As a complement of the latter contribution, ground reflection ((e) in white in Figure A4c) followed by specular scattering over the canopy ((f) in white in Figure A4c) is also

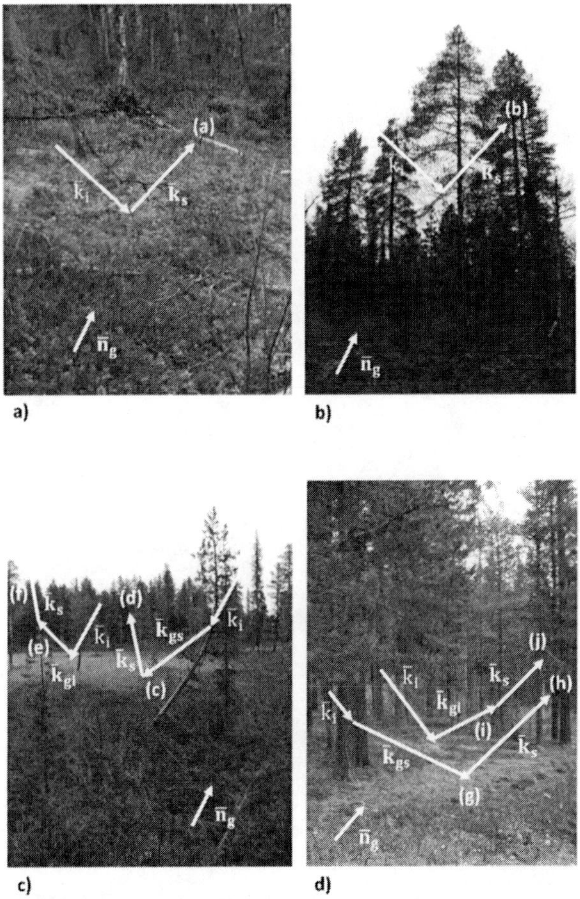

Figure A4. (**a**) Soil scattering. (**b**) Direct scattering over canopy. (**c**) Multiple scattering involving soil and canopy. (**d**) Multiple scattering involving both soil and branches.

considered. The incident wave propagates through the upper layers and is attenuated by them before it scatters over the ground. The upward reflected GPS signals penetrate the trunks and are scattered by the canopy. As the last contribution, specular scattering on the trunks ((g) in white in Figure A4d) followed by ground reflection ((h) in white in Figure A4d) is considered. This mechanism is similar to the canopy–soil one. However, the scattering process occurs in the trunks instead of the canopy, and the canopy acts as an attenuating medium. Ground reflection ((i) in white in Figure A4d) followed by trunk specular scattering ((j) in white in Figure A4d) complements the latter mechanism, being similar to the soil–canopy mechanism.

The statistics of the scattered field are approximated from a Monte-Carlo simulation in which a large number of tree structures are generated using stochastic L-systems [38] and then the scattering matrix of all generated trees are computed. The total forest element scattered field in circular polarization $\bar{E}^{s,element}_{forest}$ is obtained as the mean of 100 Monte-Carlo realizations to average the position of the scatterers. A large number of fractal-generated trees (725, 150 and 72 trees/ha) is considered during the Monte-Carlo simulations to characterize the statistics of the scattered signals. Finally, the reflectivity is defined as [39]:

$$\Gamma = \frac{1}{4\pi \sin\theta_{e_i}} \iint \sigma^0 d\Omega_s = \frac{1}{4\pi \sin\theta_{e_i}} \iint \frac{|\bar{E}^s|^2/A}{|\bar{E}^i|^2/4\pi R_0^2} \cos\theta_{e_s} d\varphi_s d\theta_{e_s} \quad \text{(A14)}$$

where θ_{ei} is the elevation angle of the incident wave, θ_{es} and φs are the elevation and azimuth angles of the scattered wave, and $\sigma 0$ is the forward scattering coefficient, A is the illuminated area, and the integration is carried out over the upper half space. Since around the direction of specular reflection the coherent component is much larger than the incoherent one, the integrand in Equation (A14) tends to a delta function. This allows reducing the integration limits around the specular direction. In our case, after inspection of the integrand, it was concluded that the contributions beyond a 4° × 4° domain had a negligible contribution to the computed reflectivity value.

REFERENCES

1. Martín-Neira, M. A PAssive Reflectometry and Interferometry System (PARIS): Application to ocean altimetry. *ESA J.* **1993**, *17*, 331–355.

2. Garrison, J.L.; Katzberg, S.J.; Hill, M.I. Effect of sea roughness on bistatically scattered range coded signals from the global positioning system. *Geophys. Res. Lett.* **1998**, *25*, 2257–2260.

3. Fabra, F. GNSS-R as A Source of Opportunity for Remote Sensing of the Cryosphere. Ph.D. Thesis, Universitat Politècnica de Catalunya, Barcelona, Spain, 2014.
4. Rodriguez-Alvarez, N. Contributions to Earth Observation Using GNSS-R Opportunity Signals. Ph.D. Thesis, Universitat Politècnica de Catalunya, Barcelona, Spain, 2011.
5. Zavorotny, V.U.; Gleason, S.; Cardellach, E.; Camps, A. Tutorial on remote sensing using GNSS bistatic radar of opportunity. *IEEE Geosci. Remote Sens. Mag.* **2014**, *2*, 8–45.
6. Garrison, J.L.; Komjathy, A.; Zavorotny, V.; Katzberg, S.J. Wind speed measurement using forward scattered GPS signals. *IEEE Trans. Geosci. Remote Sens.* **2002**, *40*, 50–65.
7. Cardellach, E.; Rius, A.; Martín-Neira, M.; Fabra, F.; Ribó, S.; Kainulainen, J.; Camps, A.; D'Addio, S. Consolidating the precision of interferometric GNSS-R ocean altimetry using airborne experimental data.*IEEE Trans. Geosci. Remote Sens.* **2014**, *52*, 4992–5004.
8. Carreno-Luengo, H.; Camps, A.; Ramos-Pérez, I.; Rius, A. Experimental evaluation of GNSS-reflectometry altimetric precision using the P (Y) and C/A signals. *IEEE Sel. Top. Appl. Earth Obs. Remote Sens.* **2014**, *7*, 1493–1500.
9. Unwin, M.; Jales, P.; Curiel, A.S.; Brenchley, M.; Sweeting, M.; Gommenginger, C.; Roselló, J. Sea state determination with GNSS reflectometry on TechDemoSat-1. In Proceedings of the 2014 ESA Small Satellites, Systems and Services Symposium, Mallorca, Spain, 26–30 May 2014.
10. Unwin, M. TechDemoSat-1 and the GNSS reflectometry experiment. Available online http://www.merrbys.co.uk:8080/CatalogueData/Documents/TDS-1%20SGR-ReSI%20Experiment.pdf (accessed on 5 June 2015).
11. Rose, R.; Wells, W.; Rose, D.; Ruf, C.; Ridley, A.; Nave, K. Nanosat technology and managed risk: An update of the CYGNSS microsatellite constellation mission development. In Proceedings of the 28th AIAA/USU Conference on Small Satellites, Logan, UT, USA, 4–7 August 2014; pp. 1–12.
12. Wickert, J.; Beyerle, G.; Cardellach, E.; Förste, C.; Gruber, T.; Helm, A.; Hess, M.P.; Hoeg, P.; Jakowski, N.; Kern, M.; *et al.* GEROS-ISS-GNSS rEflectometry, Radio Occultation and Scatterometry on-

board the International Space Station. In Proceedings of 4th International Colloquium Scientific and Fundamental Aspects of the Galileo Programme, Espoo, Finland, 28–31 October 2013.

13. Martín-Neira, M.; D'Addio, S.; Buck, C.; Floury, N.; Prieto-Cerdeira, R. The PARIS ocean altimeter in-orbit demonstrator. *IEEE Trans. Geosci. Remote Sens.* **2011**, *49*, 2209–2237.

14. Carreno-Luengo, H.; Camps, A.; Jové, R.; Alonso, A.; Olivé, R.; Amèzaga, A.; Vidal, D.; Munoz, J.F. The 3Cat-2 project: GNSS-R In-Orbit demonstrator for earth observation. In Proceedings of the 2014 ESA Small Satellites, Systems and Services Symposium, Mallorca, Spain, 26–30 May 2014.

15. Zavorotny, V.U.; Voronovich, A.G. Bistatic GPS signal reflections at various polarizations from rough land surface with moisture content. In Proceedings of the 2000 IEEE International Geoscience and Remote Sensing Symposium, Honolulu, Hawaii, USA, 24–28 July 2000; pp. 2852–2854.

16. Ferrazzoli, P.; Guerriero, L.; Pierdicca, N.; Rahmoune, R. Forest biomass monitoring with GNSS-R: Theoretical simulations. *Adv. Space Res.* **2011**, *47*, 1823–1832.

17. Egido, A.; Paloscia, S.; Motte, E.; Guerriero, L.; Pierdicca, N.; Caparrini, M.; Santi, E.; Fontanelli, G.; Floury, N. Airborne GNSS-R soil moisture and above ground biomass observations. *IEEE Sel. Top. Appl. Earth Obs. Remote Sens.* **2014**, *7*, 1522–1532.

18. Wu, X.R.; Jin, S.G. GPS-Reflectometry: Forest canopies polarization scattering properties and modelling.*Adv. Space Res.* **2014**, *54*, 863–870.

19. Martínez-Vazquez, A. Emisividad Polarimétrica del Terreno Efecto de la Vegetación. Master's Thesis, Universitat Politècnica de Catalunya, Barcelona, Spain, 2001.

20. Ledesma-Galera, I. Estudio Experimental del Comportamiento Radiométrico de LAS Superfícies naturales. Master's Thesis, Universitat Politècnica de Catalunya, Barcelona, Spain, 2002.

21. Kinnaird, A. BEXUS User Manual. Available online: http: //www.rexusbexus.net /index.php? option=com_content&view= article&id=51&Itemid=63 (accessed on 5 December 2014).

22. Cardellach, E.; Ribó, S.; Rius, A. Technical Note on Polarimetric Phase Interferometry (POPI). Available online: http://

arxiv.org/pdf/physics/ 0606099.pdf (accessed on 1 March 2015).

23. Beckmann, P.; Spizzichino, A. *The Scattering of Electromagnetic Waves from Rough Surfaces*; Artech House Inc.: Norwood, MA, USA, 1963; p. 125.

24. Smyrnaios, M.; Schon, S. Multipath Propagation, Characterization and Modelling in GNSS. Available online: http: //cdn.intechopen.com/pdfs-wm/43710.pdf (accessed on 9 September 2015).

25. Woo, T.K. Optimum semi-codeless carrier phase tracking of L2. In Proceedings the 12th International Technical Meeting of the Satellite Division of the Institute of Navigation, Nashville, TN, USA, 14–17 September 1999; pp. 82–99.

26. Yi-Cheng, L.; Sarabandi, K. A Monte Carlo coherent scattering model for forest canopies using fractal-generated trees. *IEEE Trans. Geosci. Remote Sens.* **1999**, *37*, 440–451.

27. Prusinkiewicz, P.; Lindenmayer, A. *The Algorithmic Beauty of Plants*; Springer-Verlag: New York, NY, USA, 1990.

28. Tsang, L.; Kong, J.A.; Shin, R.T. *Theory of Microwave Remote Sensing*; Wiley Interscience: New York, NY, USA, 1985.

29. Martínez-Vazquez, I.; Camps, A.; Lopez-Sanchez, J.M.; Vall-llosera, M.; Monerris, A. Numerical simulation of the full-polarimetric emissivity of vines and comparison with experimental data. *Remote Sens.* **2009**, *1*, 300–317.

30. Caicoya, A.T.; Kugler, F.; Hajnsek, I.; Papathanassiou, K. Boreal forest biomass classification with TANDEM-X. In Proceedings of the 2012 IEEE International Geoscience and Remote Sensing Symposium, Munich, Germany, 22–27 July 2012; pp. 3439–3442.

31. Global map of forest height produced from NASA's ICESAT/GLAS, MODIS and TRMM sensors. Available online: http:// www.nasa.gov/ topics/ earth/features/earth20120217map. html (accessed on 1 March 2015).

32. Imhoff, M.L. Radar backscattering and biomass saturation: Ramifications for global biomass inventory.*IEEE Trans. Geosci. Remote Sens:* **1995**, *33*, 511–518.

33. Lin, Y.-C.; Sarabandi, K. Electromagnetic scattering model for a tree trunk above a tilted ground plane.*IEEE Trans. Geosci. Remote Sens.* **1995**, *33*, 1063–170.

34. Karam, M.A.; Fung, A.K.; Antar, Y.M.M. Electromagnetic wave scattering from some vegetation samples.*IEEE Trans. Geosci. Remote Sens.* **1998**, *26*, 799–808.

35. Choudhury, B.J.; Schmugge, T.J.; Chang, A.; Newton, R.W. Effect of surface roughness on the microwave emission from soils. *J. Geophys. Res.* **1979**, *84*, 5699–5706.

36. Rodriguez-Alvarez, N.; Camps, A.; Vall-llosera, M.; Bosch-Lluis, X.; Monerris, A.; Ramos-Pérez, I.; Valencia, E.; Marchán-Hernandez, J.F.; Martinez-Fernandez, J.; Baroncini-Turricchia, G.; *et al.* Land geophysical parameters retrieval using the interference pattern GNSS-R technique. *IEEE Trans. Geosci. Remote Sens.* **2011**, *49*, 71–84.

37. Ulaby, F.T.; Moore, R.K.; Fung, A.K. Radar remote sensing and surface scattering and emission theory. In*Microwave Remote Sensing: Active and Passive*; Addison-Wesley: Reading, MA, USA, 1982; Volume II, pp. 1540–1541.

38. Lindermayer, A. Developmental algorithms for multicellular organisms: A survey of L-systems.*J. Theor. Bio.* **1975**, *54*, 3–22.

39. Ulaby, F.T.; Moore, R.K.; Fung, A.K. Microwave remote sensing fundamentals and radiometry. In*Microwave Remote Sensing: Active and Passive*; Addison-Wesley: Reading, MA, USA, 1982; Volume I, pp. 249–251.

5
CHAPTER
MULTICORE SOFTWARE-DEFINED RADIO ARCHITECTURE FOR GNSS RECEIVER SIGNAL PROCESSING

Heikki Hurskainen Jussi Raasakka, Tapani Ahonen and Jari Nurmi

Department of Computer Systems, Tampere University of Technology

ABSTRACT

We describe a multicore Software-Defined Radio (SDR) architecture for Global Navigation Satellite System (GNSS) receiver implementation. A GNSS receiver picks up very low power signals from multiple satellites and then uses dedicated processing to demodulate and measure the exact timing of these signals from which the user's position, velocity, and time (PVT) can be estimated. Three GNSS SDR architectures are discussed. (1) A hardware-based SDR that is feasible for embedded devices but relatively expensive, (2) a pure SDR approach that has high level of flexibility and low bill of material, but is not yet suited for handheld applications, and (3) a novel architecture that uses a programmable array of multiple processing cores that exhibits both flexibility and potential for mobile devices. We present the CRISP project where the multicore architecture will be realized along with numerical analysis of application requirements of the platform's processing cores and network payload.

1. INTRODUCTION

Global navigation has been a challenge to mankind for centuries. However, in the modern world it has become easier with the help from Global Satellite Navigation Systems (GNSSs). NAVSTAR Global Positioning System (GPS) [1] has been the most famous implementation of GNSS and the only fully operational system available for civilian users, although this situation is changing.

Galileo [2] is emerging as a competitor and complement for GPS, as they both are satellite navigation systems based on Code Division Multiple Access (CDMA) techniques. CDMA is a technique that allows multiple transmitters to use same carrier simultaneously by multiplying pseudorandom noise (PRN) codes to the transmitted signal. The PRN code rate is higher than data symbol rate which divides the energy of a data symbol to a wider bandwidth. The used PRN codes are unique to each transmitter and thus transmitter can be identified in reception when received signal is correlated with a replica of the used PRN code.

The Russian GLONASS system, originally based on Frequency Division Multiple Access (FDMA), is adding a CDMA feature to the system with GLONASS-K satellites [3]. China has also shown interest in implementing its own system, called Compass, during the following decade [4]. The GPS modernization program [5] introduces additional signals with new codes and modulation. Realization of the new navigation systems and modernization of GPS produce updates and upgrades to system specifications.

Besides changing specifications, GNSS is also facing challenges from an environmental point of view. The resulting multipath effects make it more difficult to determine exact signal timing crucial for navigation algorithms. Research around multipath mitigation algorithms is active since accurate navigation capability in environments with heavy multipaths is desired. Among interference issues multipath mitigation is also one of the biggest drivers for the introduction of new GNSS signal modulations.

Designing a true GNSS receiver is not a trivial task. A true GNSS receiver should be reconfigurable and flexible in design so that the posibilities of new specifications and algorithms can be exploited, and the price should be low enough to enable mass market penetration.

2. GNSS PRINCIPLES AND CHALLENGES

2.1. Navigation and Signal Processing

Navigation can be performed when four or more satellites are visible to the receiver. The pseudoranges from receiver to satellites and navigation data (containing ephemeris parameters) are needed [1, 6, 7].

When pseudoranges (ρ) are measured by the receiver, they can be used to solve unknowns, the users location $(x, y, z)_u$ and clock bias b_u with known positions of satellites $(x, y, z)_i$. The relation between pseudorange, satellite position, and user position is illustrated in

$$\rho_i = \sqrt{(x_i - x_u)^2 + (y_i - y_u)^2 + (z_i - z_u)^2} + b_u. \tag{1}$$

The transmitted signal contains low rate navigation data (50 Hz for GPS Standard Positioning Service (SPS)), repeating PRN code sequence (1023 chips at 1.023 MHz for GPS SPS) and a high rate carrier (GPS SPS is transmitted at L1 band which is centered at 1575.42 MHz) [1]. For Galileo E1 Open Service (OS) and future GPS L1C it also contains a Multiplexed Binary Offset Carrier (MBOC) modulation [8, 9]. These signal components are illustrated in Figure 1.

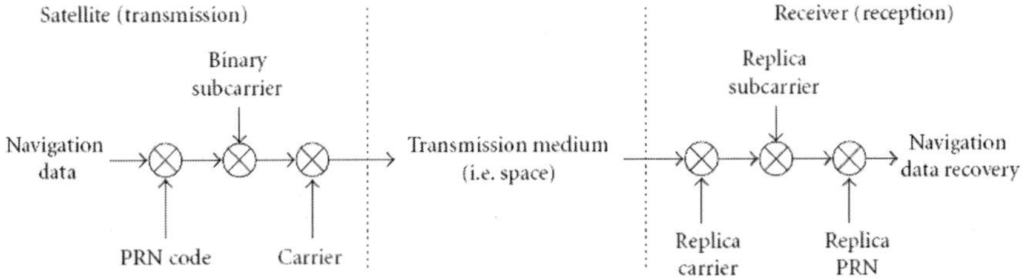

Figure 1. Principles for gnss signal modulation in transmission and demodulation in reception.

The signal processing for GNSS can be divided into analog and digital parts. Since the carrier frequencies of the GNSS are high ($>$ 1 GHz) it is impossible to perform digital signal processing on it. In the analog part of the receiver, which is called the radio front-end, the received signal is amplified, filtered, downconverted, and finally quantized and sampled to digital format.

The digital signal processing part (i.e., baseband processing) has two major tasks. First, the Doppler frequencies and code phases of the satellites need to be acquired. The details of the acquisition process are well explained in literature, for example, [1, 7]. There are a number of ways to implement acquisition, with parallel methods being faster than serial ones, but at the cost of consuming more resources. The parallel methods can be applied either as convolution in the time domain (matched filters) or as multiplication in the frequency domain (using FFT and IFFT).

Second, after successful acquisition the signals found are being tracked. In tracking, the frequency and phase of the receiver are continously fine-tuned to keep receiving the acquired signals. Also, the GNSS data is demodulated and the precise timing is formed from the signal phase measurements. A detailed description of the tracking process can be found, for example, in [1, 7]. The principles for data demodulation are also illustrated in Figure 1.

2.2. Design Challenges of GNSS

The environment we are living in is constantly changing in topographic, geometric, economic, and political ways. These changes are driving the GNSS evolution.

Besides new systems (e.g., Galileo, Compass), the existing ones (i.e., GPS, GLONASS) are being modernized. This leads to constantly evolving field of specifications which may increase the frustration and uncertainty among receiver designers and manufacturers.

The signal spectrum of future GNSS signals is growing with the new systems. Currently the GPS L1 (centered at 1575.42 MHz) is the only commercially exploited GNSS frequency band. Galileo systems E1 OS signal will be sharing the same band. Another common band of future GPS and Galileo signals will be centered at 1176.45 MHz (GPS L5 and Galileo E5a).

The GPS modernization program is also activating the L2 frequency band (centered at 1227.60 MHz) to civilian use by implementing L2C (L2 Civil) signal [10]. This band has already been assigned for navigation use, but only for authorized users via GPS Precise Positioning Service (PPS) [1].

To improve the signal code tracking and multipath performance new Binary Offset Carrier (BOC) modulation was originally introduced as baseline for Galileo and modern GPS L1 signal development [11]. The later agreement between European and US GNSS authorities further specified the usage of Multiplexed BOC (MBOC) modulation in both systems. In MBOC modulation two different binary subcarriers are added to the signal, either as time multiplexed mode (TMBOC), or summed together with predefined weighting factors as Composite BOC (CBOC) [8,9, 12].

Like any other wireless communication, satellite navigation also suffers from multipaths in environments prone to such (e.g., urban canyons, indoors). The problem caused by multipaths is even bigger in navigation than in communication since precise timing also needs to be resolved. The field of multipath mitigation is actively researched and new algorithms and architectures are presented frequently, for example, in [13–15].

Besides GNSS there are also other wireless communication technologies that are developing rapidly and the direction of development is driven towards multipurpose low cost receivers (user handsets) with enhanced capabilities [16].

3. OVERVIEW OF SDR GNSS ARCHITECTURES

In this section we present three architectures for Software-Defined Radio (SDR) GNSS receiver. A simplified definition of SDR is given in [17]. "*Radio in which some or all of the physical layer functions are software defined*."

The root SDR architecture was presented in [18]. Figure 2 illustrates an example of GNSS receiver functions mapped on to this canonical architecture. Only the reception part of the architecture is presented since current GNSS receivers are not transmitting. Radio Frequency (RF) conversion handles the signal processing before digitalization. The Intermediate Frequency (IF) processing block transfers the frequency of the received signal from IF to baseband and may also take care of Doppler removal in GNSS. The baseband processing segment handles the accurate timing and demodulation, thus enabling the construction of the navigation data bits. The division into IF and baseband sections can vary depending on the chosen solution since the complex envelope of the received signal can be handled in baseband also. Desired navigation output (Position, Velocity, and Time (PVT)) is solved in the last steps of the GNSS receiver chain.

Current state-of-the-art mass market receivers are based on a chipset or single-chip receiver [19]. The chipset or single-chip receiver is usually implemented as an Application Specific Integrated Circuit (ASIC). ASICs have high Nonrecurring Engineering (NRE) costs, but when produced in high volumes they have very low price per unit. ASICs can also be optimized for small size and to have small power consumption. Both of these features are desired in handheld, battery operated devices. On the other hand, ASICs are fixed solutions and impossible to reconfigure. Modifications in design are also very expensive to realize with ASIC technology.

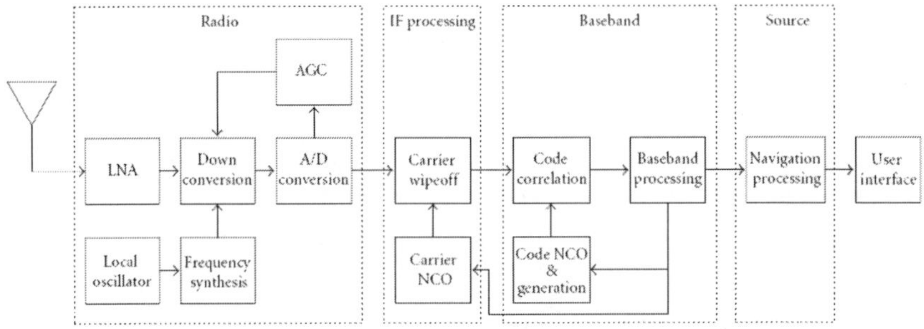

Figure 2. Canonical sdr architecture adapted to gnss. **It is modified from [18].**

This approach has proven to be successful in mass market receivers because of price and power consumption advantages although it may not hold its position with growing demand for flexibility and shortened time to market.

3.1. Hardware Accelerated SDR Receiver Architecture

The first SDR receiver architecture discussed in this paper is the approach where the most demanding parts of the receiver are implemented on a reconfigurable hardware platform, usually in the form of a Field Programmable Gate Array (FPGA) progammed with a Hardware Description Language (HDL). This architecture, comprised of radio front-end circuit, reconfigurable baseband hardware, and navigation software is well known and presented in numerous publications, for example, [16, 20–22]. FPGAs have proved to be suitable for performing GNSS signal processing functions [23]. The building blocks for hardware accelerated SDR receivers are illustrated in Figure 3.

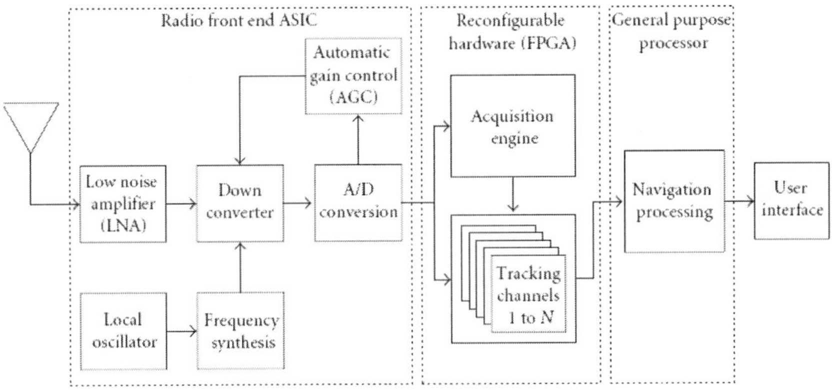

Figure 3. Hardware accelerated baseband architecture. From left to right: analog radio part, reconfigurable baseband hardware, and navigation software running on GPP.

In this architecture the RF conversion is performed by analog radio. The last step of the conversion transforms the signal from analog to digital format. IF processing and baseband functionalities are performed in accelerating hardware. The source, PVT for the GNSS case, is constructed in navigation processing.

The big advantage for reconfigurable FPGAs in comparison to ASIC technologies is the savings in design, NRE and mask costs due to shorter development cycle. The risk is also smaller with FPGAs, since the possible bugs in design can be fixed by upgrades later on. On the other hand FPGAs are much higher in unit price and power consumption.

A true GNSS Receiver poses some implementation challenges. The specifications are designed to be compatible (i.e., systems do not interfere with each other too much) and the true interoperability is reached at receiver level. One example of interoperative design challenges is the selection of the number of correlators and their spacing for tracking, since different modulations have different requirements for the correlator structure.

3.1.1. Challenges with Radio Front End

Although the focus of this paper is mainly on baseband functions, the radio should not be forgotten. The block diagram for a GNSS single frequency radio front end is given on the left-hand side of Figure 3. In the radio the received signal is first amplified with the Low Noise Amplifier (LNA) and then after necessary filtering it is downconverted to low IF, for example, to 4 MHz [24]. The signal is converted to a digital format after downconversion.

The challenges for GNSS radio design come from the increasing amount of frequency bands. To call a receiver a true GNSS receiver and also to get the best performance, more than one frequency band should be processed by the radio front-end. Dual- and/or multifrequency receivers are likely choices for future receivers, and thus it is important to study potential architectures [25].

Another challenge comes from the increased bandwidth of new signals. With increased bandwidth the radio becomes more vulnerable to interference. For mass market consumer products, the radio design should also meet certain price and power consumption requirements. Only solutions with reasonable price and power consumption will survive.

3.1.2. Baseband Processing

The fundamental signal processing for GNSS was presented in Figure 1. The carrier and code removal processes are illustrated in more detail in Figure 4. The incoming signal is divided into in-phase and quadrature-phase components by multiplying it with the locally generated sine and cosine waves. Both phases are then correlated in identical branches with several closely delayed versions (for GPS; early, prompt, and late), of the locally generated PRN code [1]. Results are then integrated and fed to discriminator computation and feedback filter. Numerically Controlled Oscillators (NCOs) are used to steer the local replicas.

An example of the different needs for new GNSS signals is the addition of 2 correlator fingers (bump-jumping algorithm) due to Galileo BOC modulation [26]. In Figure 4 additional correlator components needed for Galileo tracking are marked with darker shade. In most parts the GPS and Galileo signals in L1 band are using the same components. The main difference is that due to the BOC family modulation Galileo needs additional correlators; it is very-early (VE) and very-late (VL) to remove the uncertainty of main peak location estimation [27]. The increasing number of correlators is related to the increase in complexity, measured by the number of transistors in the final design [13].

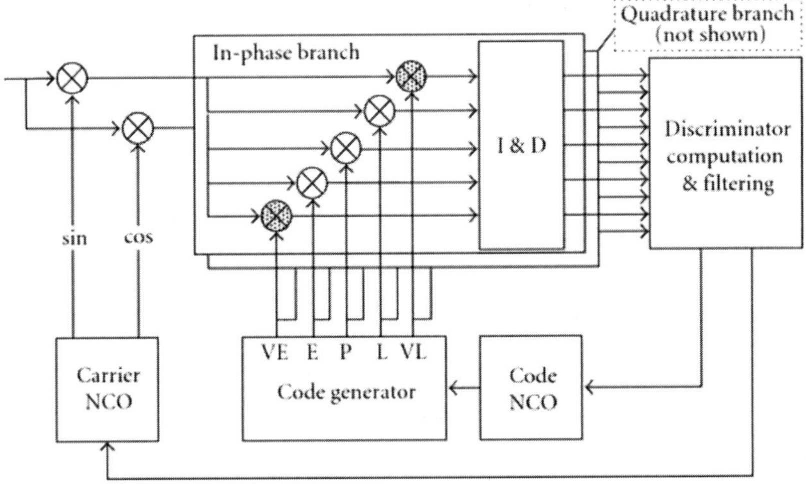

Figure 4. Gps/galileo tracking channel.

The level of hardware acceleration depends on the selected algorithms. Acquisition is rarely needed compared to tracking and thus it is more suitable for software implementation. FFT-based algorithms are more desirable for designer to implement in software since hardware languages are usually lacking direct support for floating-point number calculus. Tracking on the other hand is a process containing mostly multiplication and accumulation using relatively small word lengths. The thing that makes it more suitable for hardware implementation is that the number of these relatively simple computations is high, with a real-time deadline.

3.2. Ideal SDR GNSS Receiver Architecture

The ideal SDR is characterized by assigning all functions after the analog radio to a single processor [18]. In the ideal case all hardware problems are turned to software problems.

A fundamental block diagram of a software receiver is illustrated in Figure 5 [28]. The architecture of the radio front-end is the same that was illustrated in Figure 3. After radio the digitized signals are fed to buffers for software usage. Then all of the digital signal processing, acquisition, and tracking functions are performed by software.

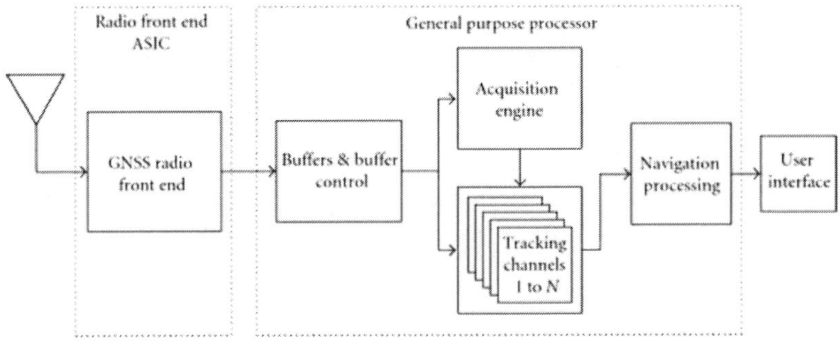

Figure 5. Software receiver architecture. On left-hand side: analog radio part, and on right-hand side: baseband and navigation implemented as software running on a GPP.

In the literature, for example, [28, 29], the justification and reasoning for SDR GNSS is strongly attributed to the well-known Moores law which states that the capacity of integrated circuits is doubling every 18–24 months [30]. Ideal SDR solutions should become feasible if and when available processing power increases. Currently reported SDR GPS receiver implementations are working in realtime only if the clock speed of the processor is from 900 MHz [31] to 3 GHz [29], which is too high for mobile devices but not, for example, a laptop PC.

In the recent years, the availability of GNSS radio front ends with USB has improved, making the implementation of a pure software receiver on a PC platform quite straightforward. The area where pure software receivers have already made a breakthrough is postprocessing applications. Postprocessing with software receivers allows fast algorithm prototyping and signal analysis. Typical postprocessing applications are ionospheric monitoring, geodetic applications, and other scientific applications [21, 32].

Software is definitely more flexible than hardware when compared in terms of time to market, bill of materials, and reconfigurable implementation. But with a required clock frequency of around 1 GHz or more, the generated heat and battery life will be an issue for small handheld devices.

3.3. SDR with Multiple Cores

What about having an array of reconfigurable cores for baseband processing? In a multicore architecture baseband processing is divided among multiple processing cores. This reduces the clock frequency needed to a range achievable by embedded devices and provides an increased level of parallelism which also eases the work load per processing unit.

An example of the GNSS receiver architecture with reconfigurable baseband approach is illustrated in Figure 6. In this example one of the four cores is acting as an acquisition engine and the remaining three are performing the tracking functions. A fixed set of cores is not desirable since the need for acquisition and tracking varies over time. For example, when receiver is turned on, all cores should be performing acquisition to guarantee the fastest possible Time To First Fix (TTFF). After satellites have been found more of the acquisition cores are moved to the tracking task.

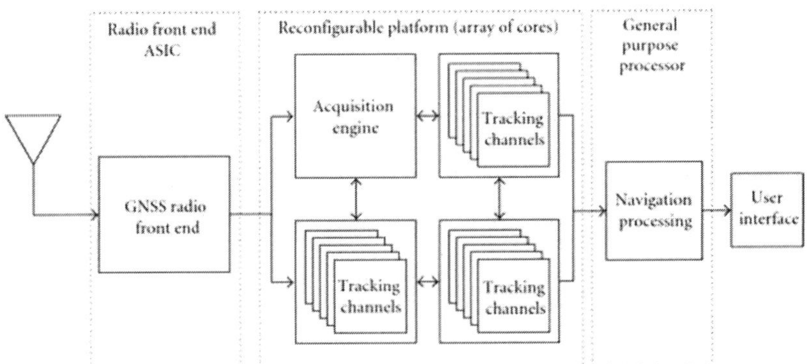

Figure 6. Software reconfigurable baseband receiver architecture. From left: analog radio part, baseband implemented on an array of reconfigurable cores, and navigation software running on GPP.

If (and when) manufactured in large volumes the (properly scaled) array of processing cores can be eventually implemented in an ASIC circuit. This lowers the per unit price and makes this solution more appealing for mass markets, while still being reconfigurable and having high degree of flexibility.

In the next section we present one future realization of this architecture.

4. CRISP PLATFORM

Cutting edge Reconfigurable ICs for Stream Processing (CRISP) [33] is a project in the Framework Programme 7 (FP7) of the European Union (EU). The objectives of the CRISP are to research the optimal utilization, efficient programming, and dependability of a reconfigurable multiprocessor platform for streaming applications. The CRISP consortium is a good mixture of academic and industrial know-how with partners; Recore (NL), University of Twente (NL), Atmel (DE), Thales Netherlands (NL), Tampere University of Technology (FI), and NXP (NL). The three-year project started in the beginning at 2008.

The reconfigurable CRISP platform, also called General Streaming Processor (GSP), designed and implemented within the project, will consist of two separate devices: General Purpose Device (GPD) and Reconfigurable Fabric Device (RFD). The GPD contains off-the-shelf General Purpose Processor (GPP) with memories and peripheral connections whereas the RFD consists of 9 reconfigurable cores. The array of reconfigurable cores is illustrated in Figure 7 [34], with "R" depicting a router.

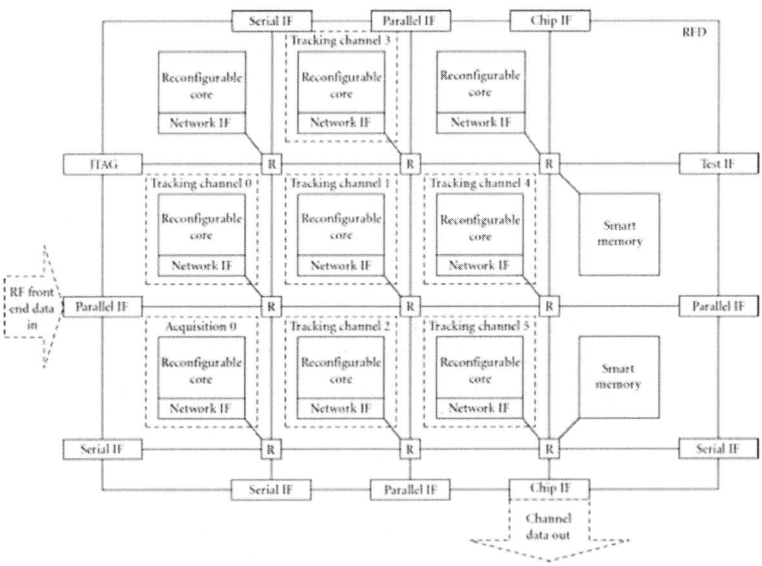

Figure 7. Array of 9 reconfigurable cores [**34**] with example mapping of gnss application illustrated, the selection of cores is random. "r" depicts router and "if" interface.

The reconfigurable cores are Montium cores (it was recently decided to use Xentium processing tile as Reconfigurable Core in the CRISP GSP. The Xentium has at least similar performance to the Montium (with respect to cycle count), but is designed for better programmability (e.g., hardware supporting optimal software pipelining)). Montium [35] is a reconfigurable processing core. It has five Arithmetic and Logical Units (ALUs), each having two memories, resulting in total of 10 internal memories. The cores communicate via a Network-on-Chip (NoC) which includes two global memories. The device interfaces to other devices and outer world via standard interfaces.

Within the CRISP project the GNSS receiver is one of the two applications designed for proof of concept for the platform. The other is a radar beamforming application which has much higher demands on computation than a standalone GNSS receiver.

4.1. Specifying the GNSS Receiver for the Multicore Platform

In the CRISP project our intention is to specify, implement, and integrate a GNSS receiver application supporting GPS and Galileo L1 Open Service (OS) signals on the multicore platform. In this case, the restriction for L1 band usage comes from the selected radio [24], but in principle the multicore approach can be extended to multifrequency receivers if a suitable radio front-end is used.

4.1.1. Requirements for Tile Processor

The requirements of GNSS L1 application have been studied in [36]. The results, restated in Table 1, indicated that a single Montium core running at 200 MHz clock speed is barely capable of executing the minimum required amount of acquisition and tracking processes. This analysis did not take into account the processing power needed for baseband to navigation handover nor navigation processing itself. With this it is evident that an array of cores (more than one) is needed for GNSS L1 purposes. The estimations given in Table 1 are based on reported [35] performance of the Montium core. The acquisition figures are computed for a search speed of one satellite per second and the tracking figures are for a single channel.

Table 1. Estimation of GNSS baseband process complexity for Montium Tile Processor running at 200 MHz, max performance of 1 GMAC/s [36].

Process	Usage (MMAC/s)	Usage of TP (%)
Acquisition (GPS)	43.66	4.4
Acquisition (Galileo)	196.15	19.6
Tracking (GPS)	163.67	16.4
Tracking (Galileo)	229.14	22.9

The results presented in Table 1 reflect the complexity of the processes when the input stream is sampled at 16.368 MHz, which is the output frequency of the selected radio front end for CRISP platform [24]. This is approximately 16 times the navigation signal fundamental frequency of 1.023 MHz.

The GNSS application can also be used with a lower rate input stream without a significant loss in application performance. For this paper, we analyzed the effect of the input stream decimation to the complexity of the main baseband processes. The other parameters, such as acquisition time and number of frequency bins for acquisition and number of active correlators per channel for tracking, remained the same as in [36].

Figures 8 and 9 illustrate the effect of decimation by factors 1, 2, 4, 8, and 16 to the utilization of the Montium Tile processor. Decimation factor 1 equates to the case where no decimation is applied, that is, results shown in Table1. The presented figures show how the complexity of both processes, measured as Montium Tile Processor utilization percentage, decreases exponentially as decimation factor increases. The behavior is the same for GPS and Galileo signals, except that utilization with Galileo signals is a bit larger than with GPS in all studied cases.

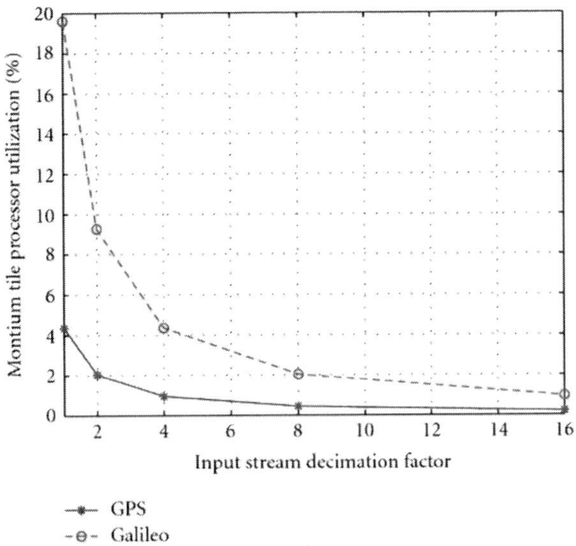

Figure 8 Acquisition process utilization of montium tile processor resources as a function of the decimation factor of the input stream.

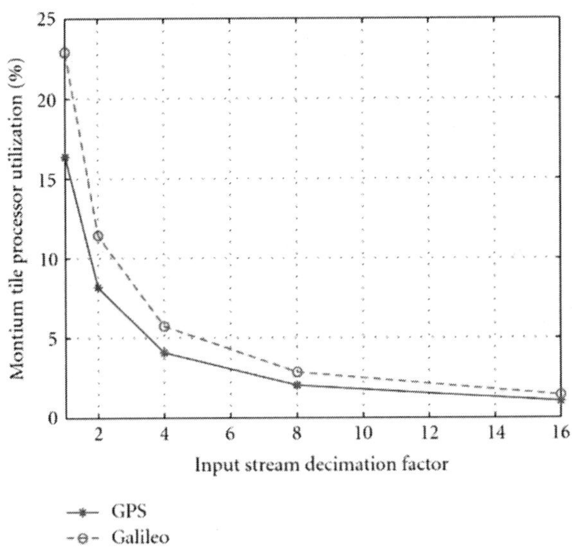

Figure 9. Tracking process utilization of montium tile processor resources as a function of the decimation factor of the input stream.

To ease the computational load of the Tile Processor the decimation of the input stream seems to be a feasible choice. The amount of decimation should be sufficient to effect meaningful savings in TP utilization without significantly degrading performance of the application. For the current GPS SPS signal, decimation by factor 4 (4.092 MHz) is feasible without significant loss in receiver performance. Factor of 8 (2.046 MHz) is equal to the Nyqvist rate for 1.023 MHz, which is the PRN code rate used in GSP SPS signal.

In the Galileo case, decimation factor 4 is the maximum decimation factor. This is because with a sampling frequency of approximately 4 MHz the BOC(1,1) component of the Galileo E1 OS signal can be still received with a maximum loss of only −0.9 dB, when compared with the reception of the whole MBOC bandwidth [12]. (This applies also to the modern GPS L1C signals, but they are not specified in our application [36].)

In the ideal case the decimation of the input stream should be changing with the receiver mode (GPS/Galileo). Since in CRISP the decimation of the radio stream will be implemented as hardware in FPGA, which is connecting the radio to the parallel interface of the final CRISP prototype platform, the run time configuration of the decimation factor is not feasible. For this reason, in the rest of the paper we will focus on the scenario where fixed decimation factor of 4 is used, resulting in a stream sample rate of 4.092 MHz.

Table 2 shows baseband complexity estimation for the case when input stream is decimated by a factor of four. When it is compared to the original figures of complexity shown in Table 2, it can be seen that the utilization of TP is over four times smaller.

Table 2. Estimation of GNSS baseband process complexity with decimated (by factor 4) input stream. Montium Tile Processor running at 200 MHz, max performance of 1 GMAC/s.

Process	Usage (MMAC/s)	Usage of TP (%)
Acquisition (GPS)	9.57	0.96
Acquisition (Galileo)	43.66	4.37
Tracking (GPS)	40.92	4.09
Tracking (Galileo)	57.28	5.73

4.1.2. Requirements for the Network-on-Chip

To analyze the multicore GNSS receiver application we built a functional software receiver with the C++ language, running on a PC. The detailed analysis of the software receiver will be given in substantial paper [37].

In our SW receiver each process was implemented as a software thread. With approximating one process per core this approach enabled us to estimate the link payload by logging communication between the threads.

We estimated a scenario where one core was allocated to perform acquisition and six cores were mapped for the tracking process. This scenario is illustrated in Figure 7. Digitized RF front-end data is input to the NoC via an interface. A specific chip interface is used to connect the RFD to the GPD, and it is used to forward channel data (channel phase measurement data related to pseudorange measurements, and navigation data) to the GPD. The Selected mapping is a compromise between minimal operative setup (one acquisition and four tracking) and the needs of dependability testing processes, where individual cores may be taken offline for testing purposes.

The scenario was simulated with a prerecorded set of real GPS signals. Since signal sources for Galileo navigation were not available, the Galileo case was not tested. The link payloads caused by the cores communicating while the software was running for 5 seconds is illustrated in Figure 10.

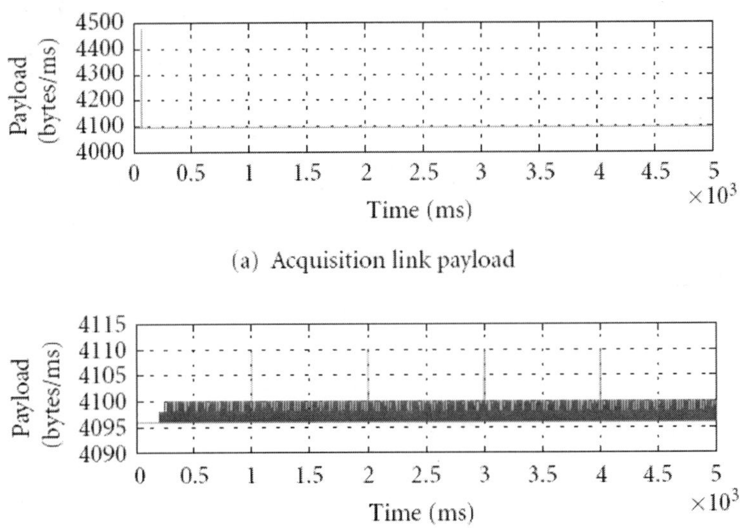

Figure 10. Link payloads for gps acquisition process (a) and average payload of gps tracking processes (b).

Acquisition link payloadAverage tracking link payload

The results show that, in GPS mode, our GNSS application causes a payload for each link/processing core with a constant baseline of 4096 Bytes/millisecond. This is caused by the radio front-end input, that is, the incoming signal. In this scenario we used real GPS front end data which was sampled at 4.092 MHz, each byte representing one sample. This sampling rate is also equal to the potential decimation scenario discussed earlier.

With a higher sampling rate the link payload baseline will be raised, but on the other hand one byte can be preprocessed to contain more than one sample, decreasing the traffic caused by radio front-end input.

The first peak in the upper part of Figure 8 is caused by the acquisition process output. When GNSS application starts, FFT-based acquisition is started and the results are ready after 60 milliseconds, which are then transmitted to tracking channels. This peak is also the largest individual payload event caused by the GNSS application.

After a short initialization period the tracking processes start to produce channel output. An Average of simulated GPS tracking link/processing core payloads is illustrated in Figure 10(b). Every 20 milliseconds a navigation data symbol (data rate is 50 Hz in GPS) is transmitted and once a second higher transmission peak is caused by the loop phase measurement data, which is transmitted to GPD for pseudorange estimation.

In Galileo mode, the payload caused by incoming signal will be equal since the same radio input will be used for both GPS and Galileo. However, the transmission of data symbols will cause a bigger payload since data rate of Galileo E1 signals is 250 symbols per second [8]. Galileo phase measurement rate will remain the same as in GPS mode.

From the results it is seen that the link payload caused by the incoming RF signal is the largest one in both operating modes, and if the link payload needs to be optimized the reduction of it is the first thing to be studied. The results also indicate that when GNSS application is running smoothly the link payloads caused by it are predictable.

Note that this estimation does not contain any overheads caused by network protocol or any other data than navigation related (dependability, real-time mapping of the processes). These issues will be studied in our future work.

4.2. Open Issues

Besides the additional network load caused by other than the GNSS application itself, there are also some other issues that remain open. There may be challenges in designing software for a multicore environment. Power consumption as well as the final bill of materials (BOMs), (i.e., final price of the multicore product) remains an open issue at the time of this writing. In future these issues will be studied and suitable optimizations performed after the prototyping and proof of concepts have been completed successfully.

5. CONCLUSIONS

In this paper we discussed three Software-Defined Radio (SDR) architectures for a Global Navigation Satellite System (GNSS) receiver. The usage of flexible architectures in GNSS receiver was justified with the need for implementing support for upcoming navigation systems and new algorithms developed, and especially for multipath mitigation. The hardware accelerated SDR architecture is quite close to the current mass market solutions. There the ASIC is replaced with a reconfigurable piece of hardware, usually an FPGA. The second architecture, ideal (or pure) SDR receiver is using a single processor to realize all necessary signal processing functions. Real-time receivers remain a challenge, but postprocessing applications are already taking advantage of this architecture.

The third architecture, SDR with multiple cores, is a novel approach for GNSS receivers. This approach benefits in both having high degree of flexibility, and when properly designed and scaled, a reasonably low unit price in high volume production. In this paper we also presented the CRISP project where such a multicore architecture will be realized along with the analysis of GNSS application requirements for the multicore platform.

We extended the previously published analysis of processing tile utilization to cover the effect of input stream decimation. Decimation by factor four seems to offer a good compromise between core utilization and application performance.

We implemented a software GNSS receiver with processes implemented as threads and used that to analyze the GNSS application communication payload for individual links. This analysis indicated that the incoming signal represents the largest part of the communication in the network between processing cores.

DECLARATIONS

Acknowledgments

The authors want to thank Stephen T. Burgess from Tampere University of Technology for his useful comments about the manuscript. This work was supported in part by the FUGAT project funded by the Finnish Funding Agency for Technology and innovation (TEKES). Parts of this research are conducted within the FP7 Cutting edge Reconfigurable ICs for Stream Processing (CRISP) project (ICT-215881) supported by the European Commission.

REFERENCES

1. Kaplan ED, Hegarty CJ (Eds): *Understanding GPS, Principles and Applications*. 2nd edition. Artech House, Boston, Mass, USA; 2006.
2. Benedicto J, Dinwiddy SE, Gatti G, Lucas R, Lugert M: GALILEO: SATELLITE SYSTEM DESIGN AND TECHNOLOGY DEVELOPMENTS. European Space Agency, November 2000
3. Revnivykh S: GLONASS STATUS AND PROGRESS. December 2008,
4. Gibbons G: INTERNATIONAL SYSTEM PROVIDERS MEETING (ICG-3) REFLECTS GNSS'S COMPETING INTEREST, COOPERATIVE OBJECTIVES. *Inside GNSS* December 2008.
5. Airforce US: GPS MODERNIZATION FACT SHEET. 2006,
6. Braasch MS, van Dierendonck AJ: GPS RECEIVER ARCHITECTURES AND MEASUREMENTS. *Proceedings of the IEEE*1999,87(1):48-64.
7. Borre K, Akos DM, Bertelsen N, Rinder P, Jensen SH: *A Software Defined GPS and Galileo Receiver—A Single-Frequency Approach*. Birkhäuser, Boston, Mass, USA; 2007.
8. GALILEO OPEN SERVICE, SIGNAL IN SPACE INTERFACE CONTROL DOCUMENT (OS SIS ICD) Draft 1, February 2008
9. INTERFACE SPECIFICATION—NAVSTAR GPS SPACE SEGMENT/USER SEGMENT L1C INTERFACES IS-GPS-800, August 2007

10. Fontana RD, Cheung W, Stansell T: THE MODERNIZED L2C SIGNAL—LEAPING FORWARD INTO THE 21ST CENTURY.*GPS World* September 2001, 28-34.

11. GALILEO JOINT UNDERTAKING—GALILEO OPEN SERVICE, SIGNAL IN SPACE INTERFACE CONTROL DOCUMENT (OS SIS ICD) GJU, May 2006

12. Hein GW, Avila-Rodriguez J-A, Wallner S, *et al.*: MBOC: THE NEW OPTIMIZED SPREADING MODULATION RECOMMENDED FOR GALILEO L1 OS AND GPS L1C. *Proceedings of the IEEE/ION Position, Location, and Navigation Symposium (PLANS '06), April 2006, San Diego, Calif, USA* 883-892.

13. Hurskainen H, Lohan ES, Hu X, Raasakka J, Nurmi J: MULTIPLE GATE DELAY TRACKING STRUCTURES FOR GNSS SIGNALS AND THEIR EVALUATION WITH SIMULINK, SYSTEMC, AND VHDL. *International Journal of Navigation and Observation* 2008, 2008:-17.

14. Kim S, Yoo S, Yoon S, Kim SY: A NOVEL UNAMBIGUOUS MULTIPATH MITIGATION SCHEME FOR BOC(KN, N) TRACKING IN GNSS. *Proceedings of the International Symposium on Applications and the Internet Workshops, 2007*-57.

15. Dovis F, Pini M, Mulassano P: MULTIPLE DLL ARCHITECTURE FOR MULTIPATH RECOVERY IN NAVIGATION RECEIVERS.*Proceedings of the 59th IEEE Vehicular Technology Conference (VTC '04), May 2004* 5: 2848-2851.

16. Dovis F, Gramazio A, Mulassano P: SDR TECHNOLOGY APPLIED TO GALILEO RECEIVERS. *Proceedings of the International Technical Meeting of the Satellite Division of the Institute of Navigation (ION GPS '02), September 2002, Portland, Ore, USA*

17. SDR FORUM January 2009, http://www.sdrforum.org/

18. Mitola J: THE SOFTWARE RADIO ARCHITECTURE. *IEEE Communications Magazine* 1995.

19. Mattos PG: A SINGLE-CHIP GPS RECEIVER. *GPS World* October 2005.

20. Mumford PJ, Parkinson K, Dempster AG: THE NAMURU OPEN GNSS RESEARCH RECEIVER. *Proceedings of the International Technical Meeting of the Satellite Division of the Institute of Navigation (ION GNSS '06), September 2006, Fort Worth, Tex, USA* 5: 2847-2855.

21. Ganguly S, Jovancevic A, Saxena DA, Sirpatil B, Zigic S: OPEN ARCHITECTURE REAL TIME DEVELOPMENT SYSTEM FOR GPS AND GALILEO. *Proceedings of the International Technical Meeting of the Satellite Division of the Institute of Navigation (ION GNSS '04), September 2004, Long Beach, Calif, USA* 2655-2666.

22. Hurskainen H, Paakki T, Liu Z, Raasakka J, Nurmi J: GNSS RECEIVER REFERENCE DESIGN. *Proceedings of the 4th Advanced Satellite Mobile Systems (ASMS '08), August 2008, Bologna, Italy* 204-209.

23. Hill J: NAVIGATION SIGNAL PROCESSING WITH FPGAS. *Proceedings of the National Technical Meeting of the Institute of Navigation, June 2004* 420-427.

24. Atmel : GPS FRONT END IC ATR0603. Datasheet, 2006

25. Detratti M, Lopez E, Perez E, Palacio R: DUAL-FREQUENCY RF FRONT END SOLUTION FOR HYBRID GALILEO/GPS MASS MARKET RECEIVER. *Proceedings of the IEEE Consumer Communications and Networking Conference (CCNC '08), January 2008, Las Vegas, Nev, USA* 603-607.

26. Fine P, Wilson W: TRACKING ALGORITHMS FOR GPS OFFSET CARRIER SIGNALS. *Proceedings of the ION National Technical Meeting (NTM '99), January 1999, San Diego, Calif, USA*

27. Hurskainen H, Nurmi J: SYSTEMC MODEL OF AN INTEROPERATIVE GPS/GALILEO CODE CORRELATOR CHANNEL.*Proceedings of the IEEE Workshop on Signal Processing Systems (SIPS '06), October 2006, Banff, Canada*327-332.

28. Akos DM: THE ROLE OF GLOBAL NAVIGATION SATELLITE SYSTEM (GNSS) SOFTWARE RADIOS IN EMBEDDED SYSTEMS.*GPS Solutions* 2003.

29. Dionisio C, Cucchi L, Marracci R: SOFTREC G3, SOFTWARE RECEIVER AND SIGNAL ANALYSIS FOG GNSS BANDS.*Proceedings of the 10th IEEE Internationl Symposium on Spread Spectrum Techniques and Applications (ISSSTA '08), August 2008, Bologna, Italy*

30. Moore GE: CRAMMING MORE COMPONENTS ONTO INTEGRATED CIRCUITS. *Proceedings of the IEEE* 1998,86(1):82-85.

31. Söderholm S, Jokitalo T, Kaisti K, Kuusniemi H, Naukkarinen H: SMART POSITIONING WITH FASTRAX'S SOFTWARE GPS RECEIVER SOLUTION. *Proceedings of the International Technical*

Meeting of the Satellite Division of the Institute of Navigation (ION GNSS '08), September 2008, Savannah, Ga, USA 1193-1200.

32. Won JH, Pany T, Hein GW: GNSS SOFTWARE DEFINED RADIO: REAL RECEIVER OR JUST A TOOL FOR EXPERTS. *Inside GNSS* July-August 2006, 48-56.

33. CRISP PROJECT December 2008, http://www.crisp-project.eu/

34. Heysters P: CRISP PROJECT PRESENTATION. June 2008, http://www.crisp-project.eu/images/publications/D6.1_CRISP_project_presentation_080622.pdf

35. Heysters PM, Rauwerda GK, Smit LT: A FLEXIBLE, LOW POWER, HIGH PERFORMANCE DSP IP CORE FOR PROGRAMMABLE SYSTEMS-ON-CHIP. *Proceedings of the IP/SoC, December 2005, Grenoble, France*

36. Hurskainen H, Raasakka J, Nurmi J: SPECIFICATION OF GNSS APPLICATION FOR MULTIPROCESSOR PLATFORM.*Proceedings of the International Symposium on System-on-Chip (SOC '08), November 2008, Tampere, Finland*128-133.

37. Raasakka J, Hurskainen H, Nurmi J: MODELING MULTI-CORE SOFTWARE GNSS RECEIVER WITH REAL TIME SW RECEIVER. *Proceedings of the International Technical Meeting of the Satellite Division of the Institute of Navigation (ION GNSS '09), September 2009, Savannah, Ga, USA*

6

CHAPTER

AN OVERVIEW OF GNSS REMOTE SENSING

Kegen Yu, Chris Rizos, Derek Burrage, Andrew G Dempster, Kefei Zhan g and Markus Markgraf

School of Geodesy and Geomatics, Wuhan University

ABSTRACT

The Global Navigation Satellite System (GNSS) signals are always available, globally, and the signal structures are well known, except for those dedicated to military use. They also have some distinctive characteristics, including the use of L-band frequencies, which are particularly suited for remote sensing purposes. The idea of using GNSS signals for remote sensing - the atmosphere, oceans or Earth surface - was first proposed more than two decades ago. Since then, GNSS remote sensing has been intensively investigated in terms of proof of concept studies, signal processing methodologies, theory and algorithm development, and various satellite-borne, airborne and ground-based experiments. It has been demonstrated that GNSS remote sensing can be used as an alternative passive remote sensing technology. Space agencies such as NASA, NOAA, EUMETSAT and ESA have already funded, or will fund in the future, a number of projects/missions which focus on a variety of GNSS remote sensing applications. It is envisaged that GNSS remote sensing can be either exploited to perform remote sensing tasks on an independent basis or combined with other techniques to address more complex applications. This paper provides an overview of the state of the art of this relatively new and, in some respects, underutilised remote sensing technique. Also addressed are relevant challenging issues associated with GNSS remote sensing services and the performance enhancement of GNSS remote sensing to accurately and reliably retrieve a range of geophysical parameters.

REVIEW

Recent developments in both ground-based and satellite-borne geospatial infrastructure have opened exciting new opportunities for geodesists to contribute to 'big issues' such as weather, climate, global warming, environment and sustainability. The ability to remotely sense the atmosphere using geodetic techniques has dramatically improved over the past decade, primarily as a result of advances in satellite-borne technologies, large-scale and dense geodetic Global Navigation Satellite System (GNSS) networks, new dedicated space missions and developments of new algorithms and innovative methodologies. Atmospheric sounding is a new area of GNSS applications based on the analysis of radio signals from GNSS satellites, which are refracted as they pass through the atmosphere and can give information on its physical properties. Depending on the way the signals (direct or reflected) are used, GNSS remote sensing can be broadly divided into two categories: GNSS reflectometry (GNSS-R) and GNSS refractometry (known as GNSS radio occultation or GNSS-RO). GNSS-R involves analysing measurements of GNSS signals reflected from the Earth's surface, while GNSS-RO utilises measurements of GNSS signals refracted by the atmosphere when the slanted propagation path is close to the Earth's limb. In the following sections more details about the theory and practice of GNSS remote sensing are provided. Interested readers are also referred to [1, 2] for more background information.

ATMOSPHERIC SENSING VIA GNSS METEOROLOGY

The use of GNSS receivers installed on low Earth orbit (LEO) satellites to measure the refracted radio signals (GNSS-RO) coupled with ground-based atmospheric sounding techniques that use continuously operating reference station (CORS) networks for the determination of water vapour (WV) variability form the basis for GNSS meteorology. The dynamics of WV have a strong influence on weather and climate due to the large energy transfers in the hydrological processes. This is particularly so during the formation and life cycle of severe mesoscale convective storm and precipitation systems. Contrary to its importance, WV remains poorly understood and inadequately measured both spatially and temporally, especially in the southern hemisphere, where meteorological data are sparse.

GNSS atmospheric remote sensing using LEO satellites such as CHAllenging Minisatellite Payload (CHAMP), Gravity Recovery and Climate Experiment (GRACE) and Constellation Observing System for Meteorology, Ionosphere, and Climate (COSMIC) is a robust technique for profiling the atmosphere from LEO satellite orbit heights down to the Earth's surface [3–5]. Many studies have demonstrated that the GNSS-RO technique is able to measure the Earth's atmospheric parameters with unprecedented high accuracy, high resolution and global coverage [6–10]. RO has been considered as a new observation type for meteorology which has the capability to provide significant information on the thermodynamic state of the atmosphere and to allow improvements in atmospheric analysis and prognosis. GPS RO has been operationally used in numerical weather prediction (NWP) models for weather forecasting by major weather forecasting centres such as those in the USA, Europe and Australia, since 2006. The beneficial impact of such data assimilation has been quantified in several studies [11–14], and it is significant particularly in data-void regions (e.g. southern hemisphere). GNSS-RO has been shown to significantly reduce both the forecast errors in the ECMWF system and NWP model biases. It has become the independent sensor technique with the fifth highest impact in operational weather forecasting, among the 24 available systems [12–15]. Most recently, Le Marshall et al. [11] have successfully assimilated MetOp, GRACE and COSMIC RO data into the Australian weather forecasting model (ACCESS) using 4DVAR, and a reliability improvement of 10 h has been achieved in a 1- to 5-day period. GPS RO has been officially used in operational Australian ACCESS system since late 2011.

Building on the success of the COSMIC RO mission, a follow-on mission (COSMIC-2) was recently approved by the USA and Taiwan that will launch six satellites into low-inclination orbits in late 2015 and another six satellites into high-inclination orbits in early 2018. Up to 12,000 high-quality profiles per day will be tracked when it is fully deployed. This revolutionary increase in the number of atmospheric and ionospheric observations will greatly benefit the research and operational communities and will open up unprecedented research opportunities for hurricane analysis and prediction, and studies of Madden-Julian Oscillation and climate processes, to name just a few. In addition, algorithm development, optimisation and refinement, error reduction, climate analysis and benchmarking, and polar applications are all frontier areas of RO research [15–21].

Vertical gradients of the atmospheric properties are retrieved from precise measurements of the radio propagation path between a LEO-based GPS receiver and an occulting GPS satellite transmitter pair. A radio occultation event (ROE) occurs when a GPS satellite is setting or rising behind the Earth's limb as viewed by a receiver onboard a LEO satellite as shown in Figure 1. During a ROE, the transmitted signals from the GPS to the LEO satellite are delayed. This delay can be converted into a profile of bending angles and consequently processed to produce a refractivity index profile. The refractivity is a function of the electron density in the ionosphere, and temperature, pressure and WV in the atmosphere. The fundamental GPS observation equation based upon which atmospheric parameters are retrieved can be expressed as follows:

$$L = \rho + c(\delta t_{\mathrm{clk}} - \delta t_{\mathrm{rel}}) - c(\delta T_{\mathrm{clk}} - \delta T_{\mathrm{rel}}) + N \cdot \lambda + \Delta\varphi_{\mathrm{excess}} + \Delta R + \varepsilon \tag{1}$$

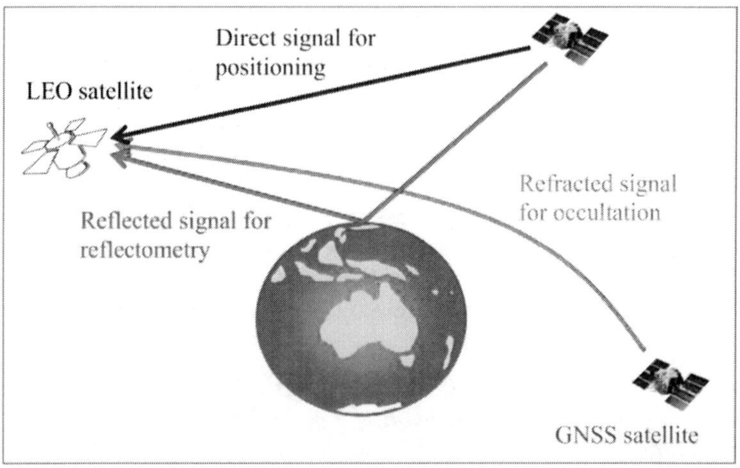

Figure 1. Illustration of the geometry of the direct, reflected and refracted gnss signals and their use for different applications.

where L is the raw GPS carrier phase measurement, ρ is the geometric distance between the LEO and GPS satellites, c is the speed of light, (δt_{clk}, δt_{rel}) are the errors related to the clock and relativistic effects of LEO, (δT_{clk}, δT_{rel}) are the errors related to GPS satellite clock and relativistic effects, N is the integer ambiguity, λ is the wavelength of the GPS L-band signal, $\Delta\varphi_{\mathrm{excess}}$ is the excess phase along the radio signal propagation path which is the phase increment due to the signal propagation delay, ΔR is the general relativistic effect, and ϵ is the measurement noise. Figure 2 shows the flowchart of a typical GNSS-RO atmospheric retrieval process, and Figure 3 shows a typical 1-day RO event distribution across the globe under a full COSMIC constellation.

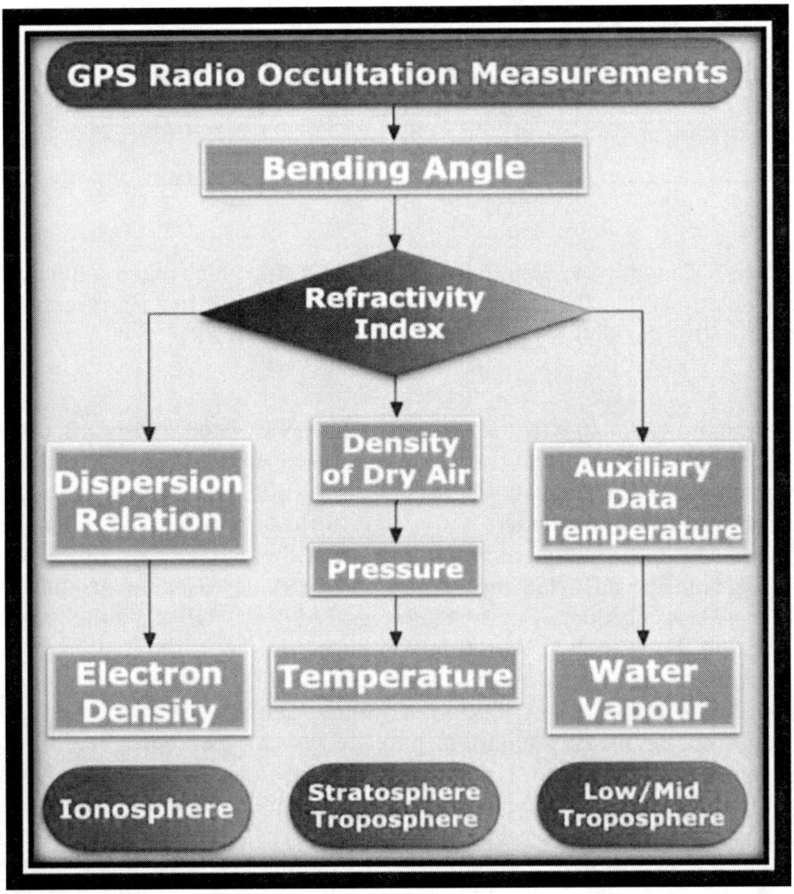

Figure 2. Typical data processing steps involved in gnss-ro atmospheric retrievals.

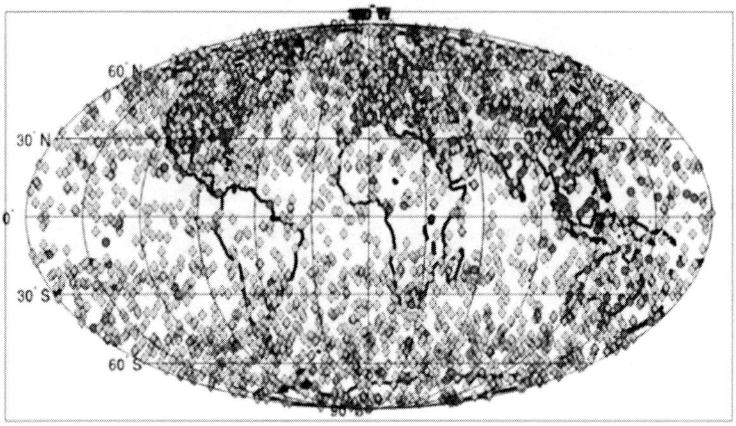

Figure 3. One-day ro event distribution across the globe under a full cosmic constellation. GPS RO sounding locations for COSMIC (in green) and radiosonde sites (in red) in a period of 24 h (after UCAR).

The L-band radio signals broadcast by the GNSS satellites are affected by both the ionospheric and tropospheric effects as they travel through the atmosphere to the permanent ground-based GNSS receivers. Using a dense CORS network, the impact of the neutral atmosphere, i.e. tropospheric delays, can be estimated as a by-product of the geodetic data processing. This ground-based GNSS meteorology is currently used to produce two different data products. Firstly, the integrated precipitable WV in the zenith direction above each GPS station can be accurately estimated using path delay measurements between a receiver and GPS satellites. Secondly, the spatial and temporal distribution of wet refractivity can be reconstructed using accurate measurements of path delays between each satellite-and-receiver link over the GNSS network. This method is called GNSS tomography which is an emerging method for 4D tropospheric reconstruction using a local GNSS CORS network. Integrated measurements of wet delays are extracted from each satellite-to-station link, which are used as the primary observable through a finite 3D grid field to estimate the time-dependent 3D structure of WV. An integrated approach is usually used in the tomographic reconstruction process by incorporating multiple sources of measurements from GNSS CORS, automated weather stations, radiosonde, RO and NWP models.

A variety of research has been conducted based on both real and simulated data. For example, [22] presented a simulation and real data investigation into 4D tropospheric tomography using GPS observations and a stochastic approach. The real data analysis used the Kilauea network in Hawaii with the voxel grid discretisation covering an area of 400 km^2 and 15 km in height. A Kalman filtering technique was used for estimating the evolution of dynamically changing parameters. An alternate method presented in [23] used raw GPS phase measurements directly for the reconstruction of WV distribution. This method was tested using simulations where the horizontal and vertical voxel resolution, observation noise and additional GNSS constellations were varied and optimised. Results indicated that high accuracy can be achieved in the modelling of wet refractivity in lower layers. However, its limitation was in resolving layers above 4,000 m altitude. Nilsson and Gradinarsky [24] investigated the benefits of GPS tomography for modelling mesoscale precipitation systems. The model was designed on the premise that the grid system would move in relation to the vector of horizontal wind velocity. It was concluded that GPS tomographic solutions can improve the initial WV field for short-term NWP predictions and forecasting. Bender and Raabe [25] extensively analysed the preconditions and constraints of ground-based GPS WV tomography and suggested that the spatial resolution is limited by the density of the ground-based GPS networks and the number of available GNSS satellites. To optimise horizontal resolution, the horizontal size of the humidity field voxels would be restricted to the mean interstation distance of the GPS network. Another important parameter presented is the elevation cutoff angle; the number of available rays increases heavily with a decreasing elevation cutoff angle and in return provides much more information on the lower troposphere. Rohm and colleagues [26, 27] presented investigations into the concept of near-real-time atmospheric model construction based on the GNSS and meteorological observations. This nowcasting approach provides accurate comparisons with NWP analysis models. Yuan et al. [28] recently investigated real-time retrieval using precise point positioning technique and good results were obtained. Choy et al. [29] investigated the propagation and distribution of WV during the Melbourne severe weather event in 2010, while Manning and colleagues [30, 31] investigated 4D methods of reconstructing wet refractivity fields using GPS tomography for the Australian region. A comprehensive investigation was carried out in terms of sensitivity analysis of multiple measurements, voxel optimisation, parameterisation method and Kalman filtering using two severe storm events in Melbourne as case studies. Bennitt and Jupp [32] studied the data assimilation of GPS zenith total delays into the Met Office NWP models, and a slight positive overall impact and significant improvements to short-range forecasts in some cases were demonstrated.

To summarise, GNSS is now an established atmospheric observing system which can accurately sense WV, the most abundant greenhouse gas, accounting for 60 to 70% of atmospheric warming. Severe weather forecasting is challenging, in part due to the high temporal and spatial variation of atmospheric WV. WV content is undersampled in the current meteorological and climate observing systems; obtaining and exploiting more high-quality humidity observations is essential to NWP data assimilation, nowcasting and climate monitoring. It is expected that the GNSS meteorology technology will significantly advance our knowledge of Earth's atmospheric structure and processes, including severe weather phenomena and climate change, in particular in data-sparse areas such as the southern hemisphere, and oceanic and polar regions.

OCEAN OBSERVATION

The complex temporal and spatial variations which shape the ocean surface are amenable to study using remote sensing methods because the sea surface both emits and strongly reflects electromagnetic radiation. Static ocean surface undulations are dominated by Earth's geoid and by its topographic response to the mean ocean circulation, while temporal departures from this surface are forced by weather and climate-related changes in ocean temperature and salinity, and surface wind stress. Lunar and solar gravitational tides, and their harmonics, contribute additional variability, on daily and longer timescales. These sea level changes, which also drive ocean currents, have been sensed remotely using nadir-viewing satellite-borne radar altimeters, almost continuously over the last four decades. Observations from these altimeters provide vital information for studies of deep-ocean circulation and boundary currents, the mid-ocean gyres, tsunamis and ocean currents on synoptic to global scales, while mesoscale (approximately 100 km) ocean eddies are only marginally resolved.

Finer-scale topographic changes associated with ocean surface roughness, or 'sea state', are dominated by locally forced wind-waves and swell from distant storms occurring on time and space scales exceeding a few hours and kilometres. In addition to their effects on shipping and other human activities, these contribute to ocean mixing, which crucially affects the exchange of momentum, heat, salt and various gases, such as CO_2, across the air-sea interface, with important implications for weather and climate studies. Sea state also affects the interaction of electromagnetic radiation with the sea surface through its impact on the effective emissivity and reflectivity of the surface. By using suitable remote sensing methods, its effects can be used to deduce sea state or mean square slope. Coupled with a wind-wave spectral model, the surface wind field can also be inferred. Alternatively, these effects can be used to derive corrections for roughness influence on retrievals of other geophysical parameters from microwave radiometers, such as sea surface salinity. Since sea state modifies the waveform of the radar reflection, nadir-viewing (monostatic) radar altimeters are also used to estimate wave height and wind speed, with high spatial resolution (approximately 7 km) along-track but poorer resolution (approximately 300 km) across-track and with an approximately 10-day repeat cycle [33]. However, over the last three decades, a new approach to measuring sea surface height, wind speed and sea state bistatic radar, offering higher spatial and temporal resolution, and utilising GNSS satellites as transmitters has been proposed and developed.

Ocean observation using GNSS signals was first proposed as the Passive Reflectometry and Interferometry System (PARIS) concept to perform mesoscale ocean altimetry [34], and an in-orbit demonstrator has been designed to implement this concept [35]. Recent developments in GNSS-R altimetry include experimental ground-based and airborne campaigns, performance studies, refinements in error budget estimates and alternative signal processing methods for retrieving sea surface height estimates from GNSS-R instrumentation. The new methods exploit a variety of techniques for enhancing performance characterised by improvements to spatial and temporal resolutions, signal to noise ratio and/or accuracy and precision. With respect to error characterisation, Martin-Neira et al. [36] studied the effects of errors in delay drift compensation on the waveform and its impact on the altimetric accuracy and precision, while Pascual et al. [37] used simulated GPS and Galileo signals to investigate the impact of bandwidth and low-power transmitters on height accuracy and precision. Rius et al. [38] investigated the scatterometric and specular delay observables and considered various sources of uncertainty including observation statistics and model parameterisation or calibration error. Methodology studies include an investigation of model-free retrieval of sea surface height using a power ratio method combined with cost function optimisation when the sea surface is rather rough [39]. A semi-codeless processing approach [40] showed how to determine the phase sign variation of encrypted P-code GNSS signals and combine it with the CA signal stream to enhance signal to noise ratio in comparison with that achieved by the interferometric method. An investigation of Doppler multilook processing [41] realised enhanced spatial and temporal resolution by intra- and inter-Doppler tracking and averaging of delay waveforms. The method demonstrated 25 to 30% improvement in altimetric precision over the interferometric method and increased flexibility in trading off spatial resolution and height estimation accuracy. In a ground-based study reported in 2001 [33], retrieval of Crater Lake surface height to 2-cm precision was achieved using GPS altimetry from relatively low satellite elevations. A year later, 5-cm precision GPS ocean altimetry was achieved from an aircraft [42]. A ground-based experiment implementing the PARIS concept in the form of an X-band instrument [43] takes advantage of high signal to noise ratio (SNR) associated with X-band signals of opportunity from digital satellite TV broadcasts. Useful overviews including summaries of these and/or other GNSS altimetry studies appear in [2, 44, 45], along with an assessment of the overall performance of GNSS-R altimetry techniques.

Using airborne data collected in 1997, Garrison et al. [46] first demonstrated the potential for retrieving sea state from the dependence of the GNSS sea surface reflection waveform width on wind speed. This waveform represented the cross-correlation between a replica of the direct signal and the signal reflected off the sea surface, as a function of delay. Coupled with knowledge of the GPS navigation code autocorrelation function (which is usually triangular in shape), this gives an estimate of the radar cross section that is closely related to the ocean surface mean square slope and hence surface roughness. During the period 2003 to 2011, the UK-Disaster Monitoring Constellation (UK-DMC) satellite carried an experimental GNSS reflectometer, which yielded sea state and wind speed estimates for representative cases of low to moderate wind strength [47]. While suitable for scatterometric measurements, this instrument was not optimised for altimetric measurements.

Since Garrison's first experiment, a variety of techniques have been developed for retrieving surface mean square slope and wind speed, and many have been tested using ground and airborne field campaigns. The following discussion of some key studies exemplifies the progress made but is not meant to be comprehensive: Retrieval of ocean wind speeds from a balloon in the stratosphere [48] produced mean square slope and corresponding wind speed estimates departing from radiosondes and satellite altimeters and a scatterometer by at most 1 m/s in relatively calm winds (less than about 9 m/s). Retrieval of hurricane winds using GPS reflectometry was first demonstrated by Katzberg et al. [49] who later calibrated retrieved hurricane winds using dropsondes [50, 51]. The GNSS-R reflectometer was found to underestimate observed winds, but a standard deviation around a linear regression fit to observations of about 5 m/s was found, for winds ranging from 10 to 40 m/s. Recently, wind speeds have been retrieved using GNSS-R instrumentation from an aircraft using scattering model-free linear regression of several Delay Doppler Map (DDM) observables [52]. The results obtained using satellite triplets, and validated using coincident dropsonde measurements, produced errors of up to 1 to 4 m/s for winds of 5 to 25 m/s, with overall accuracy and precision of the order of 1 to 2 m/s.

Methods to extract directional roughness characteristics have been developed and tested using airborne campaign data [53–55]. Wind direction retrieval is performed by accounting for differences in upwind/crosswind surface slope probability distribution, which is represented by an observable asymmetry in the glistening zone. The particular technique devised by Cardellach and Rius [55] which does not require an assigned probability distribution function can resolve the 180° directional ambiguity inherent in other GNSS-R and radar scatterometry processing methods. Retrieval of roughness characteristics using relatively powerful signals of opportunity from S-band satellite radio broadcasts produced SNR values greatly exceeding those of GPS transmissions [56]. These broadcasts can provide high-quality GNSS-R + data along the coastal margins but are not available over the deep ocean.

Useful reviews of ground-based and airborne GNSS-R campaigns, geophysical models and signal processing techniques for a variety of oceanographic applications are available [2, 44]. Campaigns performed over ocean surfaces have spanned locations in the North Sea, the Baltic Sea, the Mediterranean Sea and the Atlantic Ocean [44], and the Tasman Sea in the western South Pacific [39]. Newly developed GNSS-R signal processing techniques, exploiting favourable sampling geometries or instrument refinements, can now be used to extract such factors as significant wave height and roughness field directionality [38, 54] or to relate ocean roughness to surface brightness temperature measured at microwave frequencies [57, 58]. The possibility of retrieving ocean surface temperature and salinity from permittivity estimates derived from polarimetric GNSS-R measurements has also been investigated [59]. Such a development could complement the retrieval of sea surface salinity from current L-band radiometry missions (ESA's SMOS and NASA's Aquarius), which is strongly affected by the influence of roughness effects on emissivity.

A well-established approach to estimate sea state and wind speed using GNSS-R measurements is based on the theoretical model of [60], which solves the bistatic radar equation of the time-delayed scattered power for the radar cross section σ_0:

$$\langle |Y(\tau.f_c)|^2 \rangle = T_i^2 \iint_A \frac{G^2(\vec{\rho})|S(f_c - f_D(\vec{\rho}))|^2 \Lambda^2 (\tau - (R_0(\vec{\rho}) - R(\vec{\rho}))/c)}{4\pi R_0^2(\vec{\rho}) R^2(\vec{\rho})} \sigma_0(\vec{\rho})\, d^2\vec{\rho} \quad (2)$$

Here, τ is the time delay relative to the specular reflection; T_i is the averaging time; A is the reflecting surface area; $\vec{P}$ is the reflecting surface position vector; and R_0 and R are the distances from a surface point to the transmitter and receiver, respectively; and f_c and f_D are the frequencies of the replica signal and the reflected Doppler signal, respectively. The terms G, S and Λ represent the antenna pattern, the sinc function for the Doppler mismatch and the navigation signal correlation function, respectively. In general, this equation is solved in 2D delay Doppler space using various geometric descriptions of the corresponding DDM [61]. However, it is often simpler and faster to perform retrievals in 1D delay space by deriving the corresponding delay waveforms. In the limit set by the geometric optics theory, σ_0 is proportional to the probability density function of surface gravity wave slopes from which the mean square slope (MSS) may be derived. A semi-empirical wind-wave spectrum [62] is typically used to infer wind speed at standard meteorological height from the derived MSS.

An example of wind speed estimation, based on the model fitting method, appears in Figure 4[63]. Five theoretical 1D delay waveforms are shown corresponding to five different wind speeds (3, 4, 5, 6 and 7 m/s), with a measured delay waveform from GPS satellite PRN#13 superimposed. The data were collected by an airborne receiver at an altitude of about 3 km. The measured waveform is the result of coherent integration of 1-ms intermediate frequency (IF) signals followed by incoherent integration of 1,000 of these 1-ms integrals. The measured waveform matches well with the theoretical waveform of wind speed 4 m/s, which closely matches the observed wind speed of 4.4 m/s.

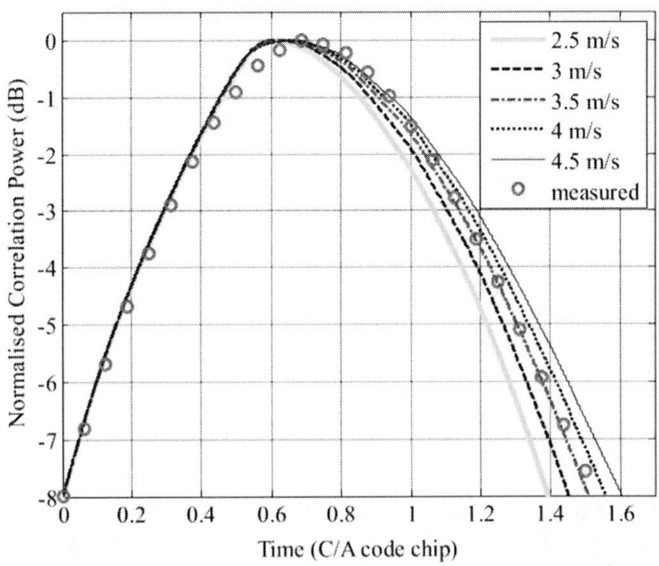

Figure 4. Wind speed estimation through matching the measured delay waveform with modelled waveforms.

Recent successes in global mapping of sea surface salinity using L-band radiometers aboard ESA's SMOS and NASA's Aquarius satellites provide additional motivation for developing techniques to deduce corrections for roughness influence on L-band emissivity. Several airborne missions have demonstrated this potential, and the planned launch of NASA's Cyclone Global Navigation Satellite System (CYGNSS) L-band reflectometer constellation in 2016, with the goal of mapping surface winds at a high temporal resolution inside tropical cyclones, will allow this concept to be tested in space. Recent extensions of GNSS-R to communications satellites operating on higher frequency bands, with improved signal strength, demonstrate enhanced precision of ocean reflectometry measurements with the possibility of expanded coverage in coastal regions [43, 56, 64].

LAND APPLICATIONS

Knowledge of soil moisture content is critical for drought and irrigation management, so as to increase crop yields and to gain a better understanding of natural processes linked to the water, energy and carbon cycles. Soil moisture can be estimated by analysing surface-scattered signals transmitted and received by active radar sensors or natural surface emission detected by microwave radiometers. Alternatively, GNSS-R signals can be employed to estimate soil moisture. Since the soil dielectric constant and reflectivity depend on soil moisture, the variation in the reflected GNSS signal power, or SNR, indicates changes in soil moisture [65]. If feasible, the empirical dielectric model based method in [66] may be applied to GNSS-based soil moisture retrieval. In this method, the transition moisture parameter (θ_t) is defined as

$$\theta_t = 0.165 + 0.49\theta_{wt} \tag{3}$$

where θ_{wt} is the wilting point moisture that is a function of the percentage of the clay and sand contents. Depending on whether the soil moisture is greater or less than the transition moisture, a specific formula can be used to calculate the soil moisture. This model employs the mixing of either the dielectric constants or the refraction indices of ice, water, rock and air and treats the transition moisture value as an adjustable parameter. The dielectric constant of the soil can be estimated by measuring the surface reflectivity. The model is quite general since it was developed by considering many different types of soils.

An alternative GNSS-R technique for soil moisture estimation is the interferometric method in which the power associated with the coherent sum of the direct and reflected GNSS signals is exploited. After continuously measuring the total power for a few hours, the location of the power notch over satellite elevation can be observed, and the soil moisture may be estimated [67]. This notch-based approach is limited to the case where a vertically polarised antenna is used and fixed as horizon-looking. It would not be applicable to standard geodetic antennas. It is an advantage to use existing GPS receivers, installed primarily for geophysical and geodetic applications, to estimate soil moisture [66]. The zenith-looking antenna at a geodetic reference receiver station captures both direct and reflected GPS signals, although the reflected signals are received at a negative elevation angle where the antenna radiation gain is non-zero. The amplitude (or SNR) and phase (or frequency) of the multipath signal can be used to infer soil moisture information [68]. These receivers, if exploited effectively, could provide a global network for services on soil moisture, vegetation and snow cover [69, 70]. GNSS-based soil moisture estimation is complicated by a number of issues including surface roughness, vegetation canopy and variation in percentage of individual soil components. To achieve reliable soil moisture estimation, these issues must be taken into account through processes such as modelling and compensation.

Forest change detection is another possible application of GNSS remote sensing. Forest change can provide useful information about global climate change impacts and carbon storage, and knowledge of forest change is vital for effective forest management. Received signal strength of the reflected GNSS signals can be employed to distinguish forest conditions from each other, since different surface covers have different reflectivities. A number of signal strength ranges may be defined to be associated with a group of surfaces such as lake/river water, typical dense forest and cleared area due to logging [71]. Figure 5 shows four ground specular reflection tracks of the reflected signals from four GNSS satellites received by a receiver on an aircraft. The tracks are colourised by the reflected signal power. The low signal power corresponds to the dense forest areas, while the high signal power occurs over three areas marked by A1 (cleared area), A2 (partially cleared area) and A3 (lake water). Accordingly, a surface can be classified by determining within which range its measured strength falls. In addition to signal strength, other signal characteristics such as those related to the observed correlation waveform may be used to enhance surface change detection or surface classification. The major issue related to surface change detection is the dependence of reflectivity on several factors such as soil moisture and surface roughness. Forest change may be quantified by evaluating forest biomass, which depends on the volume of both living and dead trees (leaves, branches and trunks). The received signal power can be used to determine the scattering coefficient which, based on simulation [72], should be a function of the biomass. However, it is a challenging problem to derive a formula that accurately describes the relationship between the scattering coefficient and the biomass.

Figure 5. Ground specular reflection tracks associated with four gps satellites, colourised by reflected signal power. The picture was generated by Google Earth and GPS Visualizer.

CRYOSPHERE APPLICATIONS

GNSS-R has been used to sense sea ice and snow, mid- to high-latitude continental snow and deep sub-surface layers of Antarctica's dry snow. Three main parameters of sea ice (i.e. thickness, surface roughness and ice permittivity) can be retrieved with GNSS-R. These parameters can be combined to help characterise different ice types including new ice, young ice, thin first-year ice, first-year ice, and multiyear ice. Sea ice thickness is a key parameter for classification and characterisation of sea ice masses, which influence the temperature and circulation pattern of both the ocean and atmosphere and thus can be used for analyses of the Earth's climate. In the case of reflections over a sea ice surface, it is possible to track phase delay precisely so as to retrieve the ice thickness through phase altimetry. ESA initiated a GPS sea-ice campaign, and a ground-based experiment was conducted in Greenland over a 7-month period in order that the complete process of sea ice formation and melting was monitored [73, 74]. The time series of the retrieved heights is in agreement with an Arctic tide model, and there is a consistency between results obtained using different antenna polarisations.

Similar to sea surface wind speed retrieval, model fitting between the theoretical and measured delay waveforms can be used to determine the roughness of an ice surface [75]. In the transition from newly formed sea ice to older and thicker ice, a gradual decrease was observed in bulk salinity and an increase in surface roughness. Roughness may also be inferred by investigating the characteristics of the phase or coherence of the interferometric signal [73, 76]. Permittivity of sea ice is another important parameter which can be estimated by determining the ratio between the co-polar and cross-polar components of the reflected signal [44, 76] or by model fitting of delay waveform [77].

Snow is a critical storage component in the hydrologic cycle and an important component of the climate system; however, current in situ observations of snow distribution are sparse. By making use of the networks of GPS stations established for geodetic applications, it is possible to monitor snow distribution on a global scale. The SNR and phase of the multipath signal are both a function of the antenna height above the ground. The presence of a snow pack around the antenna reduces the antenna height and hence changes the multipath signal SNR [78, 79]. The multipath signal SNR is a periodic function with a carrier phase given by

$$\varphi_m = \frac{4\pi h}{\lambda}\sin\theta \tag{4}$$

where h is the antenna height relative to the ground surface, λ is the wavelength, and θ is the satellite elevation angle. Differentiating the phase with respect to $\sin\theta$ will produce the signal frequency

$$f_m = \frac{2h}{\lambda} \tag{5}$$

As a consequence, the antenna height, and hence the snow depth, can be estimated by measuring the frequency of the multipath signal SNR. It is worth mentioning that GNSS signal reflection off Antarctic dry snow can be significantly different from that of typical wet snow surfaces. The L-band GNSS signals can penetrate into the dry snow down to 200 to 300 m, producing a complex interference pattern induced by the multiple reflections occurring at different layer interfaces of the sheet [80]. Radio-holographic techniques are used on each lag of the delay waveform to identify the spectral content of the signal and to identify each frequency component to different snow depths.

MISSIONS

Useful reviews of radio occultation missions can be found in [81] and [82]. The concept was first proved by the GPS/MET (GPS/Meteorology) satellite mission in 1995 to 1997. That satellite took few measurements but was followed by the more productive CHAllenging Minisatellite Payload (CHAMP) and Satellite de Aplicaciones Cientificas-C (SAC-C) satellites. CHAMP provided 8 years of radio occultation data consisting of around 440,000 measurements from February 2001 to October 2008 [83]. Those missions led to the launch in 2006 of a six-satellite constellation FORMOSAT-3/COSMIC, which provided 1,500 to 2,000 soundings per day, resulting in GPS RO becoming an operational data source for weather prediction and ionospheric monitoring. Other missions that provided significant quantities of RO data are Gravity Recovery and Climate Experiment (GRACE-A), METeorological Operations (METOP-A), Communications/Navigation Outage Forecasting System (C/NOFS), TerraSAR-X and TanDEM-X. Table 1 (taken from the COSMIC website [84]) shows the contributions made by these and other missions to both atmospheric and ionospheric sounding. With the COSMIC constellation degrading, as it has reached its design life, a new FORMOSAT-7/COSMIC-2 follow-on constellation is being constructed which has multifrequency, multi-GNSS receivers onboard 12 satellites to be launched - six into low-inclination orbits in 2015 and six into high-inclination orbits in 2018 [85].

Far fewer satellites have been able to perform GNSS reflectometry. The first was UK-DMC from 2003 to 2011 [86]. Others that have been proposed are Techdemosat-1 launched July 2014 [87] with a custom reflectometry payload; PARIS, a dedicated mission for GNSS-R [35,88]; CYGNSS, a constellation of eight microsatellites [89]; GEROS-ISS (GNSS REflectometry, Radio Occultation and Scatterometry onboard International Space Station) [90]; and Cat-2 [91]. Several of the authors are also from the University of New South Wales (UNSW), which is preparing a CubeSat to carry both GNSS occultation and reflectometry experiments as part of the QB50 mission [92]. As illustrated in Figure 1, a LEO satellite to be launched for the QB50 mission can receive the direct, reflected and refracted GNSS signals to enable positioning, reflectometry and occultation applications.

Table 1. Contributions to atmospheric and ionospheric sounding by missions

MISSION	Total atm occs	Total ion occs
CHAMP	399,968	303,291
CNOFS	120,588	0
COSMIC	4,039,311	3,707,966
GPSMET	5,002	0
GPSMETSA	4,666	0
GRACE	273,013	132,817
METOPA	993,084	0
SACC	353,808	0
TSX	276,549	0
Total	6,465,989	4,144,074

Data obtained from the COSMIC website [84] (last updated: Sat Dec 7 23:25:02 MST 2013).

INSTRUMENTS

Even though primarily designed for precise navigation, surveying and geodesy, high-end dual-frequency GPS receivers have been used for more than two decades for a diverse set of remote sensing applications. Ground-based receiver networks such as that maintained by the International GNSS Service (IGS) are routinely used to monitor the tropospheric vapour content and the ionospheric electron content. Multifrequency, multiconstellation receivers that are now available will further improve the spatial and temporal resolution in the near future. With slight modifications of standard geodetic receivers (higher data rate, special observation types), the so-called scintillation receivers can be used to study the short-term variability of the atmosphere, including scintillation, ionospheric storms or travelling ionospheric disturbances [93].

For satellite-borne GNSS applications, various modifications of terrestrial receivers are required, including hardware and software modifications to enhance survivability in a space environment (temperature, vacuum, radiation) and to ensure a proper acquisition of signals despite the high orbital velocity and the associated Doppler shifts. On the other hand, special hardware and software is required to support non-navigation remote sensing applications in space, such as radio occultation measurements. Well-known examples include the BlackJack and Integrated GPS Occultation Receiver (IGOR) flown on numerous research satellites [94], or the GRAS instrument flown on MetOp as shown in Figure 6[95]. These offer a high-rate sampling (50 Hz to 1 KHz) of raw phase observations or correlator outputs and open-loop tracking techniques. A model of the expected Doppler shift is used during occultations in the deep layers of the atmosphere to facilitate tracking at very low signal levels. Radio occultation receivers are also operated with special beam-forming antennas to increase the antenna gain in directions close to the Earth's limb [95].

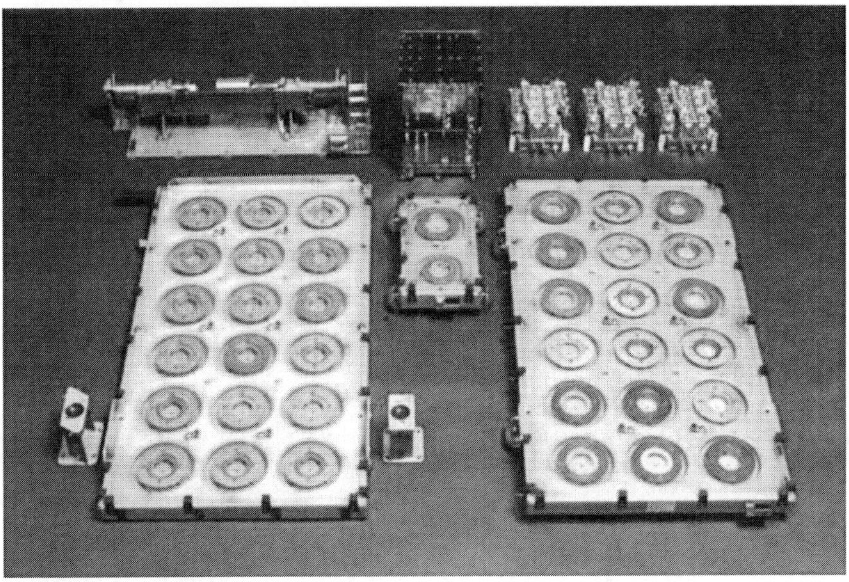

Figure 6. IMAGE OF THE GRAS INSTRUMENT FLOWN ON METOP-A. Images showing the GRAS instrument with both occultation antennas (front left and front right), the Zenith antenna (front middle) and the R/F front ends and main electronics unit (in the back).

While the applications discussed so far can largely be fulfilled with receivers using traditional correlator application-specific integrated circuits (ASICs), GNSS-R calls for a higher flexibility in the receiver design. As discussed in the previous sections, GNSS-R may be used for a variety of remote sensing applications such as studies of the sea surface height and roughness or salinity and soil moisture from the properties of passively reflected GNSS signals. Besides special antennas (down-looking, high-gain or beam-steering antennas with left-hand polarisation) [35], such receivers commonly employ a large number of slaved correlators and often exceed the capabilities of traditional ASICs. This is even truer for new advanced signal processing techniques. Recently, two novel correlation schemes have been introduced: the so-called interferometric processing [35] and the direct-signal-enhanced semi-codeless processing [96]. Both methods do not correlate the reflected signal with a clean replica signal, but directly cross-correlate the line-of-sight signal received from a zenith-looking antenna either with the reflected signal from a nadir-looking antenna for the first method or with a modelled replica of the encrypted signal obtained in a semi-codeless way for the second method. Key advantages of these new schemes are that no prior knowledge of the incoming signal is required and there is a significant improvement in the measurement precision and the carrier to noise ratio. For altimetric applications in particular, which are probably the most demanding applications in terms of the required instrument hardware and processing capabilities, this is of crucial benefit.

The disadvantage of these novel methods is that the correlation requires a substantially increased signal processing effort. To cope with these needs, a software-defined radio approach is widely applied. In the extreme case, pure software receivers have been employed by various research groups for offline processing of the pre-recorded radio frequency (RF) signals, while field-programmable gate arrays (FPGA) or fast digital signal processors (DSPs) are commonly used in real-time-capable reflectometry receivers. Examples of the latter type of receivers include the GOLD-RTR [97] and SSTL's SGR-ReSI [98] instruments developed by IEEC, Barcelona, and SSTL, Surrey, respectively.

CHALLENGING ISSUES AND FUTURE DIRECTIONS

GNSS has a global space-based positioning, navigation and timing (PNT) capability. There are many technological challenges for GNSS remote sensing to come close to matching the extraordinary success of GNSS-PNT, especially in the case of the GNSS-R technique. Satellite-borne GNSS remote sensing must demonstrate its value as a reliable, continuous, high-quality sensing technology. With the launch of multiple LEO satellites with GNSS-RO- and GNSS-R-capable instrumentation, it seems we are at last close to a renaissance. But are more LEO missions and better algorithms for geophysical parameter extraction sufficient? In the case of GNSS-RO, the answer is yes because the meteorological/climate community now assimilates GNSS-RO products into their operational and research systems and is eager for more [11–15]. However, GNSS-R is still very much a novelty technology as far as the geosciences community is concerned because current results are generated by one or more isolated academic group and not organised into independently verifiable products. Nevertheless, as mentioned earlier, it is a fact that the Plate Boundary network is being used in the USA for a range of GNSS-R applications.

The International Association of Geodesy (IAG) has been very successful in launching technique-specific services (see [99]). These services aid other geoscientists in their research, as well as supporting important applications in the wider community. Examples of the latter are the contribution of the International GNSS Service (IGS) to precise positioning [100] and the International Earth Rotation and Reference Systems Service (IERS) to the International Terrestrial Reference Frame (ITRF) [101]. The IGS does generate troposphere and ionosphere parameter products [102]; however, they are based on observations made from the global ground-based GNSS tracking network (Figure 7), not from satellite platforms. There are no geodetic services producing GNSS remote sensing products on a continuous, synoptic basis. From the IAG's perspective, the challenging issues are related to 'operationalising' GNSS remote sensing, in all its forms, so that the remotely sensed geophysical parameters (atmospheric, oceanic, wind, soil moisture, biomass, etc.) are included in the suite of geodetic outputs of the Global Geodetic Observing System (GGOS) [103].

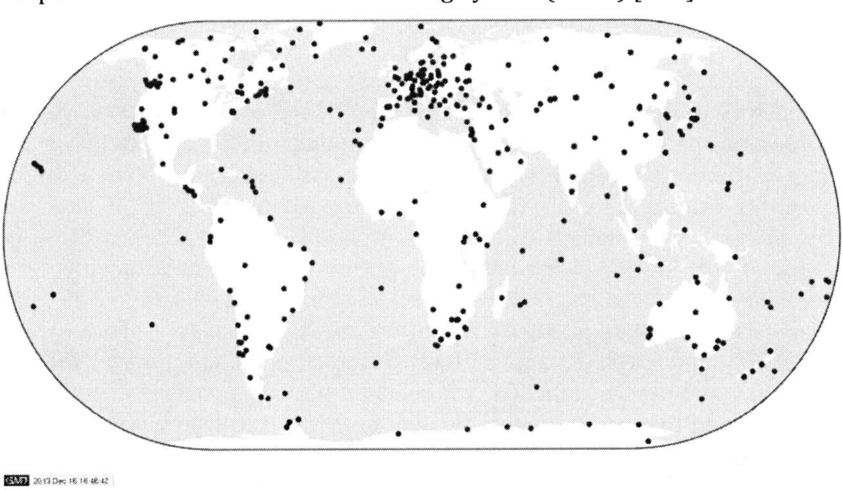

Figure 7. IGS OBSERVATIONS MADE FROM THE GLOBAL GROUND-BASED GNSS TRACKING NETWORK. One of the reasons for the success of IAG

services are the permanent geodetic observing networks such as IGS's global GNSS tracking network, providing accurate, reliable data products for science and society.

The non-technical challenges therefore include the evolution of current GNSS remote sensing science missions to operational services by*increasing* the number of LEO satellites equipped with GNSS receivers to satisfy coverage requirements in space and time; *standardising*data formats, instrumentation, and calibration, pre-processing and analysis procedures; and establishing a *coordinating* agency or authority so as to ensure continuous, high-quality product generation and dissemination. The IAG has considerable experience with geometry technique-based services (such as the IGS, ILRS, IVS and IDS) [104]; however, it has not yet established any 'geodetic imaging' services based on technologies such as synthetic aperture radar (SAR), satellite radar altimetry, lidar or GNSS remote sensing (GNSS-RO, GNSS-R). One of the future challenges is to address this shortcoming, so that GNSS remote sensing can be recognised as a geodetic technique that is making critical contributions to science and society.

CONCLUSIONS

This paper provided an overview of GNSS remote sensing (GNSS-RO and GNSS-R) with a focus on six different issues. Atmospheric sensing using GNSS-RO measurements has been extensively investigated for determining atmospheric parameters such as water vapour, temperature, pressure and humidity. Significant benefits have been shown for the southern hemisphere where there is limited access to ground-based observations, and GNSS-RO has been operationally used by major international weather forecasting models. GNSS-R, on the other hand, which has recently gained more attention from research organisations and universities, can be used to perform remote sensing tasks related to ocean observation and land applications, retrieving a wide range of geophysical parameters including sea surface wind speed and direction, sea surface height and roughness, soil moisture, biomass, snow/ice depth, coverage and density. The most important past and planned satellite missions primarily associated with GNSS-RO and GNSS-R applications were summarised. More GNSS-RO missions have been launched, while a few GNSS-R missions are planned. The state-of-the-art characteristics of the GNSS receiver were discussed, which is the key equipment for GNSS remote sensing. Finally, a number of challenging issues related to GNSS remote sensing products and services were highlighted.

REFERENCES

1. Gleason S, Gebre-Egziabher D: *GNSS Applications and Methods.* Boston: Artech House; 2009.
2. Jin S, Cardellach E, Xie F: *GNSS Remote Sensing: Theory, Methods and Applications.* Dordrecht: Springer; 2014.
3. Liou Y-A, Pavelyev AG, Liu S-F, Pavelyev AA, Yen N, Huang C-Y, Fong G-J: FORMOSAT-3/COSMIC GPS radio occultation mission: preliminary results. *IEEE Trans Geosci Rem Sens* 2007, 45(11):3813-3826.
4. Pavelyev AG, Zhang K, Liou Y-A, Pavelyev AA, Wang C-S, Wickert J, Schmidt T, Kuleshov Y: Principle of locality and analysis of radio occultation data. *IEEE Trans Geosci Rem Sens* 2013, 51(6):3240-3249.
5. Pavelyev AG, Zhang K, Matyugov SS, Liou YY-A, Wang C-S, Yakovlev OI, Kucherjavenkov IA, Kuleshov Y: Analytical model of bistatic reflections and radio occultation signals. *Radio Sci* 2011., 46(RS1009): doi:10.1029/2010RS004434
6. Rius A, Ruffini G, Romeo A: Analysis of ionospheric electron density distribution from GPS/MET occultations. *IEEE Trans Geosci Rem Sens*1998, 36(2):383-394.
7. Yoon YT, Eineder M, Yague-Martinez N, Montenbruck O: TerraSAR-X precise trajectory estimation and quality assessment. *IEEE Trans Geosci Rem Sens* 2009, 47(5):1859-1868.
8. Norman RJ, Dyson PL, Yizengaw E, Marshall JL, Wang C-S, Carter BA, Wen D, Zhang K: Radio occultation measurements from the Australian microsatellite FedSat. *IEEE Trans Geosci Rem Sens* 2012, 50(11):4832-4839.
9. Li H, Yuan Y, Li Z, Huo X, Yan W: Ionospheric electron concentration imaging using combination of LEO satellite data with ground-based GPS observations over China. *IEEE Trans Geosci Rem Sens* 2012, 50(3):1728-1735.
10. Zhang K, Fu E, Silcock D, Wang Y, Kuleshov Y: An investigation of atmospheric temperature profiles in the Australian region using collocated GPS radio occultation and radiosonde data. *Atmos Meas Tech* 2011, 4: 2087-2092.

11. Le Marshall J, Xiao Y, Norman R, Zhang K, Rea A, Cucurull L, Seecamp R, Steinle P, Puri K, Fu E, Le T: The application of radio occultation observations for climate monitoring and numerical weather prediction in the Australian Region. *Aust Meteorol Oceanogr J* 2012, 62: 323-334.

12. Cardinali C: Forecast sensitivity to observation (FSO) as a diagnostic tool. *ECMWF Tech Memo* 2009., 599:

13. Healy SB, Jupp AM, Marquardt C: Forecast impact experiment with GPS radio occultation measurements. *Geophys Res Lett* 2005. doi:10.1029/2004GL020806

14. Cucurull L, Derber JC: Operational implementation of COSMIC observations into NCEP's global data assimilation system. *Weather Forecast*2008, 23: 702-711.

15. Anthes RA, Bernhardt PA, Chen Y, Cucurull L, Dymond KF, Ector D, Healy SB, Ho S-P, Hunt DC, Kuo Y-H, Liu H, Manning K, McCormick C, Meehan TK, Randel WJ, Rocken C, Schreiner WS, Sokolovskiy SV, Syndergaard S, Thompson DC, Trenberth KE, Wee T-K, Yen NL, Zeng Z: The COSMIC/FORMOSAT-3 Mission: early results. *Bull Am Meteorol Soc* 2008, 89: 313-333.

16. Norman R, LeMarshall J, Rohm W, Carter B, Liu C, Zhang K: The impact of severe weather events on GPS signal propagation. *IEEE Int J Appl Earth Obs Rem Sens* 2014., 7(8): in press

17. Li Y, Kirchengast G, Scherllin-Pirscher B, Wu S, Schwaerz M, Fritzer J, Zhang S, Carter B, Zhang K: A new dynamic approach for statistical optimization of GNSS radio occultation bending angles for optimal climate monitoring utility. *J Geophys Res* 2014, 118(23):13,022-13,040.

18. Wang CS, Norman R, Yeh TK, Zhang K, Choy SL, Tseng TP: Investigation into the atmospheric parameters retrieved from ROPP and CDAAC using GPS radio occultation measurements over the Australian area. *Aust J Earth Sci* 2014.

19. Liu CL, Kirchengast G, Zhang K, Norman R, Li Y, Zhang SC, Carter B, Fritzer J, Schwaerz M, Choy SL, Wu SQ, Tan ZX: Characterisation of residual ionospheric errors in bending angles using GNSS RO end-to-end simulations. *Adv Space Res* 2013, 52: 821-836.

20. Steiner AK, Hunt D, Ho S-P, Kirchengast G, Mannucci AJ, Scherllin-Pirscher B, Gleisner H, von Engeln A, Schmidt T, Ao C, Leroy SS, Kursinski ER, Foelsche U, Gorbunov M, Heise S, Kuo Y-H, Lauritsen

KB, Marquardt C, Rocken C, Schreiner W, Sokolovskiy S, Syndergaard S, Wickert J: Quantification of structural uncertainty in climate data records from GPS radio occultation. *Atmos Chem Phys* 2013, 13: 1469-1484.

21. Cardellach E, Tomas S, Oliveras S, Padullés R, Rius A, De M, la Torre-Juárez FJ, Turk COA, Kursinski ER, Schreiner B, Ector D, Cucurull L: Sensitivity of PAZ LEO polarimetric GNSS radio-occultation experiment to precipitation events. *IEEE Trans Geosci Rem Sens* 2015, 53(1):190-2016.

22. Flores A, Ruffini G, Rius A: 4D tropospheric tomography using GPS slant wet delays. *Ann Geophys* 2000, 18: 223-234.

23. Seko H, Shimada S, Nakamura H, Kato T: Three-dimensional distribution of water vapor estimated from tropospheric delay of GPS data in a mesoscale precipitation system of the Baiu front. *Earth Planets Space* 2000, 52: 927-933.

24. Nilsson T, Gradinarsky L: Water vapor tomography using GPS phase observations: simulation results. *IEEE Trans Geosci Rem Sens* 2006, 44(10):2927-2941.

25. Bender M, Raabe A: Preconditions to ground-based GPS water vapour tomography. *Ann Geophys* 2007, 25(8):1727-1734. 10.5194/angeo-25-1727-2007

26. Rohm W, Zhang K, Bosy J: Limited constraint, robust Kalman filtering for GNSS troposphere tomography. *Atmos Meas Tech* 2014, 7: 1475-1486.

27. Rohm W, Yuan Y, Biadeglgne B, Zhang K, JLe M: Ground-based GNSS ZTD/IWV estimation system for numerical weather prediction in challenging weather conditions. *Atmos Res* 2014, 138: 414-426.

28. Yuan Y, Zhang K, Rohm W, Choy S, Norman R, Wang C: Real time retrieval of atmospheric water vapour from GPS precise point positioning. *J Geophys Res* 2014.

29. Choy S, Wang C, Zhang K, Kuleshov Y: GPS sensing of precipitable water vapour during the March 2010 Melbourne storm. *Adv Space Res*2013, 52(9):1688-1699.

30. Manning T: *Sensing the dynamics of severe weather using 4D GPS tomography in the Australian region*. PhD thesis, RMIT University; 2014:196.

31. Manning T, Zhang K, Rohm W, Choy S, Hurter F: Detecting severe weather in Australia using GPS tomography. *J GPS* 2012, 11(1):58-70.

32. Bennitt GV, Jupp A: Operational assimilation of GPS zenith total delay observations into the Met Office Numerical Weather Prediction Models. *Mon Weather Rev* 2012, 140(8):2706-2719.

33. Treuhaft R, Lowe ST, Zuffada C, Chao Y: 2-cm GPS altimetry over Crater Lake. *Geophys Res Lett* 2001, 28(23):4343-4346.

34. Martin-Neira M: A passive reflectometry and interferometry system (PARIS): application to ocean altimetry. *ESA J* 1993, 17: 331-335.

35. Martin-Neira M, D'Addio S, Buck C, Floury N, Prieto-Cerdeira R: The PARIS ocean altimeter in-orbit demonstrator. *IEEE Trans Geosci Rem Sens* 2011, 49(6):2209-2237.

36. Martin-Neira M, D'Addio S, Vitulli R: Study of delay drift in GNSS-R altimetry. *IEEE Int J Appl Earth Obs Rem Sens* 2014, 7(5):1473-1480.

37. Pascual D, Camps A, Martin F, Park H, Alonso Arroyo A, Onrubia R: Precision bounds in GNSS-R ocean altimetry. *IEEE Int J Appl Earth Obs Rem Sens* 2014, 7(5):1416-1423.

38. Rius A, Cardellach E, Martın-Neira M: Altimetric analysis of the sea-surface GPS-reflected signals. *IEEE Trans Geosci Rem Sens* 2010, 48(4):2119-2127.

39. Yu K, Rizos C, Dempster A: GNSS-based model-free sea surface height estimation in unknown sea state scenarios. *IEEE Int J Appl Earth Obs Rem Sens* 2014, 7(5):1424-1435.

40. Lowe S, Meehan T, Young L: Direct-signal enhanced semi-codeless processing of GNSS surface-reflected signals. *IEEE Int J Appl Earth Obs Rem Sens* 2014, 7(5):1469-1472.

41. D'Addio S, Martin-Neira M, di Bisceglie M, Galdi C, Martin Alemany F: GNSS-R altimeter based on Doppler multi-looking. *IEEE Int J Appl Earth Obs Rem Sens* 2014, 7(5):1452-1460.

42. Lowe ST, Zuffada C, Chao Y, Kroger P, Young LE, LaBrecque JL: 5-cm precision aircraft ocean altimetry using GPS reflections. Geophys.*Res Lett* 2002, 29(10):1375.

43. Ribó S, Arco JC, Oliveras S, Cardellach E, Rius A, Buck C: Experimental results of an X-Band PARIS receiver using digital

satellite TV opportunity signals scattered on the sea surface. *IEEE Trans Geosci Rem Sens* 2014., 52(2014)

44. Cardellach E, Fabra F, Nogues-Correig O, Oliveras S, Ribo S, Rius A: GNSS-R ground-based and airborne campaigns for ocean, land, ice and snow techniques: application to the GOLD-RTR data sets. *Radio Sci* 2011, 46(6):1412-1415

45. Garrison JL, Cardellach E, Gleason S, Katzberg S: Foreword to special issue on reflectometry using global navigation satellite systems and other signals of opportunity (GNSS + R). *IEEE Int J Appl Earth Obs Rem Sens* 2014, 7(5):1412-1415.

46. Garrison JL, Katzberg SJ, Hill MI: Effect of sea roughness on bistatically scattered range coded signals from the Global Positioning System.*Geophys Res Lett* 1998, 25(13):2257-2260.

47. Gleason S, Hodgart SS, Sun Y, Gommenginger C, Mackin S, Adjrad M, Unwin M: Detection and processing of bistatically reflected GPS signals from low earth orbit for the purpose of ocean remote sensing. *IEEE Trans Geosci Rem Sens* 2005, 43(6):1229-1241.

48. Cardellach E, Ruffini G, Pino D, Rius A, Komjathy A, Garrison JL: Mediterranean balloon experiment: ocean wind speed sensing from the stratosphere using GPS reflections. *Remote Sens Environ* 2003, 88(3):351-362.

49. Katzberg SJ, Walker RA, Roles JH, Lynch T, Black PG: First GPS signals reflected from the interior of a tropical storm: preliminary results from Hurricane Michael. *Geophys Res Lett* 2001, 28(10):1981-1984.

50. Katzberg SJ, Dunion J: Comparison of reflected GPS wind speed retrievals with dropsondes in tropical cyclones. *Geophys Res Lett* 2009., 36(L17602)

51. Katzberg SJ, Torres O, Ganoe G: Calibration of reflected GPS for tropical storm wind speed retrievals. *Geophys Res Lett* 2006., 33(L18602): doi:10.1029/2006GL026825

52. Rodriguez-Alvarez N, Akos DM, Zavorotny VU, Smith JA, Camps A, Fairall CW: Airborne GNSS-R wind retrievals using delay-Doppler maps.*IEEE Trans Geosci Rem Sens* 2013, 51(1):626-641.

53. Germain O, Ruffini G, Soulat F, Caparrini M, Chapron B, Silverstrin P: The Eddy Experiment: GNSS-R speculometry for directional sea-roughness retrieval from low altitude aircraft. *Geophys Res Lett* 2004, 31(L21307):1-4.

54. Komjathy A, Armatys M, Masters D, Axelrad P: Retrieval of ocean surface wind speed and wind direction using reflected GPS signals. *J Atmos Ocean Technol* 2004, 21: 515-526.

55. Cardellach E, Rius A: A new technique to sense non-Gaussian features of the sea surface from L-band bi-static GNSS reflections. *Rem Sens Environ* 2008, 112(6):2927-2937.

56. Shah R, Garrison J, Grant M: Demonstration of bistatic radar for ocean remote sensing using communication satellite signals. *IEEE Geosci Remote Sens Lett* 2012, 9(4):619-623.

57. Marchan-Hernandez JF, Rodriguez-Alvarez N, Camps A, Bosch-Lluis X, Ramos-Perez I, Valencia E: Correction of the sea state impact in the L-band brightness temperature by means of delay Doppler maps of Global Navigation Satellite signals reflected over the sea surface. *IEEE Trans Geosci Rem Sens* 2008, 46(10):29142923.

58. Valencia E, Camps A, Rodriguez-Alvarez N, Ramos-Perez I, Boxch-Lluis X, Park H: Improving the accuracy of sea surface salinity retrieval using GNSS-R data to correct the sea state effect. *Radio Sci* 2011, 46(RS0C02):1-11.

59. Cardellach E, Ribó S, Rius A: *Technical Note on POlarimetric Phase Interferometry (POPI)*. CSIC-IEEC; 2006. arXiv:physics/0606099 2

60. Zavorotny VU, Voronovich AG: Scattering of GPS signals from the ocean with wind remote sensing application. *IEEE Trans Geosci Rem Sens* 2000, 38(2):951-963.

61. Elfouhaily T, Thompson DR, Linstrom L: Delay-Doppler analysis of bistatically reflected signals from the ocean surface: theory and application. *IEEE Trans Geosci Rem Sens* 2002, 40(3):560-573.

62. Elfouhaily T, Chapron B, Katsaros K, Vandermark D: A unified directional spectrum for long and short wind-driven waves. *J Geophys Res*1997, 102(C7):15781-15796.

63. Yu K, Rizos C, Dempster AG: Sea surface wind speed estimation based on GNSS signal measurements. In *Proceedings of international geoscience and remote sensing symposium (IGARSS)*. Munich, Germany; 2012:2587-2590.

64. Garrison JL, Larson KM, Burrage D: Advancing reflectometry, workshop on reflectometry using GNSS and other signals of opportunity (GNSS + R). *Eos* 2012, 94(21):193.

65. Masters D, Axelrad P, Katzberg SJ: Initial results of land-reflected GPS bistatic radar measurements in SMEX02. *Remote Sens Environ* 2004, 92: 507-520.

66. Wang JR, Schmugge TJ: An empirical model for the complex dielectric permittivity of soils as a function of water content. *IEEE Trans Geosci Rem Sens* 1980, 18(4):288-295.

67. Rodriguez-Alvarez N, Bosch-Lluis X, Camps A, Vall-Llossera M, Valencia E, Marchan-Hernandez JF, Ramos-Perez I: Soil moisture retrieval using GNSS-R techniques: experimental results over a bare soil field. *IEEE Trans Geosci Rem Sens* 2009, 47(11):3616-3624.

68. Zavorotny VU, Larson KM, Braun JJ, Small EE, Gutmann ED, Bilich AL: A physical model for GPS multipath caused by land reflection: toward bare soil moisture retrievals. *IEEE Int J Appl Earth Obs Rem Sens* 2010, 3(1):100-110.

69. Colorado GPS Reflections Research Group , viewed 5 Aug 2014 http://xenon.colorado.edu/portal/

70. Larson KM, Small EE: Using GPS to study the terrestrial water cycle. *Eos* 2013, 94(52):505-512.

71. Yu K, Rizos C, Dempster A: Forest change detection using GNSS signal strength measurements. In *Proceedings of international geoscience and remote sensing symposium (IGARSS)*. Melbourne, Australia; 2013:1003-1006.

72. Ferrazzoli P, Guerriero L, Pierdicca N, Rahmoune R: Forest biomass monitoring with GNSS-R: theoretical simulations. *Adv Space Res* 2011, 47: 1823-1832.

73. Semmling M, Beyerle G, Stosius R, Dick G, Wickert J, Fabra F, Cardellach E, Ribó S, Rius A, Helm A, Yudanov SB, D'Addio S: Detection of arctic ocean tides using interferometric GNSS-R signals. Geophys. *Res Lett* 2011., 38(4): doi:10.1029/2010GL046005

74. Fabra F, Cardellach E, Rius A, Ribó S, Oliveras S, Nogués-Correig O, Rivas MB, Semmling M, D'Addio S: Phase altimetry with dual polarization GNSS-R over sea ice. *IEEE Trans Geosci Rem Sens* 2012, 50(6):2112-2121.

75. Belmonte M: *Bistatic scattering of global positioning system signals from Arctic sea ice*. Boulder, CO, USA: Ph.D. thesis, University of Colorado at Boulder; 2007.

76. Fabra F: *GNSS-R as a Source of Opportunity for Remote Sensing of the Cryosphere*. Barcelona: PhD Dissertation, UPC; 2013.

77. Belmonte M, Maslanik JA, Axelrad P: Bistatic scattering of GPS signals off Arctic sea ice. *IEEE Trans Geosci Rem Sens* 2009, 48(3):1548-1553.

78. Larson KM, Gutmann E, Zavorotny V, Braun J, Williams M, Nievinski FG: Can we measure snow depth with GPS receivers? Geophys. *Res Lett* 2009., 36(L17502).

79. Larson KM, Nievinski FG: GPS snow sensing: results from the EarthScope Plate Boundary Observatory. *GPS Solutions* 2012, 17(1):41-52.

80. Cardellach E, Fabra F, Rius A, Pettinato S, DAddio S: Characterization of dry-snow sub-structure using GNSS reflected signals. *Remote Sens Environ* 2012, 124: 122-134.

81. Anthes RA: Exploring Earth's atmosphere with radio occultation: contributions to weather, climate and space weather. *Atmos Meas Tech*2011, 4: 1077-1103. doi:10.5194/amt-4-1077-2011 10.5194/amt-4-1077-2011

82. Jin S, Feng GP, Gleason S: Remote sensing using GNSS signals: current status and future directions. *Adv Space Res* 2011, 47(10):1645-1653.

83. Heise S, Wickert J, Arras C, Beyerle G, Faber A, Michalak G, Schmidt T, Zus F: Reprocessing and application of GPS radio occultation data from CHAMP and GRACE. In *Observation of the System Earth from Space - CHAMP, GRACE, GOCE and Future Mission*. Edited by: By Frank F, Nico S, Wolf-Dieter S. Heidelberg: Springer; 2014. GEOTECHNOLOGIEN Science Report No. 20

84. COSMIC , viewed 9 Dec 2013 http://www.cosmic.ucar.edu/

85. FORMOSAT-7/COSMIC-2 (COSMIC-2) Science Mission , viewed 9 Dec 2013 http://www.cosmic.ucar.edu/cosmic2/

86. Unwin MJ, Gleason S, Brennan M: The space GPS reflectometry experiment on the UK disaster monitoring constellation satellite. In*Proceedings of ION-GPS/GNSS*. Portland, OR; 2003.

87. Missions , viewed 9 Dec 2013 http://www.sstl.co.uk/Missions/TechDemoSat-1

88. Martin-Neira M, D'Addio S, Buck C, Floury N, Prieto-Cerdeira R: The PARIS in-orbit demonstrator. *Proceedings of IEEE IGARSS* 2009, 2: II–322-II–325.

89. CYGNSS , viewed 9 Dec 2013 http://aoss-research.engin.umich.edu/missions/cygnss/

90. GEROS: GNSS REflectometry, Radio Occultation and Scatterometry onboard International Space Station (GEROS-ISS). , viewed 9 Dec 2013 http://www.ice.csic.es/en/view_project.php?PID=155

91. Carreno-Luengo H, Camps A, Perez-Ramos I, Forte G, Onrubia R, Diez R: 3Cat-2: a P(Y) and C/A GNSS-R experimental nano-satellite mission. In *Proceedings of international geoscience and remote sensing symposium (IGARSS)*. Melbourne, Australia; 2013:843-846.

92. ACSER QB50 project , viewed 9 Dec 2013 http://www.acser.unsw.edu.au/QB50/index.html

93. Sreeja VV, Aquino M, Forte B, Elmas Z, Hancock C, Franceschi GD, Alfonsi L, Spogli L, Romano V, Bougard B, Monico JFG, Wernik AW, Sleewaegen J-M, Canto A, Silva EFD: Tackling ionospheric scintillation threat to GNSS in Latin America. *J Space Weather Space Clim* 2011, 1(1):1-9. A05

94. Wickert J, Michalak G, Schmidt T, Beyerle G, Cheng C-Z, Healy SB, Heise S, Huang C-Y, Jakowski N, Kohler W, Mayer C, Offiler D, Ozawa E, Pavelyev AG, Rothacher M, Tapley B, Arras C: GPS radio occultation: results from CHAMP, GRACE and FORMOSAT-3/COSMIC.*Terrestrial Atmos Oceanic Sci* 2009, 20(1):35-50.

95. Bonnedal M, Christensen J, Carlstrom A, Berg A: Metop-GRAS in-orbit instrument performance. *GPS Solutions* 2010, 14(1):109-120.

96. Lowe S, Meehan T, Young L: Direct-signal enhanced semi-codeless processing of GNSS surface-reflected signals. *IEEE Int J Appl Earth Obs Rem Sens* 2014, 7(5):1269-1472.

97. Nogués-Correig O, Gali EC, Campderrós JS, Rius A: A GPS-reflections receiver that computes Doppler/delay maps in real time. *IEEE Trans Geosci Rem Sens* 2007, 45(1):156-174.

98. Steenwijk RV, Unwin M, Jales P: Introducing the SGR-ReSI: A next generation spaceborne GNSS receiver for navigation and remote-sensing. In *Proceedings of ESA workshop on satellite navigation technologies and European workshop on GNSS signals and signal processing (NAVITEC)*. Noordwijk, Netherlands; 2010.

99. International Association of Geodesy (IAG) , viewed 11 Dec 2013 http://www.iag-aig.org

100. International GNSS Service (IGS) , viewed 11 Dec 2013 http://igs.org

101. International Terrestrial Reference Frame (ITRF) , viewed 11 Dec 2013 http://itrf.ensg.ign.fr

102. IGS troposphere and ionosphere parameter products , viewed 11 Dec 2013 http://www.igs.org/products

103. Global Geodetic Observing System (GGOS) , viewed 11 Dec 2013

104. IAG services , viewed 11 Dec 2013 http://www.iers.org/nn_10880/IERS/EN/Organization/TechniqueCentres/TC.html?__nnn=true

7

CHAPTER

GNSS-BASED BISTATIC SAR: A SIGNAL PROCESSING VIEW

Michail Antoniou and Mikhail Cherniakov

School of Electronic, Electrical and Computer Engineering, University of Birmingham

ABSTRACT

This article presents signal processing algorithms used as a new remote sensing tool, that is passive bistatic SAR with navigation satellites (e.g. GPS, GLONASS or Galileo) as transmitters of opportunity. Signal synchronisation and image formation algorithms are described for two system variants: one where the receiver is moving and one where it is fixed on the ground. The applicability and functionality of the algorithms described is demonstrated through experimental imagery that ultimately confirms the feasibility of the overall technology.

Keywords: bistatic SAR signal processing SS-BSAR GNSS-based SAR

1. INTRODUCTION

Over the last decade, bistatic SAR (BSAR) has seen a substantial growth. Various BSAR topologies have been proposed, using spaceborne, airborne or even fixed platforms [1–4], and their relative merits and drawbacks have been investigated.

A special BSAR topology is called space-surface BSAR (SS-BSAR), or hybrid BSAR [5]. In this configuration, one of the platforms is spaceborne, while the other is located on or near the surface of the Earth (Figure 1). Even though this topology is a special BSAR case, it encompasses a variety of possible transmitter/receiver combinations. The transmitter could be spaceborne, while the receiver could be mounted on an aircraft, a ground moving vehicle, or it could be fixed on the ground, or vice versa. The only requirement is that at least one platform should move relative to an observation area, such that a synthetic aperture can be formed. The spaceborne platform can be a radar satellite [6–8], which is the most common case. Alternatively, it can be a transmitter of opportunity, such as a communications or digital television satellite.

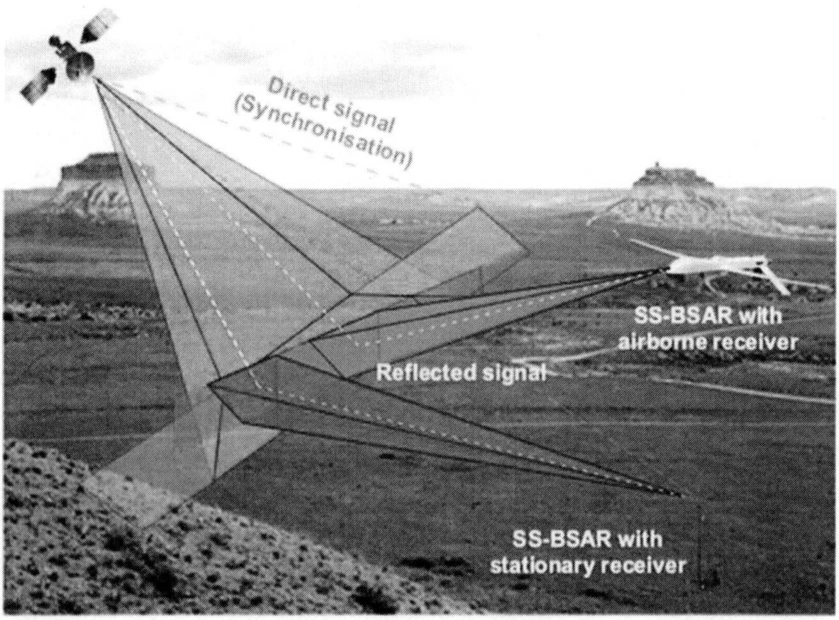

Figure 1. SS-BSAR CONCEPT.

Our research focuses on SS-BSAR using global navigation satellite systems (GNSS) as transmitters of opportunity [9–13]. There are a number of reasons for this choice. First of all, GNSS are satellite constellations consisting of at least 24 satellites, and are designed for global and permanent coverage. From the radar perspective, this enables persistent area monitoring. In addition, GNSS are designed in such a way that there are at least six to eight satellites illuminating the same point on Earth from multiple angles. This enables simultaneous image acquisition from multiple angles which may provide further scene information. The presence of multiple satellites also provides the possibility to choose the optimal bistatic acquisition geometry, minimising shadowing effects due to local landscape and maximising resolution performance.

In terms of the passive operation, use of GNSS is an ideal choice. First of all, GNSS exhibit relatively large signal bandwidths compared to other sources of opportunity, and can therefore provide sufficient range resolutions (the aggregate signal bandwidth for the Galileo E5 band is approximately 20 MHz). More importantly, however, the receiving hardware for SS-BSAR is very similar, if not identical, to a standard GPS receiver used for navigation purposes. On one hand, this implies that SS-BSAR based on GNSS is cost-effective, as standard GPS chipsets can nowadays be purchased at extremely low costs.

On the other hand, this implies that changing the signal processing algorithms within the standard GPS chipset can convert the device into a low-cost radar system, with all the advantages mentioned above. This raises the question of which signal processing algorithms can be used in this system.

All BSAR systems require two types of algorithms to provide imagery. The first type is signal synchronisation, aiming to maintain coherence for image formation. This step is necessary, since the transmitter and receiver are separated, and hence have independent clocks and local oscillators. To provide synchronisation, an extra receiving channel is usually added, which records the direct signal from the transmitter to the receiver. Extraction and analysis of the parameters of the direct signal, such as delay and phase, is sufficient to provide the necessary delay and phase references required for imaging. In traditional BSAR, where the transmitter is a radar platform, the transmitted power is high and the transmitted signal is a chirp waveform. In this case, a number of solutions have been proposed [14, 15]. In SS-BSAR with GNSS, none of the traditional solutions are applicable and signal synchronisation is a complex issue. The reason for that is that GNSS signals have a low power density near the surface of the Earth, resulting in signal-to-noise ratios (SNR) of down to −40 dB at the receiver input [16]. The other reason is that GNSS are communication signals with complex modulation schemes which should be cancelled before extraction of direct signal parameters.

The second type of algorithms required is of course image formation algorithms. The derivation of image formation algorithms has been a major challenge in BSAR since its conception. There are a number of image formation algorithms operating in the frequency domain for special BSAR topologies, such as parallel/linear transmitter and receiver trajectories or equal platform velocities [17–20]. Fast back-projection algorithms (BPAs) have been proposed for the general BSAR case [21]. However, to the best of the authors' knowledge, a generic closed-form solution for a frequency-based algorithm without any approximation is yet to be presented. The reason behind this is that the different transmitter/receiver trajectories result in azimuth signals which are difficult to convert to the frequency domain analytically. This problem is more acute for SS-BSAR. A satellite and an aircraft cannot follow parallel trajectories, nor have equal velocities. Moreover, in the moving receiver case, a suitable motion compensation algorithm is required to correct for trajectory deviations of the aircraft only.

This article describes signal synchronisation and image formation algorithms employed in SS-BSAR with GNSS transmitters of opportunity. The functionality of the algorithms, and ultimately the proposed technology, is demonstrated using a variety of experimental data, obtained from both moving receivers (roof top railways, ground moving vehicle and a helicopter) as well as a stationary one. Section 2 presents an overview of the signal synchronisation algorithm used, along with experimental results. Section 3 describes image formation algorithms developed for the moving and stationary receiver cases. Finally, Section 4 presents experimental imagery using the configurations described above.

2. SIGNAL SYNCHRONISATION

2.1. Introduction

Whether the receiver is stationary or not, SS-BSAR with GNSS uses one receiving channel to record the direct signal from the satellite for synchronisation, while another channel collects satellite signal reflections from an observation area for imaging. The direct signal channel is called the heterodyne channel (HC), and the reflected signal channel is called the radar channel (RC). Both the HC and RC have the same clocks and local oscillators, and therefore clock slippage and local oscillator drift between the transmitter and the receiver is common to both channels. As a result, clock slippage and oscillator drift can be deducted through the delay and phase of the direct signal at the HC and then compensated at the RC.

The reason for having the HC is that an antenna can be pointed towards the satellite to maximise the direct signal strength. Even in this case, the SNR at the input of the receiver can be as low as −40 dB, so the HC signal cannot be used for synchronisation directly. This implies that in order to estimate the direct signal delay and phase, and extract from them the required clock slippage and oscillator drift, a tracking algorithm based on matched filtering is required to maximise SNR. An additional problem is the structure of all GNSS signals. They are communication signals consisting of two ranging codes which are pseudorandom sequences, which generally (but not necessarily) modulate a navigation message. From these three sequences, only the primary ranging code is desired as the transmitted signal for imaging, while the other two act as interference. Hence, a method of tracking the parameters of the primary code and cancelling the interfering codes is required.

2.2. Synchronisation algorithm description

One possible method of providing signal synchronisation for SS-BSAR with GNSS is to modify signal processing algorithms used for GNSS signal tracking for navigation purposes. The purpose of this class of algorithms is to cancel all ranging codes and extract the navigation message required for positioning. However, in order to do so, they need to track the time delay, Doppler and phase of the ranging codes. In SS-BSAR with GNSS, the ranging code is the transmitted signal and hence its tracked parameters on the HC are the required synchronisation outputs.

The particular algorithm adapted from GNSS signal tracking to SS-BSAR synchronisation is the well-known block adjustment of synchronising signal (BASS) algorithm [22]. Its block diagram for signal synchronisation is shown in Figure 2.

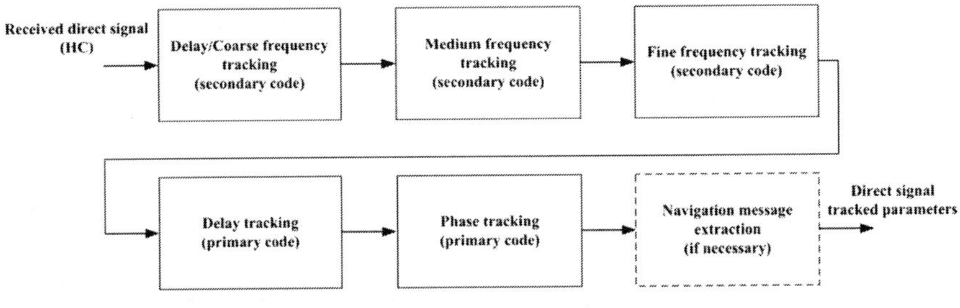

Figure 2. BASS-BASED ALGORITHM FOR SS-BSAR SYNCHRONISATION.

An algorithm for SS-BSAR signal synchronisation based on this technique was proposed at the first development stages of the system [23]. While highly accurate, it did not meet efficiency requirements, and therefore it was replaced by the algorithm described below. The generic form of a transmitted GNSS signal is:

$$Y(t) = P(t)M_P(t)\cos(\omega_c t + \varphi) + D(t)M_D(t)\sin(\omega_c t + \varphi), \quad (1)$$

where t is time, $P(t)$ and $D(t)$ are the primary and secondary GNSS ranging code envelopes, $Mp(t)$ and $M_D(t)$ are the associated navigation messages, ω_c is the signal carrier frequency and φ is the initial signal phase.

After quadrature demodulation and SAR data formatting, the direct signal received at the HC can be written as follows:

$$s(t_n, u) = P[t_n - \tau_{dP}(u)]M_P[t_n - \tau_{dP}(u)]\exp[j(\omega_d(u)t_n + \varphi_{dP}(u))] + j \times D[t_n - \tau_{dD}(u)]M_D[t_n - \tau_{dD}(u)]\exp[j(\omega_d(u)t_n + \varphi_{dD}(u))] \quad (2)$$

where $t_n \epsilon [0, PRI]$ denotes fast-time and PRI is the pulse repetition interval, $u \epsilon [-T/2, T/2]$ is slow-time and T is the dwell time, and $\tau_{dP/D}(u)$, $\omega_d(u)$ and $\varphi_{dP/D}(u)$ are the instantaneous direct signal time delay, Doppler and initial phase associated with each code, respectively, all of which are varying with slow-time. Note that the primary and secondary codes are different in structure and length, and therefore their time delays and phases are different by a constant value. However, their Doppler, defined as the derivative of their phases, are approximately equal.

As mentioned before, the signal required for SS-BSAR image formation is the primary code $P(t)$, and hence all of its parameters should be tracked for synchronisation. However, according to (2), the secondary code $D(t)$ is acting as a deterministic interference. The BASS algorithm first tracks the secondary code parameters prior to its compensation, and then proceeds to track the required parameters of the primary code.

The first stage in the algorithm combines the delay $\tau_{dD}(u)$ and coarse Doppler frequency tracking of the secondary code. These parameters are provided at every PRI, which for GNSS is usually 1 ms, resulting in a pulse repetition frequency (PRF) of 1 kHz. The tracking process consists of a bank of matched filters. The envelope of each filter is a locally generated replica of $D(t)$, modulated with a different Doppler frequency. The Doppler separation between filters is equal to the PRF (1 kHz), varied incrementally between −20 and 20 kHz which is the maximum Doppler expected from a

GNSS satellite [22]. Therefore, this stage is a 2D search algorithm in delay and Doppler, with a Doppler resolution of 1 kHz, hence the term "coarse frequency". Figure 3 shows this processing step for a single Doppler frequency, ω_{di}, and Figure 4 presents a typical 2D delay/coarse frequency estimate for one PRI, obtained from experimental data. The location of the peak indicates the estimated delay and coarse frequency.

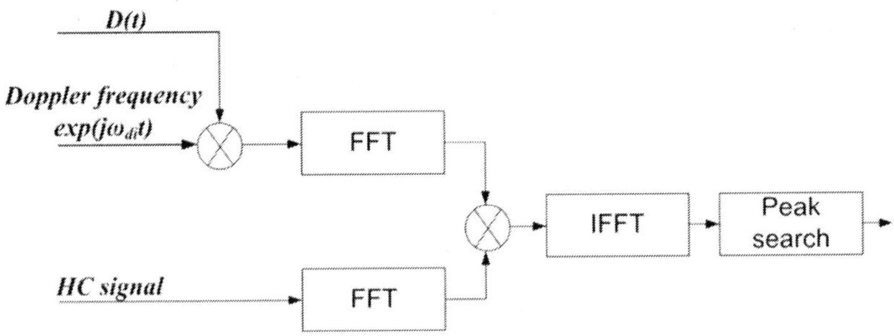

Figure 3. DELAY/COARSE FREQUENCY TRACKING BLOCK DIAGRAM.

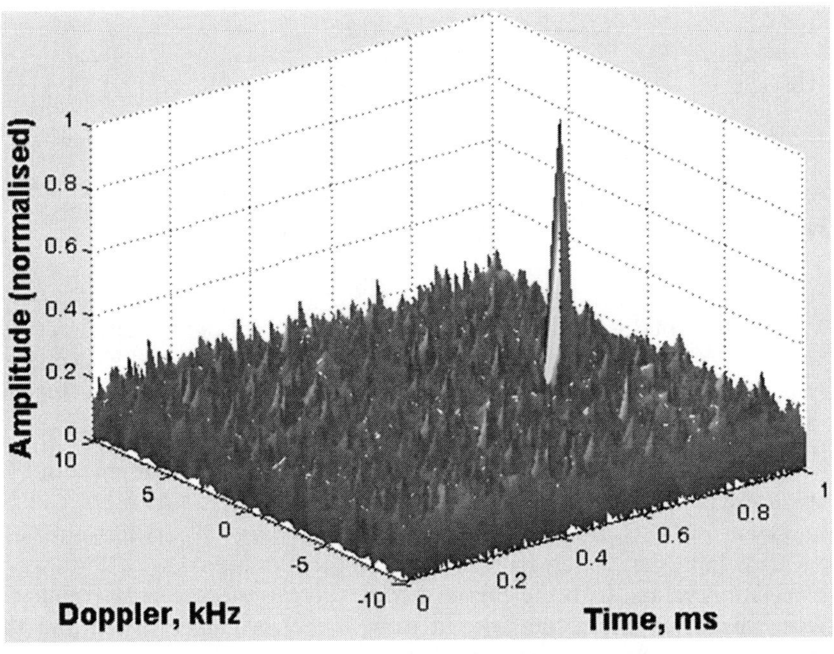

Figure 4. OUTPUT OF DELAY/COARSE FREQUENCY TRACKER FOR 1-MS DATA.

The medium and fine frequency tracking steps complete the Doppler tracking process. Medium frequency tracking provides Doppler estimates with 200 Hz resolution through fast Fourier transform (FFT) processing, whereas fine frequency tracking operates on the phase difference between signals at adjacent PRIs and can therefore track Doppler with an accuracy of up to two decimal places in practice. For these tasks to be carried out effectively, the Doppler waveform [the exp() factor in (2)] needs to be recovered without phase transitions due to the presence of the secondary code. This is achieved by stripping $D[t_n - \tau_{dD}(u)]$ from the data (Figure 5).

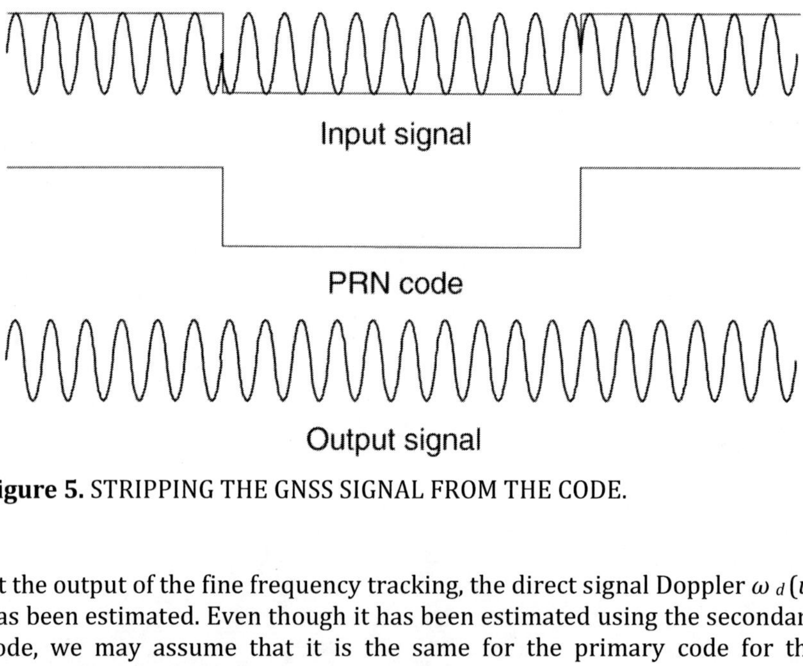

Figure 5. STRIPPING THE GNSS SIGNAL FROM THE CODE.

At the output of the fine frequency tracking, the direct signal Doppler $\omega_d(u)$ has been estimated. Even though it has been estimated using the secondary code, we may assume that it is the same for the primary code for the reasons described above.

Following the fine frequency tracking, the time delay and Doppler of $D(t)$ are known. Therefore, the secondary code with its parameters can be removed from (2). Note that the navigation message $M_D(t)$ has not been tracked; however, it can be viewed as a random signal with low cross-correlation values with the primary code and therefore can be neglected. With this observation, the second term in (2) is compensated and the remaining received signal may be written as follows:

$$s(t_n, u) = P[t_n - \tau_{dP}(u)] M_P[t_n - \tau_{dP}(u)] \exp[j(\omega_d(u) t_n + \varphi_{dP}(u))] \quad (3)$$

In order to track $\tau_{dP}(u)$, matched filtering is used. The reference signal is the envelope of $P(t)$, shifted in Doppler by $\omega_d(u)$ which was estimated in the previous step. Finally, the phase and the navigation message (if one exists) can be extracted after the time-delayed and Doppler shifted primary code have been stripped from (3), as shown in Figure 6. The navigation message is a BPSK signal, and can be regarded as a phase transition of ±π on $\varphi_{dP}(u)$. Therefore, using a phase transition detector, both the navigation message and the phase can be found.

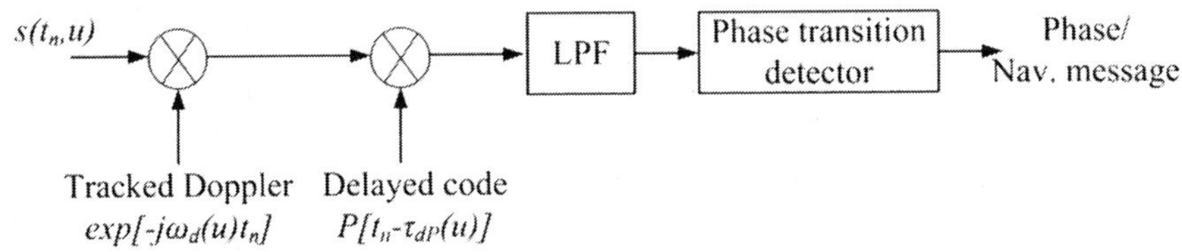

Figure 6. PHASE/NAVIGATION MESSAGE EXTRACTION.

At the output of the signal synchronisation algorithm, the direct signal time delay $\tau_{dP}(u)$, Doppler $\omega_d(u)$, phase $\varphi_{dP}(u)$ and navigation message $M_d(t)$ have been estimated. In the following section, the derived algorithm is confirmed using various experimental data.

2.3. Experimental confirmation

The proposed algorithm was tested using two sets of experimental data. The first set was obtained using a GLONASS transmitter and a fixed receiver, and the second was obtained with a Galileo transmitter and an airborne receiver. These variants were selected to test the validity of the algorithm, as well as its functionality for different configurations and topologies.

2.3.1. GLONASS transmitter and fixed receiver

A data acquisition dedicated to the verification of the synchronisation algorithm was conducted with a GLONASS transmitter and a fixed receiver (Figure 7). The direct signal from the satellite was captured by a stationary low-gain antenna, and the synchronisation algorithm was applied to extract its delay, Doppler, phase and navigation message. The experimental parameters are shown in Table 1.

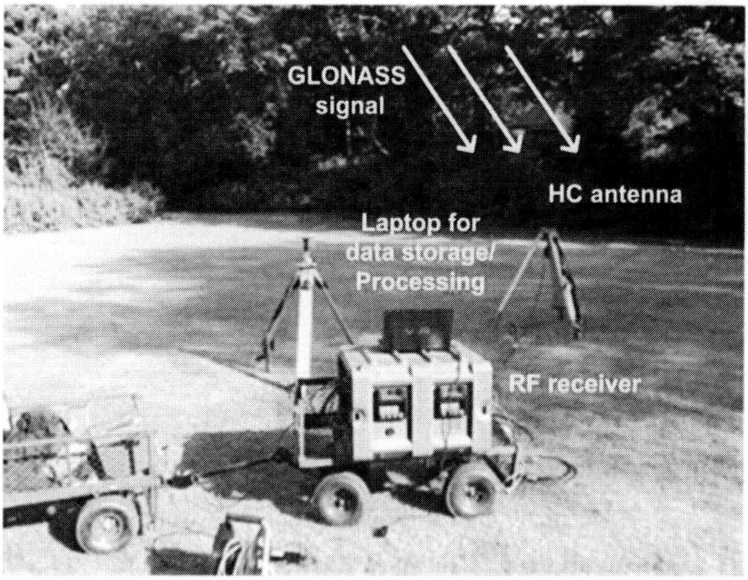

Figure 7. SS-BSAR EXPERIMENTAL SETUP WITH FIXED RECEIVER FOR SYNCHRONISATION TESTING.

Table . 1 EXPERIMENTAL PARAMETERS

Parameter	Value
Satellite used	GLONASS COSMOS 732
Operating frequency	1603.6875 MHz (L1)
Satellite signal used	P-code
Signal bandwidth	5.11 MHz
Dwell time	300 s
Satellite elevation	48°–52°
Satellite azimuth	298°–301°
PRF	1 kHz

The primary GLONASS code is the P-code, while the C/A-code is the secondary, or interfering code. Figure 8a shows the tracked P-code delay, while Figure 8b presents the Doppler frequency at the output of the fine frequency tracker. Note that in practice, a least-mean squares algorithm is applied to the tracking output to smooth Doppler variation due to receiver noise. However, both delay and Doppler outputs are clear and without significant errors.

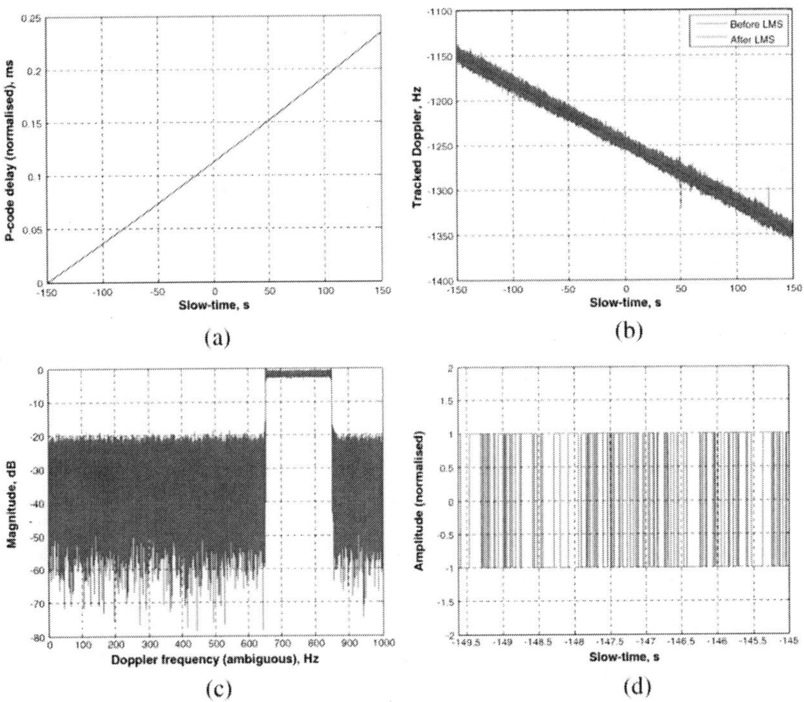

Figure 8. TRACKED DIRECT SIGNAL: (A) delay, **(B)** Doppler, **(C)** phase spectrum, **(D)** decoded navigation message.

The tracked phase spectrum is shown in Figure 8c. This result was generated by taking the complex exponential of the tracked phase, followed by an FFT. Effectively, this is the azimuth spectrum of the direct signal. The obtained results shows a near-perfect chirp signal spectrum, which is as expected from the instantaneous phase history of the satellite. Finally, Figure 8d presents the decoded navigation message in the first 5 s of the data for better visualisation.

The results presented in Figure 8 demonstrate the functionality and high performance level of the proposed method. Note that the tracked outputs contain both the true time delay and Doppler variation, as well as receiver artefacts such as clock slippage and local oscillator drift. In the next chapter, methods of cancelling them out for effective image formation will be described.

2.3.2. Galileo transmitter and airborne receiver

Experiments with a Galileo transmitter and an airborne receiver were conducted in order to verify the imaging capability of the system. The imaging experiment will be described in more detail in Section 3. Prior to image formation, signal synchronisation was required. Experimental parameters related to synchronisation are shown in Table 2. The difference in this case was that the HC was onboard the helicopter, and therefore the direct signal delay and Doppler variations were much more dynamic and affected by receiver motion errors. The Galileo E5bQ signal was used for imaging, which is free of a navigation message, and therefore navigation message extraction was not required.

Table 2. AIRBORNE EXPERIMENT PARAMETERS RELATED TO SYNCHRONISATION

Parameter	Value
Satellite used	Galileo GIOVE-B
Operating frequency	1207.14 MHz
Satellite signal used	E5bQ
Signal bandwidth	10.23 MHz
Dwell time	30 s
Equivalent PRF	1 kHz

Figure 9 shows the tracked direct signal parameters. Comparing Figures 9 and 8, it is clear that the direct signal Doppler in the moving receiver case is influenced by the trajectory deviations of the helicopter during its flight. The same effect is also visible in the tracked azimuth phase spectrum, which is no longer representative of a chirp signal (Figure 9c).

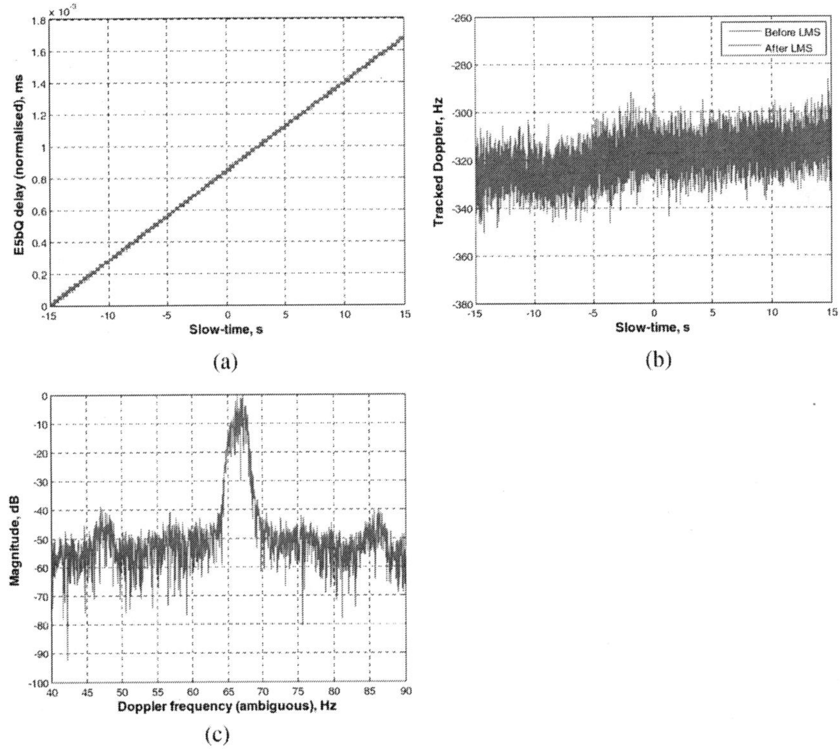

Figure 9. TRACKED DIRECT SIGNAL: (A) **delay,** (B) **Doppler,** (C) **phase spectrum.**

A number of conclusions can be derived from the experimental results in this section. First of all, the proposed algorithm can operate irrespective of the GNSS transmitter used, and irrespective of the topology, even in a dynamic environment such as an airborne receiver where trajectory deviations affect the direct signal parameters. In terms of the performance, all tracked outputs are obtained with sufficiently high SNR. In the fixed receiver case, the tracked Doppler curve is linear, implying a stable signal Doppler history that resembles a chirp signal. In the moving receiver case, effects of trajectory deviations are visible, an issue which should be dealt with at the image formation stage.

3. IMAGE FORMATION

Following signal synchronisation, an image formation algorithm is required to generate imagery of an interrogated scene. In the general BSAR case, image formation algorithms need to take two things into account. The first is the outputs of the synchronisation algorithm to maintain coherence, and the second is the topology of the bistatic acquisition. The reason for the latter is that different acquisition geometries imply different bistatic range/Doppler histories, which may or may not allow efficient processing algorithms in the frequency domain. This is particularly true for the SS-BSAR case, where there is little control in the choice of acquisition geometry due to the spaceborne transmitter. Another factor to consider is the receiver configuration. A fixed receiver on the ground may generally be simpler in terms of the processing; however, this is not necessarily the case. On the other hand, a moving receiver has the added complexity of motion compensation.

For SS-BSAR with GNSS transmitters, one of the requirements was system operation independent of the acquisition geometry for the airborne receiver case. To accomplish this task, a range-Doppler algorithm with built-in motion compensation was originally built for this specific topology. This algorithm was tested on both simulated and experimental data under various geometries, and its full details can be found in [13, 24].

However, due to the complexity of the algorithm, the requirement for system operation with both stationary and moving receivers, and the fact that the overall system study was at the proof-of-concept stage, a simpler bistatic BPA was built for global operation. In the next sections, the details of the algorithm will be provided for the moving and stationary receiver cases.

3.1. BPA for GNSS-based SAR with a moving receiver

A three-dimensional GNSS-based SAR geometric model operating in strip map mode with a moving receiver is shown in Figure 10. The receiver could be airborne or a ground moving vehicle simply by adjusting the receiver altitude.

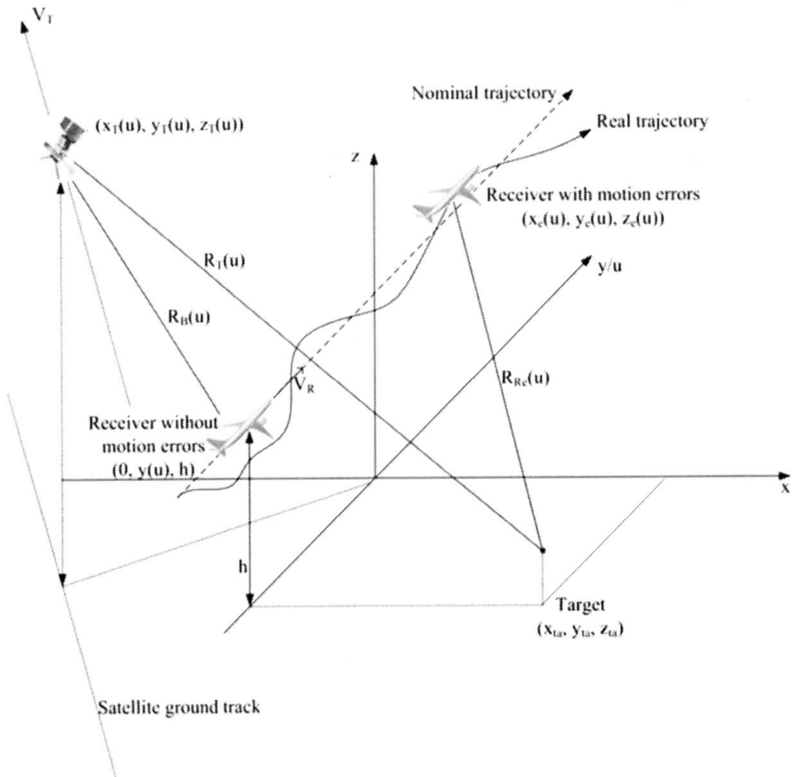

Figure 10. AIRBORNE SS-BSAR GEOMETRY.

The coordinate system obeys the right-hand rule. The receiver moves along the y-axis and for simplicity, but without loss of generality, operates in broadside mode. This axis represents cross-range (y), and is linearly related to slow-time u by the speed of the receiver V_R. The$x$- and z-axes represent ground range and height, respectively. The origin of the coordinate system is the nadir of the midpoint of the synthetic aperture of the receiver. In the absence of motion errors, the instantaneous coordinates of the SAR platform are given by $(0,y(u),h)$, where $h$is the receiver altitude and $y(u) = V_R U$. If motion errors are present, the platform is shifted from its nominal coordinates. In the airborne case, this could be due to atmospheric turbulence during flight, whereas in the ground moving vehicle case, these errors may refer to road anomalies. As a result, the true coordinates of the platform are given by $(x_e(u), y_e(u), z_e(u))$. The totality of across- and along-track errors will be referred to as motion errors hereafter. The satellite is assumed to fly in a straight line, which is approximately true for a relatively short observation time. Its instantaneous co-ordinates are given by $(x_T(u), y_T(u), z_T(u))$.

Both the HC and RC are onboard the receiver. The instantaneous range from the transmitter to the receiver is the transmitter–receiver baseline, and given by

$$R_B(u) = \sqrt{[x_T(u)-x_e(u)]^2 + [y_T(u)-y_e(u)]^2 + [z_T(u)-z_e(u)]^2} \quad (4)$$

On the other hand, for a point target located at co-ordinates (x_{Ta}, y_{Ta}, z_{Ta}), the instantaneous transmitter-target and receiver-target ranges$R_T(u)$ and $R_R(u)$ are equal to

$$R_T(u) = \sqrt{[x_T(u)-x_{Ta}]^2 + [y_T(u)-y_{Ta}]^2 + [z_T(u)-z_{Ta}]^2} \quad (5)$$

$$R_R(u) = \sqrt{[x_e(u)-x_{Ta}]^2 + [y_e(u)-y_{Ta}]^2 + [z_e(u)-z_{Ta}]^2} \quad (6)$$

The HC records the direct signal, with a time delay and phase related to $R_B(u)$. The RC records the reflected signal from the point target, with a time delay and phase related to the sum $R_T(u)$ and $R_R(u)$. However, all of these parameters are severely affected by external, time varying factors which make image focusing impossible without their compensation. These are related with time delay and phase errors due to receiver artefacts, denoted by t_{eRx}, φ_{eRx} such as clock slippage and local oscillator drift, as well as errors due to atmospheric propagation, teatm,φ_{eatm}. In the presence of these errors, the direct and reflected signals at the HC and RC may be written as (after quadrature demodulation):

$$s_{HC}(t,u) = p\left\{t-\left[\frac{R_B(u)}{c} + t_{e_{Rx}} + t_{e_{atm}}\right]\right\}$$
$$exp\left\{-j\left[\frac{2\pi}{\lambda}R_B(u) + \varphi_{e_{Rx}} + \varphi_{e_{atm}}\right]\right\} \quad (7)$$

$$s_{\mathrm{RC}}(t,u) = p\left\{t - \left[\frac{R_T(u) + R_R(u)}{c} + t_{e_{Rx}} + t_{e_{\mathrm{atm}}}\right]\right\}$$
$$\times \exp\left\{-j\left[\frac{2\pi}{\lambda}(R_T(u) + R_R(u)) + \varphi_{e_{Rx}} + \varphi_{e_{\mathrm{atm}}}\right]\right\} \quad (8)$$

where $p(t)$ is the transmitted signal envelope (the primary GNSS code in this case), c is the speed of light and λ is the radar wavelength.

Inspection of (7) and (8) shows that the HC and RC have been modelled as having the same errors. That is due to the fact that they are identical channels of the same receiver, and therefore they share common receiver errors and approximately equal atmospheric errors. Therefore, errors tracked during the HC synchronisation can be used to compensate errors in the RC.

An additional problem for image formation is the trajectory deviations of the receiver. In traditional monostatic SAR, where the same platform is used for signal transmission and echo reception, motion compensation (MoComp) deals with the correction of trajectory deviations and is a well-known problem. In SS-BSAR, the situation is quite different. MoComp is required due to trajectory deviations of the receiving platform only, but the range and phase history of the signal are composed of the sum of the transmitter-target and receiver-target ranges.

A block diagram of a BPA built to tackle all the problems mentioned above is shown in Figure 11.

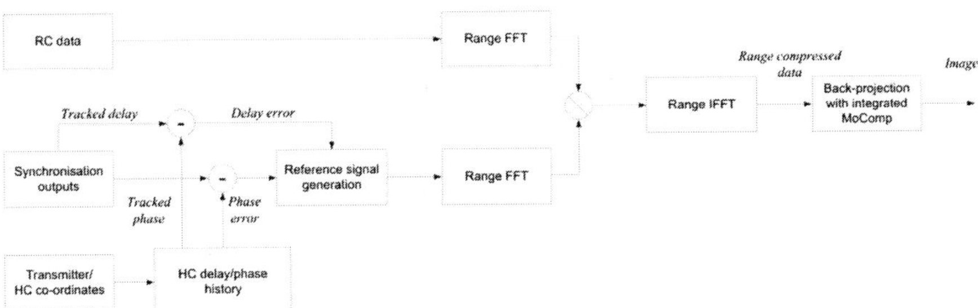

Figure 11. BPA BLOCK DIAGRAM FOR AIRBORNE SS-BSAR.

The BPA consists of two major steps in general: range compression and the calculation of the back-projection integral. For the GNSS-based SAR case, the major peculiarity lies in the generation of the reference signal for range compression, and correction factors for MoComp.

The signal synchronisation algorithm tracks the direct signal time delay and phase, that is $\frac{R_B(u)}{c} + t_{e_{Rx}} + t_{e_{\text{atm}}}$ and $\frac{2\pi}{\lambda} R_B(u) + \varphi_{e_{Rx}} + \varphi_{e_{\text{atm}}}$. In order to compensate for receiver and atmospheric errors, it is necessary to separate them from the terms associated with the baseline length $R_B(u)$. This is possible if the transmitter and receiver co-ordinates are known. That way, terms associated with $R_B(u)$ can be calculated based on (4) and then removed from the synchronisation outputs. As soon as these errors are isolated, a reference signal for range compression may be written as follows:

$$s_0(t,u) = p[t-(t_{e_{Rx}} + t_{e_{\text{atm}}})]\exp\left[-j\left(\varphi_{e_{Rx}} + \varphi_{e_{\text{atm}}}\right)\right] \quad (9)$$

Note that all the receiver and atmospheric errors change with slow-time u, but are not direct functions of this parameter.

Range compression can be conducted in the fast-time frequency domain via an FFT on (8) and (9), complex conjugate multiplication and an inverse FFT. At the output of this operation, the range-compressed RC data may be modelled as follows:

$$r(t,u) = R_x\left[t - \frac{R_T(u) + R_R(u)}{c}\right]\exp\left\{-j\frac{2\pi}{\lambda}[R_T(u) + R_R(u)]\right\}, \quad (10)$$

where $R_x(t)$ is the cross-correlation function between the received and reference signals in the fast-time direction. From (10), it is clear that through the synchronisation algorithm and the modified range compression scheme, the range-compressed data are effectively free of receiver and atmospheric errors. The time delay and phase histories are solely due to propagation time delay and phase.

The final stage in the BPA is the computation of the back-projection integral, which is similar to the monostatic case. The implementation of the BPA involves the generation of a rectangular grid of points with co-ordinates (x_i, y_j). The BPA back-tracks signal returns at the time delays (transmitter-target and receiver-target) associated with each target based on (5) and (6), and integrates the data over slow-time [25].

The challenge in this case is to remove phase artefacts associated with motion errors of the receiver, since they corrupt the received signal phase history, by amounts that differ for different point targets.

This is one of the stronger advantages in using the BPA. Since it operates on individual point targets, an integrated MoComp scheme can be applied during the integration stage, without the need for special blocks in the algorithm. In addition, the MoComp is tailored to suit each point target individually, and therefore motion errors are fully removed throughout the target scene.

The MoComp proposed is based on altering the signal phase history from every point target, so that it corresponds to that of a motion error free receiver trajectory.

The receiver-target range without motion errors can be calculated as follows:

$$R_{Rn}(u) = \sqrt{x_{Ta}^2 + [y(u) - y_{Ta}]^2 + [h - z_{Ta}]^2} \quad \textbf{(11)}$$

The instantaneous range difference can then be calculated between the ideal range history for the target at $(x_{Ta}(u), y_{Ta}(u), z_{Ta}(u))$, given by (11), and the real range history in (6):

$$\Delta R_R(u) = R_R(u) - R_{Rn}(u) \quad (12)$$

The back-projection integral with the integrated MoComp can be written as follows:

$$f(x_i, y_j) = \int_u r[t_{ij}(u), u] \exp\left[j\frac{2\pi}{\lambda}\Delta R_{ij}(u)\right] du \quad (13)$$

where $r(\cdot)$ is the range-compressed data given by (10), and $t_{ij}(u) = \frac{R_{Tij(u)} + R_{Rij(u)}}{c}$, $\Delta R_{ij}(u)$ are the time delay history and MoComp phase factor for the target at (x_i, y_j). When dealing with discrete data, the integration process can be approximated using complex summation.

It is important to note at this point that both the reference signal generation and the MoComp are done based on theoretical calculations of the appropriate range and phase histories. Therefore, the accuracy of the algorithms and the quality of the final image depend heavily on the accuracy to which the transmitter and receiver co-ordinates are known, as we will see later on.t

3.2. BPA for GNSS-based SAR with a fixed receiver

Even though the fixed and moving receiver cases may appear to be similar, they possess some fundamental differences. In the moving receiver case, the dwell time on target is defined by the receiving antenna beam width and the speed of the receiving platform. This is because GNSS satellites have a wide antenna footprint on the ground which is enough to illuminate the same area on the ground for hours with an almost constant power density. Typically, dwell times are in the order of 30–40 s, and over this interval the satellite's trajectory may be approximated as a straight line. In addition, due to the much closer proximity of the receiver to the imaging scene, image resolution in the azimuth direction is predominantly defined by the relative motion of the receiver relative to the imaging area, while the contribution of the transmitter's motion is almost negligible [13].

On the other hand, in the fixed receiver case aperture synthesis can only be provided due to the satellite's motion only. To accomplish a significantly high azimuth resolution, the dwell time must be increased so that the transmitter forms a longer aperture. To obtain an azimuth resolution in the order of 3 m or less, dwell times longer than 5 min are required [26]. However, during this interval the satellite's trajectory can no longer be approximated as a straight line, and data acquisition is more similar to a spotlight mode of operation. Under these conditions, frequency-based BSAR image formation algorithms are difficult to derive. This is the reason why the BPA is a convenient solution.

In terms of the operation, the BPA for the fixed receiver case is very similar to the moving receiver configuration. The instantaneous transmitter-target range is the same, and given by (5), with the geometric model essentially the same as the one shown in Figure 10. Assuming the receiver location is the origin of the co-ordinate system, i.e. the point (0,0,0), the receiver-target range for a target at (x_{Ta}, y_{Ta}, z_{Ta}) is fixed throughout the dwell time and given by

$$R_R = \sqrt{x^2{}_{Ta} + y^2{}_{Ta} + z^2{}_{Ta}} \tag{14}$$

In the absence of motion errors, the same BPA can be employed as in the moving receiver case (Figure 11), but without the MoComp step. Therefore, the same reference signal as in (9) can locally be generated and used for range compression, and at the output of this operation the signal can be written as (similar to (10))

$$r(t,u) = R_x\left[t-\frac{R_T(u)+R_R}{c}\right]\exp\left\{-j\frac{2\pi}{\lambda}[R_T(u)+R_R]\right\} \tag{15}$$

Finally, the equivalent back-projection operation is equal to

$$f\left(x_i, y_j\right) = \int_u r\left[t_{ij}(u), u\right] du \tag{16}$$

which is similar to (13) but without the MoComp step and

$$t_{ij}(u) = \frac{R_{Tij(u)}+R_{Rij}}{c}.$$

4. CASE STUDIES

The signal processing algorithms described above have been applied to experimental data collected both from fixed and moving receivers. The aim of this section is to demonstrate experimental imagery obtained from these configurations. These images confirm not only the functionality of the algorithms employed, but also the feasibility of each system.

4.1. GLONASS transmitter and a fixed receiver

One of the experimental datasets was collected using a GLONASS transmitter and a fixed receiver. The experimental parameters are listed in Table 3.

It was stated in the previous section that image formation processing requires knowledge of the transmitter and receiver positions. The receiver location in the experiment was marked using a standard GPS receiver. For the transmitter, precise satellite ephemeris data were acquired (5-cm accuracy). Satellite positions were given at a 15-min interval in the WGS84 co-ordinate system. To confirm to the PRF of 1 kHz, these data were interpolated using a tenth-order Lagrange polynomial. Then the co-ordinates of the transmitter and receiver were converted from WGS84 to a local co-ordinate system with the location of the HC as the origin.

Table 3. EXPERIMENTAL PARAMETERS FOR FIXED RECEIVER TRIALS

Parameter	Value
Satellite	GLONASS COSMOS 744
Satellite signal	P-code
Signal bandwidth	5.11 MHz
Carrier frequency	1604.8125 MHz
Equivalent PRF	1 kHz
Dwell time	5 min
Satellite elevation during acquisition (relevant to HC antenna)	68°–65°

The receiver was placed at the roof of the School of Electronic, Electrical and Computer Engineering (EECE) of the University of Birmingham, which is a five-storey building. The RC antenna was overlooking the area to the north of the building. A satellite photograph of the observation area is shown in Figure 12a, taken from Google Earth. The imaging area has dimensions 1 × 0.8 km^2 in range and cross-range, respectively. The obtained radar image, after signal synchronisation and image formation, is shown in Figure 12b, superimposed on the photograph in Figure 12a. The colour scale is in dB, where 0 dB represents the intensity of the maximum signal return in the image. The dynamic range of the image was artificially clipped to 30 dB. The region 0–100 m in the image was discarded since this is the region containing the direct signal compressed response.

Figure 12 shows a good coincidence between the radar image and the satellite photograph. The locations of areas with buildings correspond to areas of high echo intensity in the image, while grassy areas exhibit a low reflectivity.

Unfortunately, verification of the images cannot be made using calibrated targets such as corner reflectors. Even though they are standard tools for monostatic SAR, their properties are less predictable for BSAR, let alone GNSS-based SAR with its highly asymmetric structure. For this reason, image verification is best provided using in-scene targets.

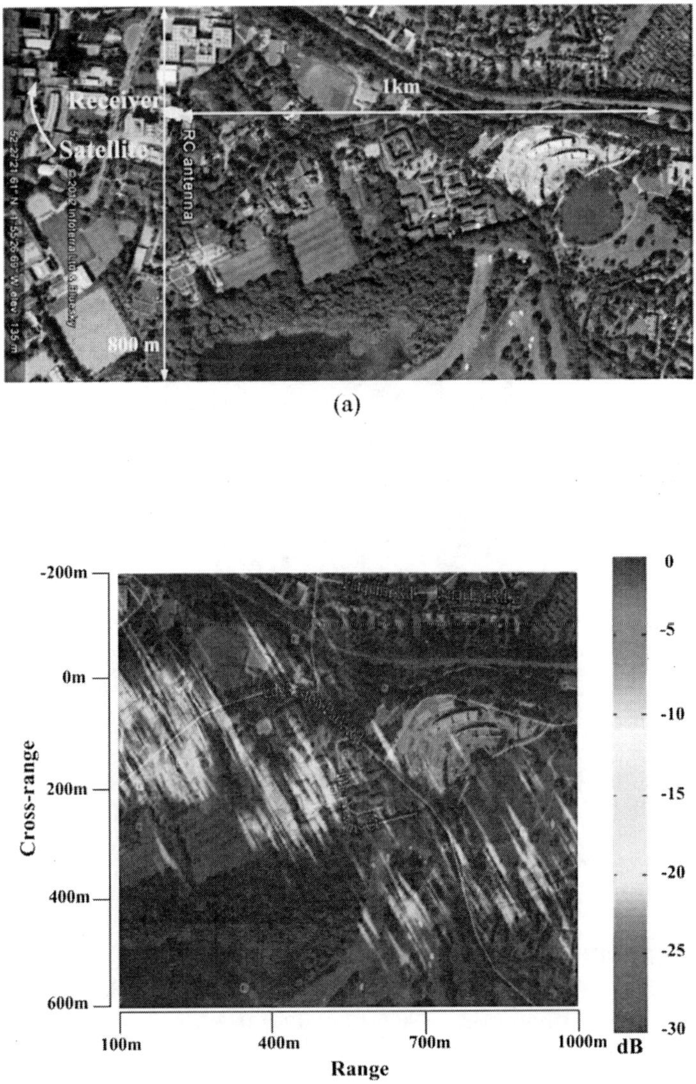

(a)

(b)

Figure 12. IMAGING WITH GLONASS AND A FIXED RECEIVER: (A) Satellite photograph of imaging area (from Google Earth), **(B)** experimental radar image super-imposed on imaging scene.

One notable area in the image is at 625 m in range and 120 m in cross-range, where an isolated compressed target return exists. This is the compressed echo from a building (shown in Figure 13a) which has recently been erected and hence does not appear in the photograph (the construction site in ranges 700 m and beyond in its place). An enlargement of this signal return is shown in Figure 13b. This return was compared to the point-spread function (PSF) of a target based on the data acquisition parameters. The PSF was calculated based on [5] and the obtained result can be seen in Figure 13c. Note that in the compressed return in Figure 13b, 0 dB represents the peak magnitude of the target in question. Moreover, the dynamic range has been clipped to 15 dB for better visualisation of the target's response compared to the returns from other targets.

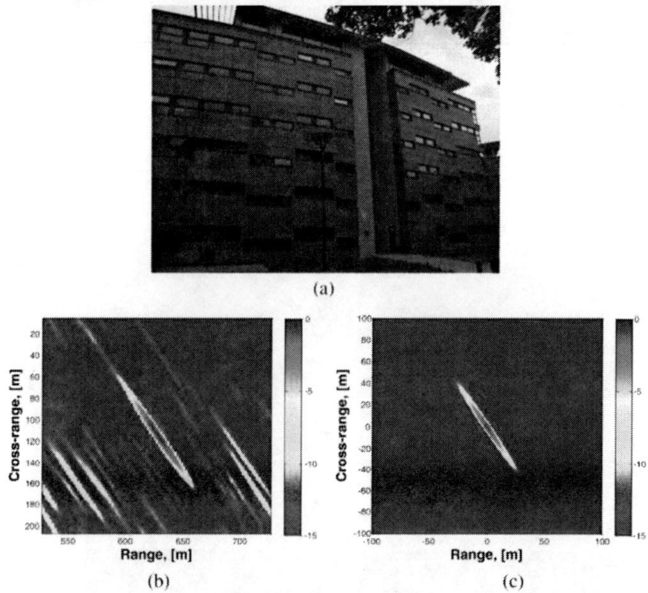

Figure 13. FIXED RECEIVER IMAGE VERIFICATION: (A) Photograph of target at (625,120)m, **(B)** radar response of **(A)**, taken from Figure 12 **(B)**, **(C)**theoretical PSF for a point target.

Comparison of the figures shows there is a high correlation between the compressed return and the theoretical expectation, which verifies the validity and accuracy of the proposed algorithms. The shape and orientation of the returns are practically identical. The size of the PSF is slightly larger in the experimental result; however, this is expected due to the comparison of the response of an extended building with that of a point target.

In addition, it should be stated that the imaging area is not the ideal testing ground for our system, since it is effectively an urban area being imaged with a medium resolution radar. However, the acquired results demonstrate the system's feasibility in practice.

4.2. Galileo transmitter and receiver on ground moving vehicle

The second dataset was collected using Galileo as the transmitter, while the receiver was mounted on a ground moving vehicle (Figure 14). Trials were taken in the area of Clee Hill, Ludlow, UK. The experimental parameters are shown in Table 4. The experimental site and the radar image are shown in Figure 15. The radar image parameters are the same as those in Figure 12. Note that the elevation of the receiver was comparable to the imaging scene, and therefore the majority of signal returns were collected from the front face of the buildings.

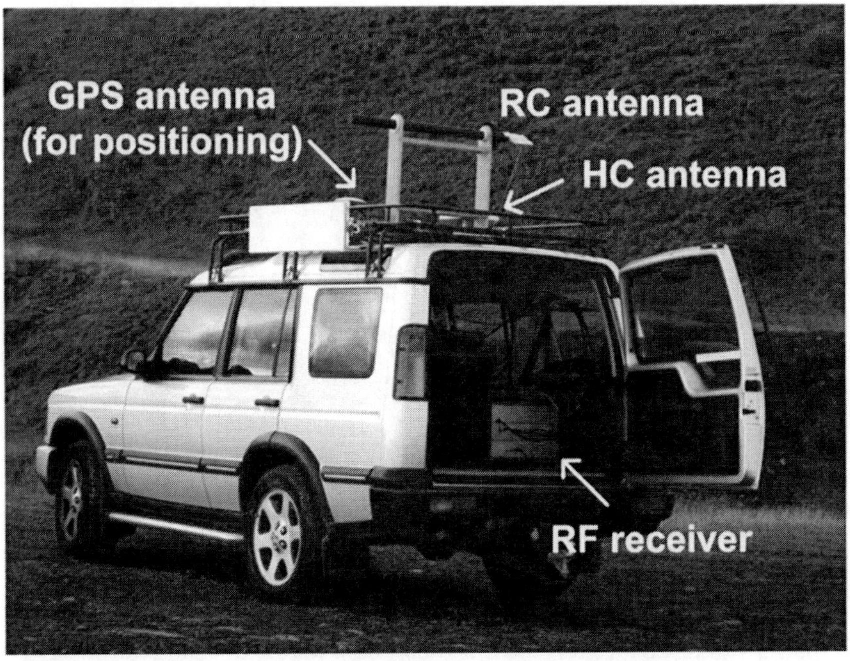

Figure 14. SS-BSAR RECEIVER ON GROUND MOVING VEHICLE.

(a) (b)

Figure 15. IMAGING WITH GALILEO AND A RECEIVER ONBOARD A GROUND MOVING VEHICLE: (A) **Satellite photograph of imaging area (from Google Earth),**(B) **experimenta; radar image super-imposed on imaging scene.**

Table 4. EXPERIMENTAL PARAMETERS FOR GROUND MOVING RECEIVER TRIALS

Parameter	Value
Satellite	Galileo GIOVE-A
Satellite signal	E5bQ
Signal bandwidth	10.23 MHz
Carrier frequency	1207.140 MHz
Equivalent PRF	1 kHz
Dwell time	30 s
Receiver speed (nominal)	30 km/h
Receiver aperture length	250 m

The observation area consists of four sparsely separated buildings which should readily be visible in the radar image. All four targets have been detected. In addition, lower intensity echoes match to the orientation of tree lines in the photograph. To verify this image, cross sections in the area around Target 1 were taken along the range and cross-range directions

(Figure 16). The expected bistatic resolutions are 27 m in range and 1 m in cross-range.

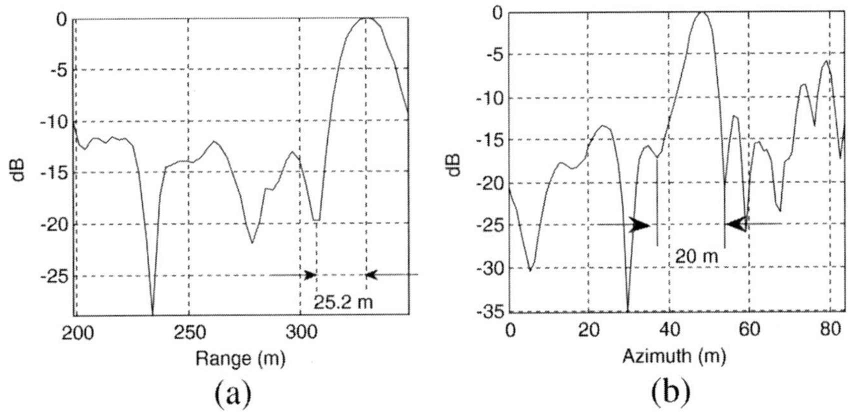

Figure 16. CROSS-SECTIONS FOR TARGET 1: (A) Range and **(B)** azimuth.

The width of the response of the target in range is 25.2 m, which is close to the expected range resolution. The theoretically predicted azimuth resolution is approximately 1 m, which is much smaller than the width of the building. However, we can compare the overall length of the target's response to its physical width. The total width of the target's response is approximately 22 m, in good accordance to its physical dimension. The smaller peaks on either side of the peak target response (the 0-dB point) are also noteworthy. They resemble the sidelobes of a sinc function, which is the form of the expected azimuth signal response. The magnitude of the rightmost peak is also approximately −12 dB. The magnitude of the leftmost peak is lower, at around −16 dB. This could be an image artefact due to inaccuracy in knowledge of the receiver's precise trajectory and velocity. Nevertheless, both the range and azimuth cross-section analyses indicate proper system functionality.

4.3. Galileo transmitter and airborne receiver

The final set of measurements was taken using Galileo as the transmitter, while the receiver was mounted on an AS355 helicopter (Figure 17a). The same receiving hardware as in the ground moving vehicle case was used.

Trials were done around the East Fortune airfield in Scotland (Figure 17b). Synchronisation results from this dataset were shown in Section 2.3.2 and the synchronisation-related parameters were listed in Table 2. The other experimental parameters are presented in Table 5.

(a) (b)

Figure 17. **EXPERIMENTAL SETUP FOR AIRBORNE TRIALS: (A)** Receiving system, **(B)** aerial photograph of imaging scene.

Table 5. EXPERIMENTAL PARAMETERS FOR AIRBORNE TRIALS

Parameter	Value
Receiver speed (nominal)	72 km/h
Receiver aperture length	~800 m
Receiver altitude (nominal)	250 m
Satellite elevation interval	70°–80°
Satellite azimuth interval	100°–130°

This set of data was the most demanding in terms of the MoComp, due to the irregular motion of the helicopter. This can readily be seen from the synchronisation results for this dataset, which were presented in Figure 9. The helicopter location was recorded with a standard GPS receiver with a 1-Hz update rate, which was not sufficient to sample trajectory deviations sufficiently. In addition, the helicopter used was not equipped with any

Inertial Navigation System (INS) and used its own GPS receiver to navigate. For these reasons, it was expected that the obtained imagery would not be as accurate as the previous two cases.

The obtained image is shown in Figure 18, superimposed on a satellite photograph of the observed area.

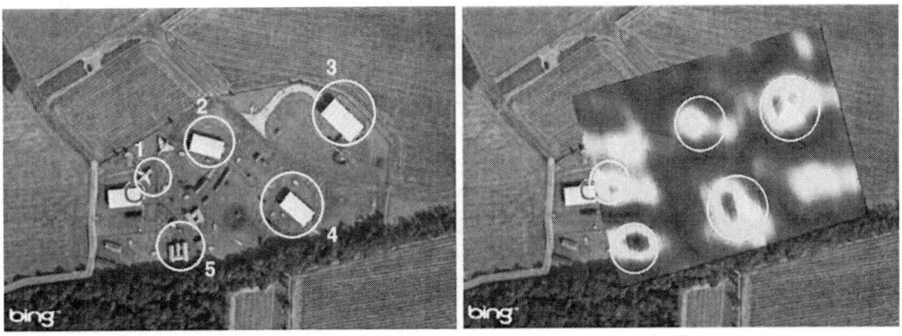

(a) (b)

Figure 18. IMAGING WITH GALILEO AND AN AIRBORNE RECEIVER: **(A)** Satellite photograph of imaging area, **(B)** experimental radar image super-imposed on imaging scene.

The image shows that five main targets have been detected. All of them correspond to buildings (such as hangars) or aircraft which could yield significantly high reflections, such as Targets 4 and 5. Higher intensity parts in the lower right part of the image are due to an occupied car park.

It is clear from the observed imagery that the image is de-focused. For example, the signal return of an aircraft above the leftmost hangar (Target 1) appears to be completely smeared. In addition, signal returns are not registered in their appropriate locations, such as Targets 2 and 4 (aircraft hangars) which appear shifted. Furthermore, Target 3 appears as multiple peaks in the image, implying asymmetric sidelobe levels in the PSF. These artefacts are due to the accuracy and update rate of the GPS receiver onboard the helicopter which are inadequate to sample trajectory deviations sufficiently, plus the absence of inertial navigation equipment which was unfortunately outside our control.

Nevertheless, the detected targets prove that such a system is feasible, and highlight the need for additional SAR post-processing methods based on in-scene targets to increase image sharpness.

5. CONCLUSIONS AND FUTURE WORK

The scope of this article is to describe the signal processing algorithms required to convert GNSS satellites into bistatic imaging radars. Two types of algorithms are needed for this purpose: signal synchronisation and image formation. The signal synchronisation algorithm is a modification of the BASS algorithm, a communications signal processing technique normally used in GNSS for navigation purposes. With regards to image formation, a bistatic BPA has been developed specifically for this topology. The algorithm is generic in the sense that it can accommodate all GNSS transmitting systems, and both moving and fixed receiver configurations, with all their peculiarities. The proposed algorithms were applied to real data collected from a number of topologies, including different GNSS transmitters and different types of receivers, including fixed, onboard a ground moving vehicle or an aircraft. The obtained results highlight not only the functionality of the proposed algorithms, but also the feasibility of the overall system itself. With the advent and progression of the GNSS market, it is believed that GNSS-based radars and imaging radars in particular are a new emerging technology.

Our future work is split into two directions. Research on fixed receiver topologies is done to establish a coherent change detector, where image de-correlation sources and new change detection algorithms are being developed. In terms of the moving receiver cases, it is sought to repeat the airborne trials with a more accurate receiver positioning information, and to develop image post-processing algorithms to correct for receiver trajectory deviations using a combination of in-scene target returns along with theoretical calculations.

DECLARATIONS

Acknowledgements

This study was supported by the Electro-Magnetic Remote Sensing Defence Technology Centre (EMRS DTC) of the UK MoD (Grant no. 1–27) and the Engineering and Physical Sciences Research Council (EPSRC) of the UK government (Grant no. EP/G056838/1). The authors thank Dr. Rui Zuo (ST Ericsson), Dr. Zhou Hong (Research Radar Institute, Beijing) and Mr. Zhangfan Zeng (University of Birmingham) for their contributions to this research.

REFERENCES

1. Rodriguez-Cassola M, Prats P, Schulze D, Tous-Ramon N, Steinbrecher U, Marotti L, Nannini M, Younis M, Lopez-Dekker P, Zink M, Reigber A, Krieger G, Moreira A: First bistatic spaceborne SAR experiments with TanDEM-X. *IEEE Geosci. Remote Sens. Lett.* 2012, 9(1):33-37.

2. Goh AS, Preiss M, Stacy NJS, Gray DA: Bistatic SAR experiment with the Ingara imaging radar. *IET Proc. Radar Sonar Navigat.* 2010, 4(3):426-437.

3. Dubois-Fernandez P, Cantalloube H, Vaizan B, Krieger G, Horn R, Wendler M, Giroux V: ONERA-DLR bistatic SAR campaign: planning, data acquisition, and first analysis of bistatic scattering behaviour of natural and urban targets. *IET Proc. Radar Sonar Navigat.* 2006, 153(3):214-223.

4. Baque R, Dreuillet P, Ruault Du Plessis O, Cantalloube H, Ulander L, Stenstrom G, Jonsson T, Gustavsson A: LORAMbis—a bistatic VHF/UHF SAR experiment for FOPEN. *IEEE Radar Conference* 2010, 832-837.

5. Cherniakov M: *Bistatic Radar—Emerging Technology*. Wiley, New York; 2008.

6. Walterscheid I, Espeter T, Brenner AR, Klare J, Ender JHG, Nies H, Wang R, Loffeld O: Bistatic SAR experiments with PAMIR and TerraSAR-X: setup, processing, and image results. *IEEE Trans. Geosci. Remote Sens.* 2010, 48(8):3268-3279.

7. Rodriguez-Cassola M, Baumgartner SV, Krieger G, Moreira A: Bistatic TerraSAR-X/F-SAR experiment: description, data processing, and results. *IEEE Trans. Geosci. Remote Sens.* 2010, 48(2):781-794.

8. Fujimura T, Totsuka H, Imai N, Matsuo S, Kimura T, Ishii T, Oura Y, Shimada M: The bistatic SAR experiment with ALOS/PALSAR and Pi-SAR-L. *Proceedings IGARSS* 2011, 4221-4224.

9. Zeng T, Cherniakov M, Long T: Generalized approach to resolution analysis in BSAR. *IEEE Trans. Aerosp. Electron. Syst.* 2005, 41(2):461-474.

10. Antoniou M, Saini R, Cherniakov M: Results of a space-surface bistatic SAR image formation algorithm. *IEEE Trans. Geosci. Remote Sens.*2007, 45(11):3359-3371.

11. Cherniakov M, Saini R, Zuo R, Antoniou M: Space-surface bistatic synthetic aperture radar with global navigation satellite system transmitter of opportunity-experimental results. *IET Proc. Radar Sonar Navigat.* 2007, 1(6):447-458.

12. Antoniou M, Zeng Z, Liu F, Cherniakov M: Experimental demonstration of passive BSAR imaging using navigation satellites and a fixed receiver. *IEEE Geosci. Remote Sens. Lett.* 2012, 9(3):477-481.

13. Antoniou M, Cherniakov M, Hu C: Space-Surface Bistatic SAR image formation algorithms. *IEEE Trans. Geosci. Remote Sens.* 2009, 47(6):1827-1843.

14. Lopez-Dekker P, Mallorqui JJ, Serra-Morales P, Sanz-Marcos J: Phase synchronisation and Doppler centroid estimation in fixed receiver bistatic SAR systems. *IEEE Trans. Geosci. Remote Sens.* 2008, 46(11):3459-3471.

15. Krieger G, De Zan F: Relativistic effects in bistatic SAR processing and system synchronisation. *Proceedings EUSAR* 2012, 231-234.

16. He X, Zeng T, Cherniakov M: Signal detectability in SS-BSAR with GNSS non-cooperative transmitter. *IET Proc. Radar Sonar Navigat.* 2005, 152(3):124-132.

17. Wang R, Loffeld O, Neo YL, Nies H, Walterscheid I, Espeter T, Klare J, Ender JHG: Focusing bistatic SAR data in airborne/stationary configuration. *IEEE Trans. Geosci. Remote Sens.* 2010, 48(1):452-465.

18. Wang R, Loffeld O, Nies H, Knedlik S, Ender JHG: Chirp scaling algorithm for the bistatic SAR data in the constant offset configuration. *IEEE Trans. Geosci. Remote Sens.* 2009, 47(3):952-964.

19. Neo YL, Wong FH, Cumming IG: Processing of azimuth-invariant bistatic SAR data using a Range-Doppler algorithm. *IEEE Trans. Geosci. Remote Sens.* 2008, 46(1):14-21.

20. Wang R, Loffeld O, Nies H, Knedlik S, Ul-Ann Q, Medrano-Ortiz A, Ender JHG: Frequency domain bistatic SAR processing for the spaceborne/airborne configuration. *IEEE Trans. Aerosp. Electron. Syst.* 2010, 46(3):1329-1345.

21. Rodriguez-Cassola M, Prats P, Krieger G, Moreira A: Efficient time-domain image formation with precise topography accommodation

for general bistatic SAR configurations. *IEEE Trans. Aerosp. Electron. Syst.* 2011, 47(4):2949-2966.

22. Tsui J: *Fundamentals of Global Positioning System Receivers*. Wiley, New York; 2004.

23. Saini R, Zuo R, Cherniakov M: Problem of signal synchronization in space-surface bistatic synthetic aperture radar based on global navigation satellite emissions-experimental results. *IET Proc. Radar Sonar Navigat.* 2010, 4(1):110-125.

24. Antoniou M, Zuo R, Plakidis E, Cherniakov M: Motion compensation algorithm for passive space-surface bistatic SAR. *IET Radar Conference*2009, 1-5.

25. Soumekh M: *Synthetic Aperture Radar Signal Processing with MATLAB Algorithms*. Wiley, New York; 1999.

26. Cherniakov M, Plakidis E, Antoniou M, Zuo R: Passive space-surface bistatic SAR for local area monitoring—primary feasibility study. *EuRad*2009, 89-92.

8

CHAPTER

A FEASIBILITY ANALYSIS OF LAND-BASED SINS/GNSS GRAVIMETRY FOR GROUNDWATER RESOURCE DETECTION IN TAIWAN

Kai-Wei Chiang, Cheng-An Lin and Chung-Yen Kuo

Department of Geomatics, National Cheng Kung University, No. 1, University Road, Tainan 701, Taiwan

ABSTRACT

The integration of the Strapdown Inertial Navigation System and Global Navigation Satellite System (SINS/GNSS) has been implemented for land-based gravimetry and has been proven to perform well in estimating gravity. Based on the mGal-level gravimetry results, this research aims to construct and develop a land-based SINS/GNSS gravimetry device containing a navigation-grade Inertial Measurement Unit. This research also presents a feasibility analysis for groundwater resource detection. A preliminary comparison of the kinematic velocities and accelerations using multi-combination of GNSS data including Global Positioning System, Global Navigation Satellite System, and BeiDou Navigation Satellite System, indicates that three-system observations performed better than two-system data in the computation. A comparison of gravity derived from SINS/GNSS and measured using a relative gravimeter also shows that both agree reasonably well with a mean difference of 2.30 mGal. The mean difference between repeat measurements of gravity disturbance using SINS/GNSS is 2.46 mGal with a standard deviation of 1.32 mGal. The gravity variation because of the groundwater at Pingtung Plain, Taiwan could reach 2.72 mGal. Hence, the developed land-based SINS/GNSS gravimetry can sufficiently and effectively detect groundwater resources.

Keywords: land-based SINS/GNSS gravimetry; gravity disturbance; groundwater resource detection

1. INTRODUCTION

According to the Taiwan Central Weather Bureau [1] and Water Resources Agency [2], the annual precipitation with an average of 62,000 mm per year in Taiwan is considered to be abundant. However, the supply of water to humans, agriculture, and industry is deficient because the water flows to the oceans directly and quickly because the topography is extremely steep. An alternative water supply source is to pump groundwater during surface water shortages. Searching for potential groundwater areas can be difficult, particularly when new or more available water resources may be stored deep underground. Gravity changes are caused directly by mass variations, and hence, fill thickness could produce subtle anomalies in the Earth's gravitational field, which can be used to estimate fill thickness and bedrock topography if properly detected. Along these lines, gravity measurements have the potential for use in the detection or monitoring groundwater resources [3].

Conventional methods for measuring gravity changes include the use of terrestrial gravimeters and astronomical techniques. Although both methods can provide accurate gravity information with high resolution, the required instruments are extremely expensive and the field work is both labor- and time-consuming. Airborne and marine gravimeters have been developed to improve the efficiency of field work, however, the instruments are still expensive. Based on the circumstances in Taiwan, not only is the topography complicated and steep but also flight altitude is restricted by Taiwanese laws, so the flight altitude cannot be low. The gravity accuracy observed by airborne gravimeters decreases because of the downward continuation of gravity from the flight altitude to ground level. At the same time, the performance of marine gravimeters was also relatively poor due to ship vibrations.

With the advances in space technology, the global gravity field can be efficiently determined using satellite missions, such as Gravity Recovery and Climate Experiment and Gravity Field and Steady-state Ocean Circulation Explorer (GOCE). However, satellite missions cannot observe short wavelength gravity fields. For example, the gravity field derived from the measurements of the GOCE mission launched in 2009 has an accuracy of 1 mGal (10−5 m/s2) with a spatial resolution of 100 km [4].

Strapdown Inertial Navigation System (SINS) and Global Navigation Satellite System (GNSS) have been integrated for navigation and kinematic positioning over the past few decades. An Inertial Navigation System (INS) contains an onboard computer and an Inertial Measurement Unit (IMU), consisting of three-axis accelerometers and three-axis gyroscopes used to determine the motion of the vehicle and provide full navigation states. Benefiting from the improvements in IMU hardware and GNSS kinematic positioning technology, the moving base SINS/GNSS gravimetry has been developed and proven successfully to perform well in estimating gravity. Schwarz [5] presented the equations for estimating gravity vectors based on GNSS-derived kinematic acceleration and IMU-measured dynamic acceleration. Airborne SINS/GNSS gravimetry can locally augment low resolutions (>150 km) to medium information (5 to 150 km), and land-based gravimetry can be increase to high resolution information (<5 km) [4]. The moving base SINS/GNSS gravimetry is an effective and mature method for surveying local gravity within a relatively short period, and can now be used in geophysical and geodetic applications.

However, because of the attenuation of the gravity field with altitudes, detecting short wavelength gravity signals has become a challenge. Considering that the power of the gravity field intensifies as the altitude decreases, the signal-to-noise ratio (SNR) should increase in the land-based system. The lower velocity of the land-based vehicle can also improve the resolution and accuracy of gravity disturbance [6], which is defined as the difference between the actual gravity and the normal gravity. In general, the high resolution gravity disturbance based on the smoothing of 180 s can be obtained with an accuracy of 3 mGal in a vertical component [6].

Land-based SINS/GNSS gravimetry, which has better mobility and flexibility, can serve as a good alternative. Hence, to increase the SNR and avoid the errors of a downward continuation process, in the current study a prototype of land-based gravimetry based on SINS/GNSS is developed for groundwater detection.

2. PROBLEM STATEMENTS

Moving base gravimetry is constructed by integrating GNSS and INS. GNSS can provide precise positioning and kinematic acceleration, whereas INS can measure the specific forces encountered by the vehicle. In general, INS has two kinds of mechanization for INS namely, gimbaled INS and SINS [7].

Gimbaled INS is a physical realization of the navigation-frame (*n*-frame) that uses a gyro-stabilized platform with three orthogonal accelerometers. Thus, the accelerometers and gyroscopes do not encounter kinematic measurement errors directly. The configurations may lead to good gravity estimates [8], but the shortcomings are that the platform has a relatively large size and is expensive. Compared with gimbaled INS, SINS can provide an analytical image of the *n*-frame, and the accelerometers and gyroscopes are installed in the unit. The advantages of SINS are its smaller size, lower cost, and more operational flexibility [9]. Therefore, the moving base SINS/GNSS gravimetry is developed; the gravity disturbance from this system can achieve an accuracy of 3 to 5 mGal [10].

In the SINS/GNSS gravimetry technique, as shown in Figure 1, the IMU accelerometers measure the specific force along the trajectory, whereas the GNSS provides the kinematic acceleration of the vehicle. In principle, the measured specific force contains gravity and the effects of the motion. According to Newton's second law of motion, in a non-rotating coordinate frame, gravity can be derived from the difference between the specific force and kinematic acceleration.

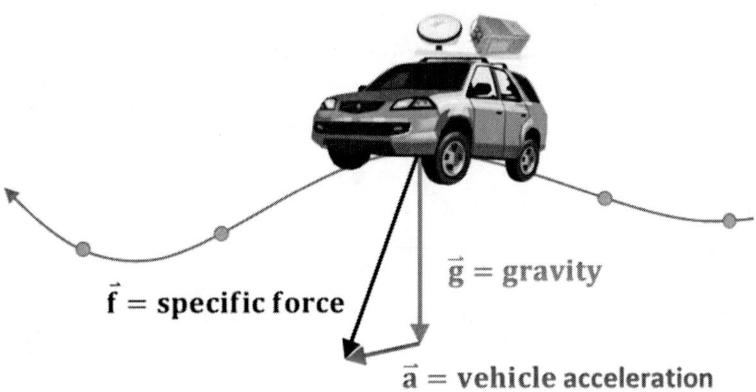

Figure 1. Concept of land-based SINS/GNSS gravimetry.

In practice, the extremely low SNR is the primary problem that needs to be addressed in determining the gravity signal. Representatively, the magnitude of gravity disturbance does not exceed 100 mGal in the vertical component over the distance of about 100 km [11]. However, non-gravitational accelerations measured by SINS/GNSS could be over 100 to 1000 times larger than the gravity disturbance [12,13]. A direct and easy method of making the gravity disturbance signal more noticeable from measurement errors is to decrease the altitude of the vehicle [6].

The first purpose of this research is to implement a land-based SINS/GNSS gravimetry method that can provide optimal estimations of gravity disturbance because of its low altitude compared with airborne SINS/GNSS gravimetry. The main idea is that gravity disturbance can be obtained directly by differencing the GNSS-derived kinematic acceleration and IMU-measured dynamic acceleration. Compared with the traditional method, the computations in this research can be carried out in both the inertial-frame (*i*-frame) and the *n*-frame, thereby avoiding a significant amount of computations related to the equations of the error dynamics applied in the traditional method.

In terrestrial tests, both IMU and GNSS measurements have much larger noises because of the rough terrain and vehicle vibrations. Thus, a low-pass filter is required to smooth the data. High resolution gravity disturbance with an accuracy of 3 mGal in a vertical component based on the smoothing of 180 s can be obtained [14]. Accuracy and resolution do not only rely on the algorithms of the data processing, but rather rely mostly on the performance characteristics of the IMU, velocity of the vehicle, altitude, and quality of the positioning solutions [15]. High-end navigation grade IMU, moderate velocity, low altitude, and cm-level positioning solutions are required to obtain accurate results of SINS/GNSS gravimetry. However, system configurations may not be always ideal during tests. Hence, detecting gravity signals with sufficient accuracies may be meaningful and worthwhile for novel studies.

The accurate determinations of a vehicle's velocity and kinematic acceleration using GNSS are often acquired in a moving base SINS/GNSS gravimetry. In turn, the differentiation of GNSS signals is a crucial procedure for obtaining estimates and is important in post-processing where the acceleration is estimated conventionally by differentiating GNSS positions [16]. High precision results can now be achieved because of the evolution of GNSS. Hence, the second purpose of this study is to evaluate kinematic velocity and acceleration using Multi-GNSS, including Global Positioning System (GPS), Global Navigation Satellite System (GLONASS), and BeiDou Navigation Satellite System (BDS).

3. GRAVITY DETERMINATION DERIVED FROM SINS/GNSS GRAVIMETRY

IMU measures the dynamic acceleration directly applied on lifting and dragging, and thus, measured gravity is retrievable from the kinematic acceleration computed from the GNSS position [15]. SINS/GNSS gravimetry can be estimated after obtaining the determination of the measured gravity. The principle equation in the *n*-frame is described in the following [17]:

$$\delta g^n = {a_u}^n - {\ddot{x}_u}^n - \gamma^n + \left(\frac{V_e}{R_N + h} + 2\omega_E \cos\varphi\right) V_e + \frac{{V_n}^2}{R_M + h} \tag{1}$$

where γ is the normal gravity, Ve and Vn are the east and north components of the vehicle's velocity, respectively, ωE is the rotation rate of Earth, h and φ are the ellipsoidal height and geodetic latitude, and RNand RM are prime and meridian curvature radii, respectively. The sum of the fourth and fifth terms at the right side is usually called the Eötvös correction, which results from the rotation of the reference coordinate frame.

In this research, all calculations are performed in the *i*-frame. Therefore, it is conceptually simpler, more straightforward and computationally less compared to the traditional approach (Equation (1)). The basic equation is based on Newton's second law of motion expressed in a non-rotating and free falling coordinate frame (*i*-frame) [7], which reads as follows:

$$g^i = {a_u}^i - {\ddot{x}_u}^i \tag{2}$$

where g is the measured gravity, xu¨ is the vertical kinematic acceleration from the upward component, and au is the dynamic acceleration.

The principal algorithms of the proposed SINS/GNSS gravimetry are divided into three parts, including the lever-arm correction, GNSS-derived kinematic acceleration, and IMU-measured dynamic acceleration.

3.1. Lever-Arm Correction

Because the phase center of the GNSS antenna does not coincide with the fiducial center of the IMU, the kinematic accelerations computed from the GNSS positions cannot be compared directly with the IMU-measured dynamic accelerations. The displacement vector between the GNSS antenna and the IMU is called the lever-arm, and the GNSS positions should be corrected to ensure that the lever-arm effect obtains reasonable gravimetry results [7]. Supposing that the GNSS position of the antenna is obtained in *i*-frame, then the position of the IMU in *i*-frame is given by:

$$X^{i}_{IMU} = X^{i}_{GNSS\ antenna} - C^{i}_{b}L^{b} \tag{3}$$

where Lb is the lever-arm displacement vector in *b*-frame and Cib is the rotation matrix transforming Lb in *b*-frame to that in *i*-frame. The rotation matrix is computed in terms of geodetic latitude φ and longitude λ:

$$C^{i}_{b} = C^{i}_{n} \times C^{n}_{b} \tag{4}$$

$$C^{i}_{n} = C^{i}_{b} \times (C^{n}_{b})^{-1} = \begin{bmatrix} -\sin(\varphi)\cos(\theta+\lambda) & -\sin(\theta+\lambda) & -\cos\varphi\cos(\theta+\lambda) \\ -\sin(\varphi)\sin(\theta+\lambda) & \cos(\theta+\lambda) & -\cos\varphi\sin(\theta+\lambda) \\ \cos\varphi & 0 & -\sin\varphi \end{bmatrix} \tag{5}$$

where θ=ωE(t−t0), and t0 is an initial epoch when the inertial and Earth-centered and Earth-fixed (ECEF) coordinate frames are assumed to be parallel. The initial epoch is usually chosen as the first epoch when both GNSS and INS data are available. The INS epoch closest to the GNSS epoch must be used because the sampling rates differ.

3.2. Kinematic Acceleration

Precise GNSS positioning solutions can be processed using a standard differential technique based on the phase observation from GNSS receivers on the moving vehicle and at a number of base stations. Conventionally, the GNSS positions Xe are given in the ECEF coordinate frame (*e*-frame). However, to reduce the computational complexity, the transformation of the GNSS positions from *e*-frame to *i*-frame should be performed before the differentiation. The equation can be expressed as:

$$X^{i} = C_{e}^{i}X^{e} = \begin{bmatrix} \cos(-\omega_{E}t) & \sin(-\omega_{E}t) & 0 \\ -\sin(-\omega_{E}t) & \cos(-\omega_{E}t) & 0 \\ 0 & 0 & 1 \end{bmatrix} X^{e} \tag{6}$$

where Cie is a matrix that rotates the coordinate frame of Xe along the third axis by the angle ωEt and t is time with respect to the reference epoch.

The kinematic acceleration of the vehicle can be determined by the numerical differentiation of the precise GNSS positions corrected by the lever-arm displacement:

$$\ddot{X}_{IMU}^{i} = \frac{d^{2}}{dt^{2}}\left(X_{GNSS\ antenna}^{i} - C_{b}^{i}L^{b}\right) \tag{7}$$

The smoothing process, which reduces the noise to the level of a few mGal, should be applied to obtain accurate kinematic accelerations.

3.3. Dynamic Acceleration

The dynamic accelerations sensed by the IMU are referenced to the platform frame because the SINS is mounted to the strapdown mechanization platform. Hence, to obtain dynamic accelerations in *i*-frame, the transformation should be applied according to:

$$a^{i} = C_{b}^{i}a^{b} \tag{8}$$

The transformation matrix Cib is calculated from the gyroscope output using the optimal estimates from Rauch-Tung-Striebel (RTS) smoother [18] with the aid of GNSS. The dynamic acceleration ab is obtained after correcting systematic errors (bias and scale factor), which are also estimated using the RTS smoother. Therefore, the measured gravity determined by the proposed SINS/GNSS gravimetry is based on the following:

$$g^n = C_i^n(a_u{}^i - \ddot{x}_u{}^i) \tag{9}$$

where the transformation matrix Cni can be obtained by transposing Cin. The starting points in the epoch for the smoothing of both GNSS and INS data must be the same and thus, the difference between kinematic and dynamic accelerations should be smoothed.

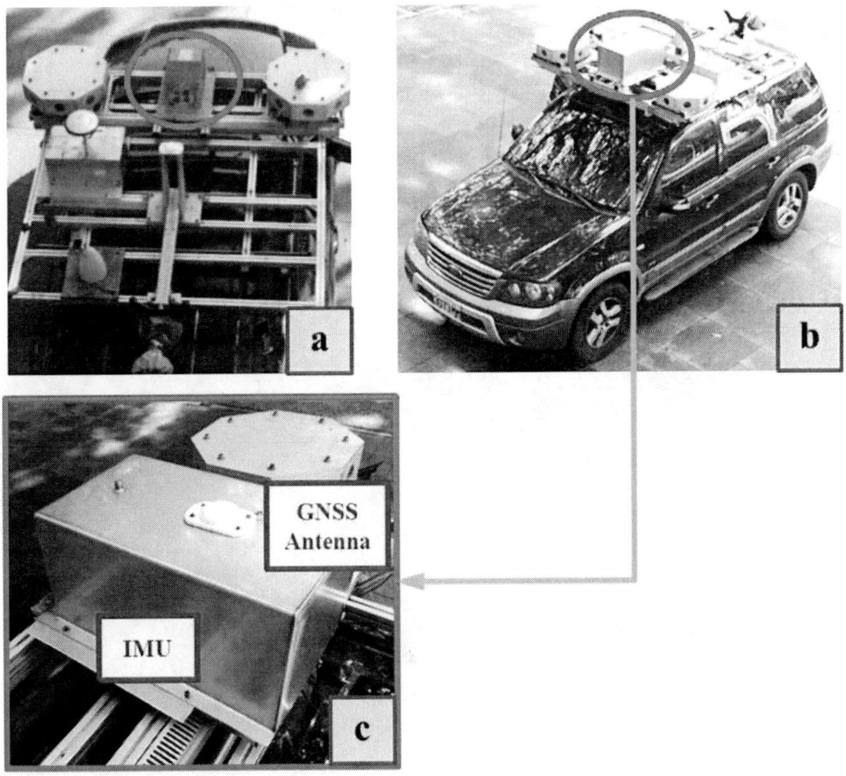

Figure 2. (**a**) Setup of sensors applied for the land-based system. (**b**) A steel box used for protecting IMU and the GNSS antenna was installed on top of its. (**c**) A zoom-in view of (b).

4. IMPLEMENTATION OF PROPOSED LAND-BASED SINS/GNSS GRAVIMETRY

Figure 2a shows our land-based system which was used for terrestrial experiments and the IMU was just mounted on top of the vehicle without the steel box. In Figure 2b, the IMU was enclosed in a steel box for protection and the GNSS antenna was installed on top of the steel box to prevent multipath effects. Figure 2c is zoom-in view of Figure 2b.

The kinematic positions of the land-based vehicle are computed using the differential GNSS (DGNSS) technique. The GNSS instrumentation of the rover on top of the land-based vehicle was a NovAtel ANT-A72GOLA-TW, which can receive L1/L2 GPS and G1/G2 GLONASS signals. The pulse per second signal is also sent to the IMU for time synchronization.

The IMU is a strapdown navigation grade unit of type iNAV-RQH containing three QA2000 accelerometers and three GG1320 ring laser gyroscopes, as shown in Figure 3. The maximum output rate of iNAV-RQH is 300 Hz. Table 1 lists the main performance characteristics of both sensors.

Figure 3. iNAV-RQH IMU.

Table 1. Performance characteristics of iNAV-RQH.

	QA2000 Accelerometer	GG1320 Ring Laser Gyroscope
Measurement Range	±2 g	±500 °/s
Non-linearity	15 μg/g^2	10 PPM
Scale Factor	70 PPM	10 PPM
Angular Random Walk	-	0.0018 °/$\sqrt{h}$
Acceleration Noise Density	8 μg/$\sqrt{Hz}$	-
Bias Repeatability	< 15 μg	0.002 °/h

The flowchart for the estimation of gravity presented in this study is shown in Figure 4. The four main steps involved are described as follows.

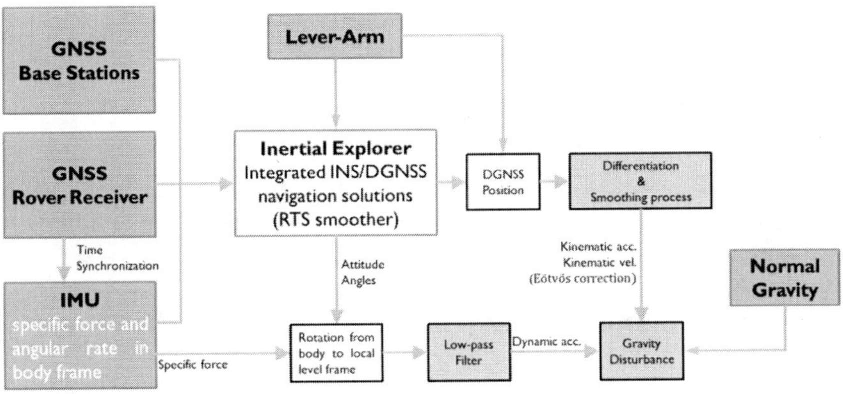

Figure 4. Data processing flowchart of the proposed SINS/GNSS gravimetry.

First, the Inertial Explorer software from NovAtel was used to process the DGNSS mode and tightly-coupled integration using the RTS smoother. The optimal results of DGNSS positions and attitude angles were obtained for further processing.

Second, the kinematic accelerations were determined using the DGNSS positions based on a numerical differentiator. Numerical time-differentiation utilizes various discrete and analytic techniques. In this study, a fifth-order B-spline differentiator [19] was used to obtain kinematic accelerations. A smoothing process is necessary in the determination of kinematic accelerations because the numerical differentiation would increase the high frequency noises. Kincaid and Cheney [20] suggested the theory and model in numerical computations of the B-spline smoother and therefore, a third-order B-spline smoother with a window length of 180 s was adopted.

The raw data of accelerometers contain gravity and the effects of vehicle motion. IMU-measured dynamic accelerations were also processed by the same smoother, *i.e.*, third order B-spline smoother to remove high frequency errors of the data in the third step.

In the final step, gravity disturbance was computed by differencing the smoothed kinematic and dynamic accelerations. The biases of the accelerometers derived from the INS/GNSS RTS smoother were used in this step to remove the systematic errors directly. The standard deviations (STD) of the estimated biases in three components (X, Y and Z) are 9.60 mGal, 14.66 mGal and 7.12 mGal, respectively. In this way, reasonable and good estimations of gravity disturbance could be achieved.

5. RESULTS AND DISCUSSION

Various scenarios and evaluations based on the purpose of this study were considered in the experiments and discussed in the following statements.

5.1. Performance of SINS/GNSS Gravimetry

The performance of navigation grade iNAV-RQH was validated by first stopping the land-based vehicle at a point to perform Zero Velocity Update (ZUPT) and collect IMU and GNSS data in static mode for 3 to 5 min. A total of 209 ZUPT points were used in the experiment. During ZUPT periods, the motion of the vehicle was zero, and does not involve either kinematic acceleration or velocity. The IMU-measured dynamic accelerations represent the gravity vector. The errors of GNSS-derived kinematic accelerations and velocities would not affect the results of the calculated gravity. Figure 5 shows the two sections containing nine and five ZUPT points, respectively, selected for analysis because altitudes in these two sections have larger changes. Figure 5 illustrates that the correlation of measured gravity and elevation is highly negative, with a correlation coefficient of 0.957. Therefore, the measured gravity from iNAV-RQH shows reasonable results, with the mean gravity gradient about 0.283 mGal/m.

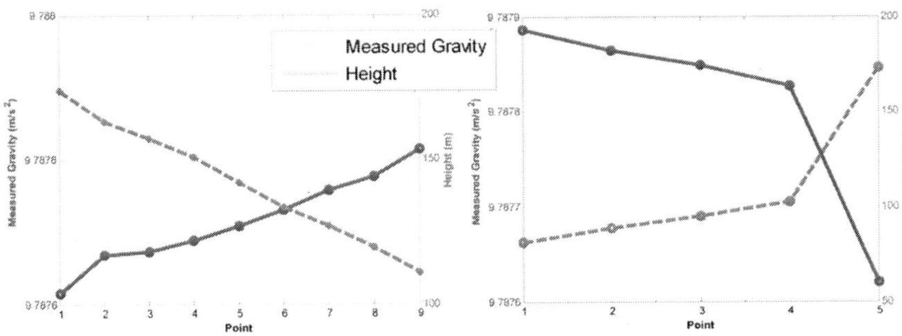

Figure 5. Relationship between the measured gravity and ellipsoidal height.

An alternative method of assessing the accuracy of iNAV-RQH measured gravity is by employing the relative gravity next to the vehicle measured by using a Graviton EG relative gravimeter, an electronic version from LaCoste & Romberg. The repeatability of the Graviton EG is 0.001 mGal and 0.003 mGal in controlled and field conditions, respectively, and the absolute drift is smaller than 1 mGal per month. The contiguous 22 ZUPT points along the trajectory were selected and the gravity difference (Delta G) of the sections computed from Graviton EG and iNAV-RQH, as illustrated in Figure 6.

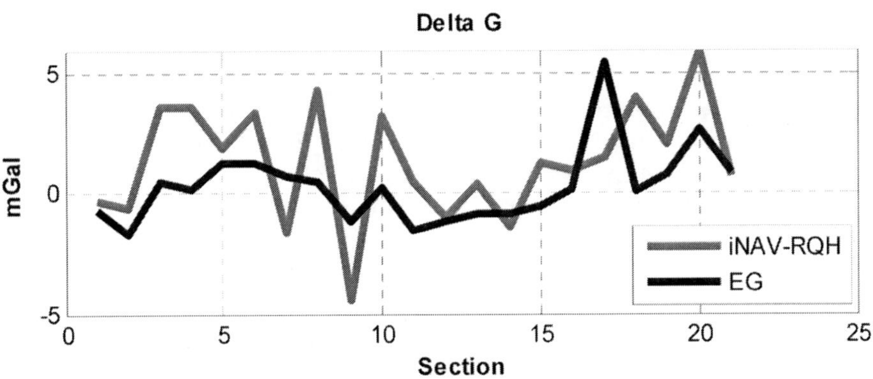

Figure 6. Comparison of the gravity at ZUPT points relative to the previous point derived from iNAV-RQH and Graviton EG.

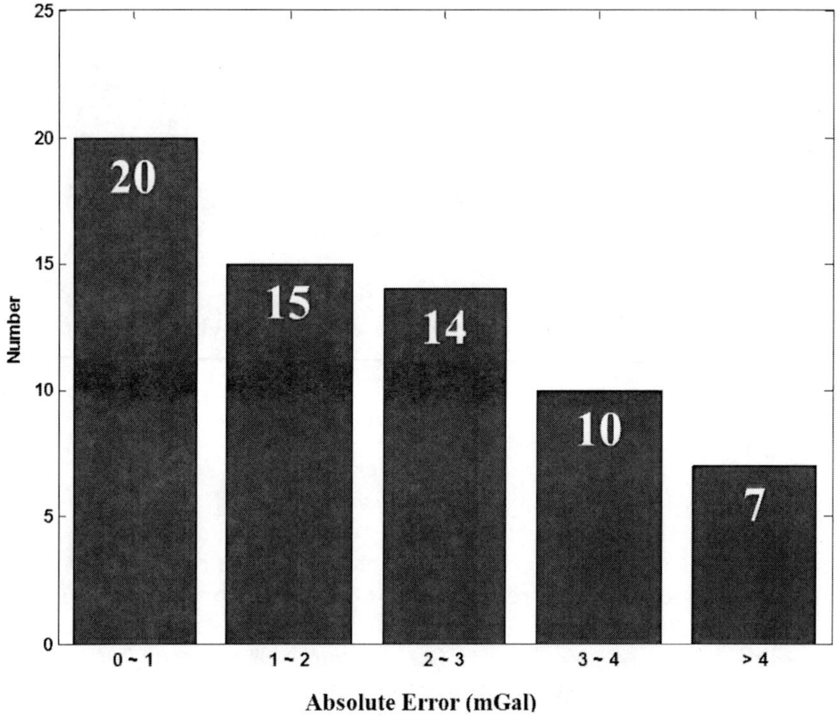

Figure 7. Statistical absolute error results.

The Delta G observed using the Graviton EG was assumed as the reference, and the absolute errors meaning the Delta G difference between the Delta G from iNAV-RQH and Graviton EG were calculated. Most of the absolute errors, as shown in Figure 7, range between 0 and 1 mGal. The mean of the absolute errors is 2.30 mGal with a standard deviation of 2.41 mGal. The comparison between the Delta G obtained through the gravimeter Graviton EG and through iNAV-RQH indicates that iNAV-RQH is capable of providing mGal-level gravity.

5.2. Determinations of Kinematic Velocity and Acceleration Using Multi-GNSS

As previously mentioned, the accuracy of the computed kinematic velocity and acceleration is crucial for estimating gravity using a moving base SINS/GNSS gravimetry. The multi-combination of GNSS data were collected at ZUPT points and processed using the DGNSS technique. The absence of errors in data and processing would be indicated by the computed kinematic velocity in horizontal and kinematic accelerations in the vertical being zero at ZUPT points.

The testing GNSS receiver was an MB5000 GNSS Sensor from COMTECH, which was active with L1/L2 GPS, G1/G2 GLONASS, and B1/B2/B3 BDS signal. The exterior view is shown in Figure 8.

Figure 8. MB5000 GNSS antenna and receiver.

Table 2. Statistics of kinematic velocity and acceleration computed using different combinations of GNSS data.

	RMSE		
	V_e (10^{-5} m/s)	V_n (10^{-5} m/s)	a_u (mGal)
GPS + GLONASS	14.05	13.01	3.91
GPS + GLONASS + BDS	11.62	7.28	3.47
	Improvement (%)		
From two systems to three systems	17.30	44.03	11.14

The DGNSS positioning results were processed using the Compass Solution software from COMTECH, and the static baseline mode was adopted. Two kinds of combined GNSS data were used in the study. The first dataset comprised a combination of GPS and GLONASS data for further computation, while the other dataset was a combination of GPS, GLONASS, and BDS. Based on the root mean square error (RMSE) of differences between the true value of zero and the computed kinematic acceleration, the kinematic acceleration derived from the data of three satellite systems indicated improvement of more than 11% compared with that from the two systems (Table 2). This improvement results from the higher availability of satellite measurements and better geometry from three satellite systems, which can elevate the accuracy of estimated gravity disturbance. The interface of the MB5000 software including the sky view and tracking satellites information is shown in Figure 9.

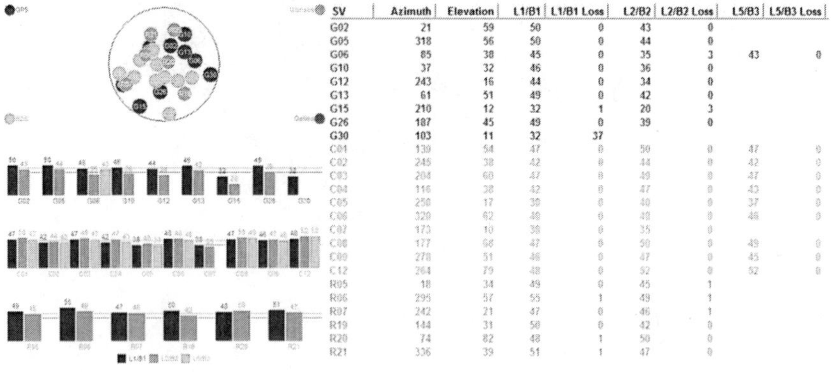

Figure 9. MB5000 software interface.

In this research, multi-combination of GNSS data observed by GPS, GLONASS, and BDS were used to evaluate the accuracies of the kinematic velocity and acceleration. The result shows obvious improvements, however, because the differential GNSS technique for kinematic data is not available so far, the multi-combination of GNSS data were not used to estimate gravity disturbances.

5.3. Feasibility Analysis for Groundwater Resource Detection

The feasibility for groundwater resource detection using SINS/GNSS gravimetry was analyzed by selecting the Pingtung Plain in Taiwan (Figure 10) as the study area based on previous civil engineering and hydrology studies by the Taiwan Central Geological Survey [21]. The yellow line represents the potential area for water storage containing groundwater resources of approximately 1.2 km3. The red zone is the experimental region, covering an area of about 30×30 km2.

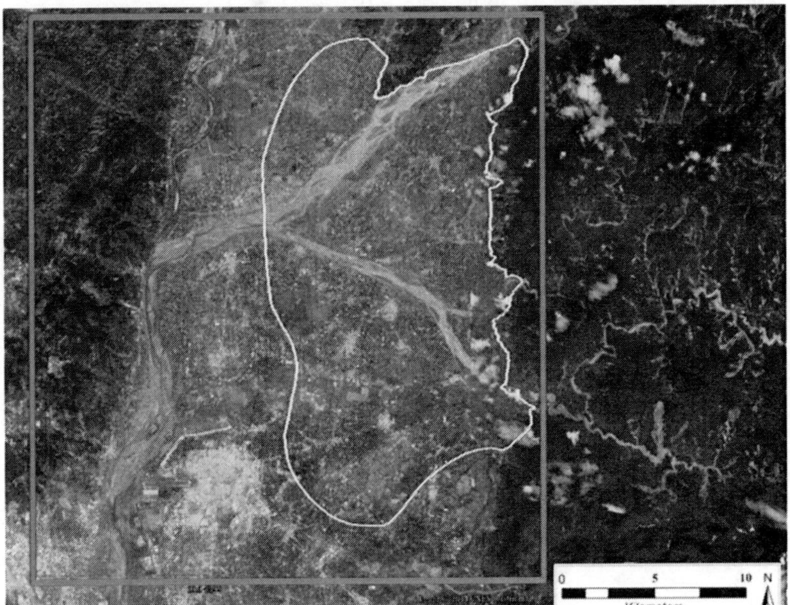

Figure 10. Experiment area at Pingtung Plain, Taiwan (background map courtesy of Google Earth).

The planned ZUPT were planted densely approximately every 1 km to ensure that better gravimetry results can be obtained and that the resolution of the test can be improved. The total number of ZUPT points was 209.Figure 11 shows the distribution of the ZUPT points.

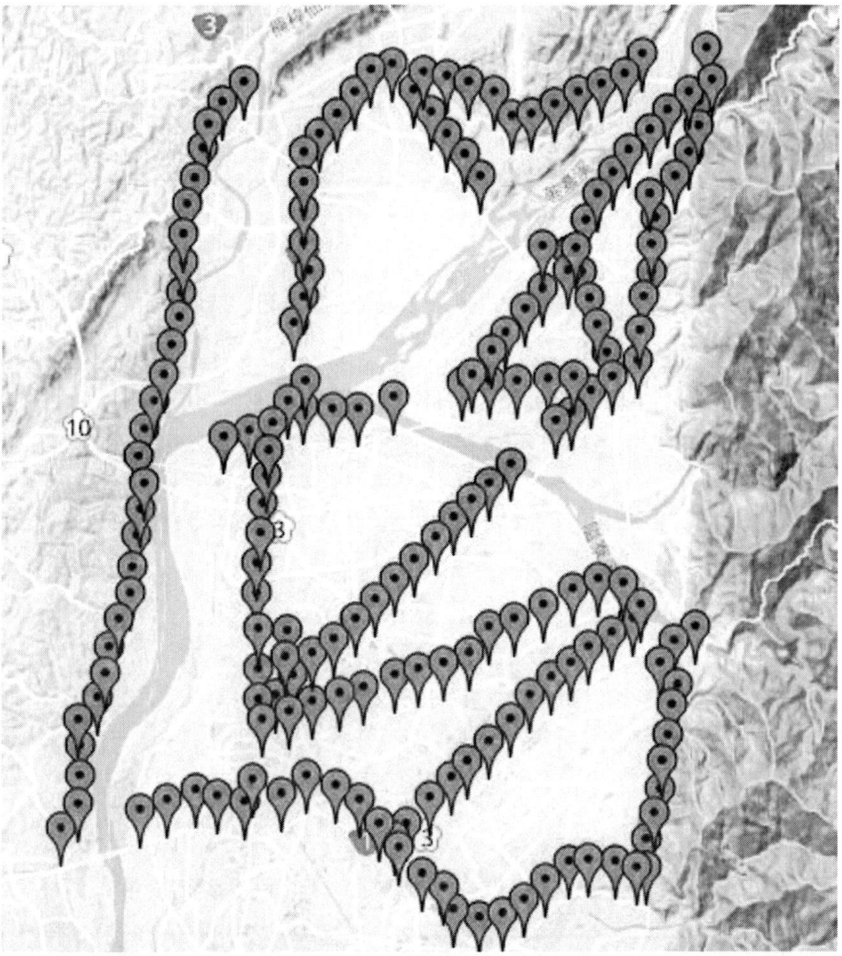

Figure 11. Distribution of ZUPT points (background map courtesy of Google Map).

Following Gerlach *et al.* [22], a fourth-order Butterworth low-pass filter with a cut-off frequency of 0.1 Hz was applied to process dynamic accelerations. In this study, both B-spline smoother and Butterworth filter were adopted to remove high frequency noises, and then the results obtained were compared. In Table 3, the averaged standard deviations (STD) show that the IMU-measured dynamic accelerations at ZUPT points processed by the B-spline smoother performed better because the high frequency noises were removed more accurately, allowing the gravity signals to stand out better. Therefore, the B-spline smoother is more effective and operative than the Butterworth filter and thus is used as the core filter of the developed SINS/GNSS gravimetry.

Table 3. Statistical results of different filters.

	Averaged STD of Dynamic Acc. (mGal)
B-spline smoother	4.84
Butterworth filter	107.70

The repeatability of the proposed land-based SINS/GNSS gravimetry can also be verified by collecting data twice from the same ZUPT point. The mean and standard deviation of the absolute difference in gravity disturbance can be calculated based on the entire repeated ZUPT points as listed in Table 4. As an indication of accuracy, the gravity disturbances at ZUPT points were examined after the smoothing process. The various extreme differences should be considered as outliers that do not fall within the three-sigma intervals. Finally, a total of 186 ZUPT points were obtained after excluding the large misfits. Table 4 also shows the significant improvement in the preliminary accuracy of the proposed land-based SINS/GNSS gravimetry.

Table 4. Accuracy of estimated gravity disturbances from the proposed system.

	Overall Result	Result After Excluding Large Misfits
Mean (mGal)	5.81	2.46
STD (mGal)	6.39	1.32

According to the size of potential area for Pingtung Plain [21], the equation used to compute gravity change caused by groundwater [23] can be expressed as follows:

$$g = 0.42 \times P_S \times H_{GW} \times 10^{-3} \tag{10}$$

where g is gravity change (mGal) and HGW=175.1 (m) is the groundwater level change of the potential area from empty to full, which is computed based on the groundwater recharge and area of Pingtung Plain. PS refers to soil porosity and the value PS=37 that we adopted for further calculation was based on [24].

By considering the parameters of the Pingtung groundwater resource area, a variation in gravity of 2.72 mGal is determined according to Equation (9). The accuracy of the proposed land-based SINS/GNSS gravimetry after excluding the large outliers can be obtained and used for groundwater resource detection. The performances are based on an effective and appropriate smoothing process, which is crucial in reducing the high frequency noises.

6. CONCLUSIONS

In this research, a land-based SINS/GNSS gravimetry consisting of a navigation grade IMU iNAV-RQH and a GNSS receiver has been successfully developed. iNAV-RQH used as a gravimeter has an accuracy of 2.30 mGal compared with the results obtained from the Graviton EG relative gravimeter in stationary state. A comparison of the kinematic velocities and accelerations indicates that the use of a multi-combination of GNSS data yielded better results in the computation than the two-system option.

The proposed land-based system was also validated through several experiments, and a preliminary accuracy of the proposed land-based SINS/GNSS gravimetry with a standard deviation of 1.32 mGal was achieved after excluding any large outliers. A comparison with the gravity variation of 2.72 mGal resulting from the Pingtung groundwater in Taiwan shows that this system has potential for groundwater resource detection. The developed SINS/GNSS gravimetry is more feasible, effective and efficient to use for geodetic applications than using gravimeters and the spatial resolution of satellite missions.

ACKNOWLEDGMENTS

The authors acknowledge the financial support from the Ministry of Science and Technology (NSC 102-2221-E-006-137-MY3) and 2014 NCKU Top-Notch Project. We also acknowledge the assistance of

Christopher Jekeli of the Ohio State University for providing us with the B-spline filter codes. We likewise thank the editor and anonymous reviewers.

AUTHOR CONTRIBUTIONS

Cheng-An Lin contributed to the implementation of land-based SINS/GNSS gravimetry including hardware integration and algorithms. He also conducted various experiments, performed data analyses, and drafted the manuscript. Chung-Yen Kuo and Kai-Wei Chiang played important roles in the conception of this research; they provided not only guidance but also the necessary resources and assisted in the revisions of the manuscript.

REFERENCES

1. Taiwan Central Weather Bureau. Available online: http://www.cwb.gov.tw/eng/index.htm (accessed on 27 April 2015).
2. Water Resources Agency, Taiwan Ministry of Economic Affairs. Available online: http://eng.wra.gov.tw/lp.asp?ctNode=6299&CtUnit=1177&BaseDSD=43 (accessed on 27 April 2015).
3. Chandler, V.W. Gravity Investigation for Potential Ground-water Resources in Rock County, Minnesota. In*Report of Investigations*; Minnesota Geological Survey: Saint Paul, MN, USA, 1994; RI-44.
4. Deurloo, R.; Bastos, L.; Bos, M. On the Use of UAVs for Strapdown Airborne Gravimetry. In *Geodesy for Planet Earth*; International Association of Geodesy Symposia; Springer-Verlag: Berlin/Heidelberg, Germany, 2012; Volume 136, pp. 255–261.
5. Schwarz, K.P. Simultaneous Determination of Position and Gravity from INS/DGPS. *Wiss. Arb. Fachricht. Geod. Geoinform. Univ. Hann.* **2006**, *258*, 141–148.
6. Li, X. Moving Base INS/GPS Vector Gravimetry on a Land Vehicle. Ph.D. Thesis, Department of Civil and Environmental Engineering and Geodetic Science, The Ohio State University, Columbus, OH, USA, 2007.

7. Jekeli, C. *Inertial Navigation System with Geodetic Application*; Walter de Gruyter: Berlin, Germany, 2001.
8. Salychev, O.S.; Bykovsky, A.V.; Voronov, V.V.; Schwarz, K.P.; Liu, Z.; Wei, M.; Peneka, J. Determination of Gravity and Deflections of the Vertical for Geophysical Applications Using the ITC-2 Platform. In Proceedings of the International Symposium on Kinematic Systems in Geodesy, Geodynamics and Navigation, University of Calgary, Calgary, AB, Canada, 30 August–2 September 1994.
9. Kwon, J.H. Airborne Vector Gravimetry Using GPS/INS. Ph.D. Thesis, Department of Civil and Environmental Engineering and Geodetic Science, The Ohio State University, Columbus, OH, USA, 2000.
10. Jekeli, C.; Kwon, J.H. Results of Airborne Vector (3-D) Gravimetry. *Geophys. Res. Lett.* **1999**, *26*, 3533–3536.
11. Kwon, J.H.; Jekeli, C. A New Approach for Airborne Vector Gravimetry Using GPS/INS. *J. Geod.* **2001**, *74*, 690–700.
12. Schwarz, K.P.; Wei, M. Some unsolved problems in airborne gravimetry. In Proceedings of International Association of Geodesy Symposium, Gravity and Geoid Joint Symposium of the International Gravity Commission and the International Geoid Commission, Graz, Austria, 11–17 September 1994; Springer: Berlin, Germany, 1994; pp. 131–150.
13. Hannah, J. *Airborne Gravimetry: A Status Report*; Department of Surveying, University of Otago: Dunedin, New Zealand, 2001.
14. Li, X.; Jekeli, C. Ground-Vehicle INS/GPS Gravimetry. *Geophysics* **2008**, *73*, 1–10.
15. Li, X. Strapdown INS/DGPS airborne gravimetry tests in the Gulf of Mexico. *J. Geod.* **2011**, *85*, 597–605.
16. Bruton, A.M.; Glennie, C.L.; Schwarz, K.P. Differentiation for High-Precision GPS Velocity and Acceleration Determination. *GPS Solut.* **1999**, *2*, 7–21.
17. Glennie, C.L. An analysis of airborne gravity by strapdown INS/DGPS. Ph.D. Thesis, Department of Geomatics Engineering, University of Calgary, Calgary, AB, Canada, 1999.
18. Rauch, H.; Tung, F.; Striebel, C. Maximum likelihood estimates of linear dynamic systems. *AIAA J.* **1965**, *3*, 1445–1450.

19. De Boor, C. *A Practical Guide to Splines*; Springer-Verlag: New York, NY, USA, 1978.

20. Kincaid, D.; Cheney, W. *Numerical Analysis*; Brooks/Cole Publishing Company: Pacific Grove, CA, USA, 1996.

21. Central Geological Survey, Taiwan Ministry of Economic Affairs. Available online: http://www.moeacgs.gov.tw/english/index.jsp (accessed on 27 April 2015).

22. Gerlach, C.; Dorobantu, R.; Rothacher, M. Results of a combined INS/GPS experiment for geodetic application. *Navigation (Paris)* **2005**, *53*, 31–47.

23. Hwang, C.; Cheng, T.C.; Cheng, C.C.; Hung, W.C. Land Subsidence Using Absolute and Relative Gravimetry: A Case Study in Central Taiwan. *Surv. Rev.* **2010**, *43*, 27–39.

24. Chen, Y.H. A Numerical Analysis on Sea Water Intrusions in the Pingtung Plain and Its Remediation Strategy. Master's Thesis, Hydraulic and Ocean Engineering Department, National Cheng Kung University, Tainan, Taiwan, 2014.

9

CHAPTER

ADAPTIVE CORRELATION SPACE ADJUSTED OPEN-LOOP TRACKING APPROACH FOR VEHICLE POSITIONING WITH GLOBAL NAVIGATION SATELLITE SYSTEM IN URBAN AREAS

Hang Ruan, Jian Li *, Lei Zhang and Teng Long

Key Laboratory of Electronic and Information Technology in Satellite Navigation (Beijing Institute of Technology), Ministry of Education, School of Information and Electronics, Beijing Institute of Technology, No. 5 Zhongguancun South Street, Haidian District, Beijing 100081, China

ABSTRACT

For vehicle positioning with Global Navigation Satellite System (GNSS) in urban areas, open-loop tracking shows better performance because of its high sensitivity and superior robustness against multipath. However, no previous study has focused on the effects of the code search grid size on the code phase measurement accuracy of open-loop tracking. Traditional open-loop tracking methods are performed by the batch correlators with fixed correlation space. The code search grid size, which is the correlation space, is a constant empirical value and the code phase measuring accuracy will be largely degraded due to the improper grid size, especially when the signal carrier-to-noise density ratio (C / N_0) varies. In this study, the Adaptive Correlation Space Adjusted Open-Loop Tracking Approach (ACSA-OLTA) is proposed to improve the code phase measurement dependent pseudo range accuracy. In ACSA-OLTA, the correlation space is adjusted according to the signal C / N_0 . The novel Equivalent Weighted Pseudo Range Error (EWPRE) is raised to obtain the optimal code search grid sizes for different C / N_0 . The code phase measuring errors of different measurement calculation methods are analyzed for the first time. The measurement calculation strategy of ACSA-OLTA is derived from the analysis to further improve the accuracy but reduce the correlator consumption. Performance simulation and real tests confirm that the pseudo range and positioning accuracy of ASCA-OLTA are better than the traditional open-loop tracking methods in the usual scenarios of urban area.

Keywords: GNSS; vehicle positioning; open-loop tracking; pseudo range accuracy

1. INTRODUCTION

Vehicle positioning with global navigation satellite system (GNSS) in urban areas is challenging for the reason that satellite signals often suffer from power attenuation and multipath. As the pilot signal is transmitted with the origin data signal in the new GNSS such as the European Galileo and the modernization of the US Global Positioning System (GPS) [1,2,3], the limitation of the coherent integration length brought by the unknown phase inversion no longer exists. As a result, it is possible to further improve the GNSS receiver sensitivity in urban area. However, for the traditional closed-loop tracking approach, the integration length cannot be extensively extended since closed loop is a feedback system and long integration length will break the loop stability. Moreover, it has been verified that lengthening the integration time has little improvement for the sensitivity of the closed-loop tracking [4]. In addition, for multipath mitigation, specially designed methods are also needed.

The block processing based open-loop tracking technique has been proposed to efficiently overcome the limitations of the traditional closed-loop sequential tracking technique for GNSS receivers [5]. The basic process of open loop tracking is detailed in [6]. As it is an open-loop system, open-loop signal tracking is absolutely stable and the power accumulation length can be greatly expanded which will tremendously improve the signal receiving sensitivity [7,8]. Moreover, the sensitivity will be further improved by the introduction of the pilot channel which eliminates the restriction to lengthen the coherent integration length [9]. Additionally, the time-frequency two-dimension correlation result of the open-loop tracking technique makes it convenient to distinguish the line-of-sight (LOS) signal and multipath signal accurately in the harsh urban environments [10,11]. The open-loop tracking method can also detect and lock the signal quickly since it combines the acquisition and tracking stage. As a result, high dynamic signal tracking and limb scan signal tracking can be achieved by it as well [12,13,14].

Hence, the open-loop signal tracking method realized by batch correlators has been applied to vehicle positioning in urban areas [15,16]. In the method, by lengthening the integration time, the sensitivity is improved so that the weak signal in urban area can be detected. With the help of the time-frequency image, the LOS signal is picked out and multipath is mitigated. In order to further keep the LOS signal peak away from multipath peaks, the strategy to choose the most accurate peak instead of the dominate peak among the correlation results is proposed [17]. In [18], amplitude threshold is used to categorize the LOS signal, multipath and noise to keep the correctness of peak detection. The time to position is shortened by the single stage quickly signal locking. However, including the traditional scheme that computes the correlation results with real correlators and the interpolation based correlation calculation method in [6], the pseudo ranges of all the previously designed open loop methods are measured by discrete code search grids with fixed empirical width. No previous study has focused on the impacts of the grid size on the measurement accuracy of open-loop tracking. In fact, the measuring accuracy is probably distorted by the inapposite grid sizes. If the grid is too small, the LOS signal energy that adjacent search grids contain is close to each other so that the most accurate peak is difficult to be distinguished from its adjacent ones and false peak detection will tremendously destroy the accuracy especially when signal power is low. While, for large grid size, it is easy to detect the correct peak but the relatively large grid size brings large uncertainty for the measurement estimation. Although early minus late discriminator (EMLD) is introduced to eliminate the uncertainty [8,19], the measurement accuracy will still be highly degraded when signal power decreases.

In this paper, Adaptive Correlation Space Adjusted Open-Loop Tracking Approach (ACSA-OLTA) is proposed to improve the code phase measurement accuracy of open-loop tracking. In the method, different from the previous open-loop tracking schemes, the space between the batch correlators, namely the code search grid size, is adjusted according to the signal carrier-to-noise density ratio ($C/N0$) which is an indicator of the signal power [20]. The code phase calculating methods are also switched according to $C/N0$. Since the grid size should not be too large or too small so as to gain the balance between the errors induced by the uncertainty within the grid and false peak detection at the same time, the new Equivalent Weighted Pseudo Range Error (EWPRE) is raised based on the false peak detection probability and mean measurement error of the search grids to gain the optimal grid size for different $C/N0$. The code phase measuring error of EMLD is analyzed and compared with other measurement extracting method of open loop tracking for the first time. The $C/N0$ to switch the pseudo range calculation methods is developed by accuracy comparison to further improve the performance but reduce the correlators needed.

The remainder of the paper is organized as follows. In Section 2, the basic concept of open-loop signal tracking method is described based on the batch architecture of GNSS signal processing. Hereafter in Section 3, the ACSA-OLTA is proposed via the calculation of the designed EWPRE and the measurement deviation. Section 4 presents the simulation and the test with the help of the collected real satellite signal data which confirm the performance improvements of ACSA-OLTA. Section 5 concludes the paper.

2. OPEN-LOOP TRACKING OF GNSS SIGNALS

This section reviews the main concept of the open-loop tracking of GNSS signals. The block diagram of the batch correlators based open-loop tracking method is illustrated in Figure 1. The input signal is transferred to multiple identical correlators at the same time. In the presence of Additive Gaussian White Noise (AGWN), the input signal is expressed as follow

$$s(t_n) = \sqrt{2P}D(t_n - \tau)c(t_n - \tau)cos(2\pi(f_{IF} + f_d)t_n + \varphi_0) + n_w(t_n) \tag{1}$$

where t_n is the discrete sampling time: $t_n = t_0 + nT_s$ and T_s is the sampling period. P is the incoming signal power. $D(\cdot)$ is the navigation message data or secondary code. $c(\cdot)$ is the Pseudo Random Noise (PRN) primary code. τ is the code phase shift of the incoming signal. f_{IF} is the signal down-conversion frequency. f_d is the instantaneous value of the Doppler frequency shift of the received satellite signal. φ_0 is the initial value of the carrier phase. n_w is the additive Gaussian white noise with the single band power density of N 0 . Since an orthogonal set of binary pseudorandom codes are generated for different satellites, the signals from different satellites can be processed independently from each other [21]. As a result, the input signal model here contains only one satellite for simplicity.

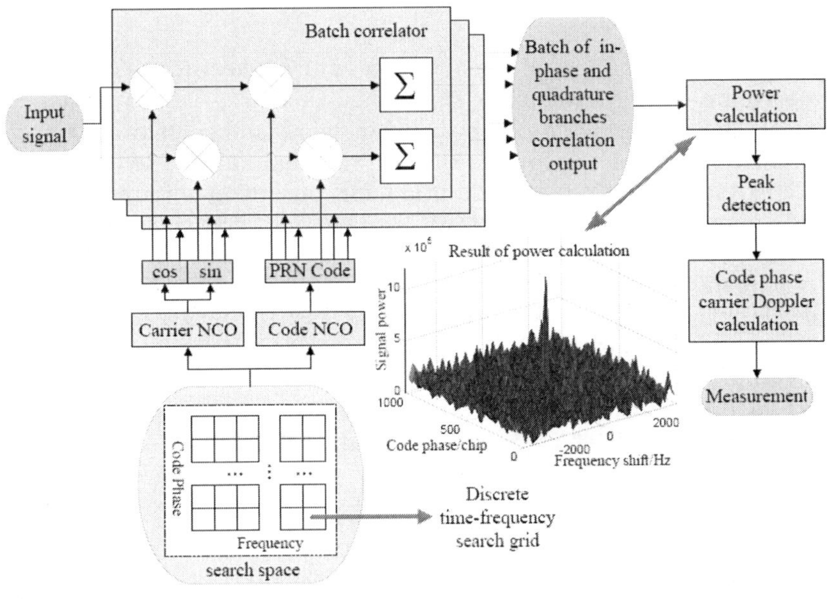

Figure 1. The block diagram of the batch correlators based open-loop tracking method.

Based on the architecture of batch correlators shown in Figure 1, open-loop tracking estimates the signal code phase and Doppler frequency shift via discrete searching. Each combination of the code phase and Doppler frequency shift makes up a discrete search grid in the time-frequency two-dimension searching space. Every correlator in the architecture corresponds to one of the search grids. The correlation space, namely the code phase delay between the adjacent correlators in code dimension is the size of code search grid. The frequency difference between those in frequency dimension is the frequency search grid size. In the correlator, the input signal is correlated with the local reference signal which is composed of the carrier and code sequence generated by the Numerically Controlled Oscillators (NCO). Correlation is performed to compute the outputs of the in-phase (I) and quadrature (Q) branches as below

$$I_{d,k} = \frac{1}{N_c} \sum_{n=0}^{N_c-1} s(t_n)c(t_n - \tau_d) \cdot \sqrt{2}cos(2\pi(f_{IF} + f_k)t_n) \quad (2)$$

$$Q_{d,k} = \frac{1}{N_c} \sum_{n=0}^{N_c-1} s(t_n)c(t_n - \tau_d) \cdot \sqrt{2}sin(2\pi(f_{IF} + f_k)t_n) \quad (3)$$

where N c is the number of signal samples during the interval of coherent integration. In order to ensure signal independence, I and Q correlation values must be accumulated over an integer number of primary code periods which is 1 ms for the GPS L1CA signal and the length cannot extend the navigation data duration for the unknown transitions of navigation data when no external assisted information is introduced. However, for new GNSS signal such as GPS L5 with pilot channel, the accumulation length can be set as long as possible when the secondary code is synchronized [22]. With the help of Assisted-GPS (AGPS), the time of code phase transition induced by navigation data is known and the coherent integration can be expanded over the data duration by taking care of the transition [23]. τ d and f k are the code phase and Doppler frequency shift of the corresponding search grid which are

$$\tau_d = d \cdot \tau_b \quad (4)$$

$$f_k = k \cdot f_b \quad (5)$$

where τ b and f b are the sizes of code and frequency search grid respectively. The correlator outputs are

$$I_{d,k} = \sqrt{P}R(\tau - \tau_d)sinc(\pi(f_d - f_k)T_c)cos\Delta\varphi + n_I \quad (6)$$

$$Q_{d,k} = \sqrt{P}R(\tau - \tau_d)sinc(\pi(f_d - f_k)T_c)sin\Delta\varphi + n_Q \quad (7)$$

where R (·) is the autocorrelation function of the PRN code, T c is the coherent integration time, Δ φ is the carrier phase error. n I and n Q are the noise components which are still AGWN. Their variances are uniformly expressed as below

$$\sigma_{n,\pm}^2 = \frac{2}{N_c^2} \sum_{n=0}^{N_c-1} E[n_w^2(t_n) \cdot \frac{1 \pm cos(4\pi(f_{IF} + f_k)t_n)}{2}] \\ = \frac{N_0}{2T_c} \tag{8}$$

where $E[x]$ is the expectation of x. "+" and "−" correspond to the variances of n I and n Q respectively and they are uniformly represented as σ n 2 in the following.

The energy of the correlator outputs is computed based on I and Q values as below which makes up the entire image of the signal

$$P(\tau_d, f_k) = I_{d,k}^2 + Q_{d,k}^2 \tag{9}$$

The satellite signal is deemed to be in the grid with the maximum power peak in the batch correlator outputs as shown in Figure 1. The measurement of the incoming signal are calculated according to the code phase and Doppler frequency shift of the reference signal of the grid as below

$$(\hat{\tau}, \hat{f_D}) = arg \max_{\tau_d, f_k} P(\tau_d, f_k) \tag{10}$$

The estimation results τ ^ and f D ^ of the signal are used to form the final signal measurements. Different from the closed loop that there is a limitation in the power accumulation length owning to the requirement of the system stability, the power accumulation length of the open-loop tracking is not limited. It can be lengthened to hundreds of micro seconds and the sensitivity can be greatly improved. In this paper, the coherent integration length of 300 ms is taken as an example and the sensitivity is sufficient for the urban application [18]. For other values, although the detail parameters may change, the basic conception is not altered and the analysis results are the same.

Nevertheless, there is a serious drawback for the open loop tracking method, which is the degraded measurement accuracy. As the measurement of open loop tracking is calculated from the parameters of the discrete search grid of the detected peak, the uncertainty in the grid will probably deteriorate the measurement accuracy. If the signal is detected in the false grid, the measurement error will be even further enlarged. Compared with the Doppler frequency measurement, the problem to code phase measurement is more serious.

For Doppler frequency, considering the long coherent integration length of open loop tracking, the width of the frequency search grid is limited to an extremely small value due to the large power loss brought by the frequency mismatch [21]. If the coherent integration is as long as 300 ms as the statement above, the frequency search grid size is at most 5 Hz. Converted to velocity in the unit of m/s, taking the center frequency 1575.42 MHz of GPS L1CA signal as an example, the uncertainty within the grid is at most 0.48 m/s. Furthermore, the false detection in the frequency dimension will almost never occur because even the neighboring grids of the right one contain little signal power. Accordingly, the effect of the discrete searching on the Doppler frequency measurement accuracy is a little.

For the code phase measurement, the situation is really different. In the autocorrelation function of the PRN code, signal power is contained within the code phase delay from −1 chip to 1 chip and it has nothing to do with the coherent integration length. Considering the code chip duration in distance is as long as hundreds of meters, the errors induced by the uncertainty in the grid and false detection will both exist and considerably degrade the accuracy. Since both the two errors are related to the code search step of open loop tracking, the code search grid size should be specially chosen to improve the measurement accuracy. Additionally, in the architecture, each combination of the parameters needs a correlator to compute the signal power. Small grid size means large numbers of correlators are needed and leads to huge computational loads. Some kinds of parallel correlation computations which use joint time-frequency domain techniques such as direct and inverse discrete Fourier transforms (DFT and IDFT, accordingly) have been utilized to reduce the correlators demanded [24]. However, it is usually used for frequency searching. For code searching, it is not suitable since the extremely long power accumulation of open loop tracking will tremendously increase the scale of FFT and it is much more complex than the architecture of batch correlators. Therefore, computational loads should be considered together with the measurement accuracy improvement.

3. ACSA-OLTA

It is known that the positioning accuracy of GNSS is directly related to the satellite pseudo range accuracy. Since pseudo range is calculated from the code phase measurement, the code phase measuring accuracy of open loop tracking is concerned in this section. Firstly, we study the code phase measurement errors of open loop tracking with different code search grid sizes and the equivalent weight pseudo range error (EWPRE) is proposed to obtain the optimal grid size. Then, the code phase measuring error of different measurement calculation methods are assessed and compared with each other. Finally, based on the analysis, the adaptive correlation space adjusted open loop tracking Algorithm (ACSA-OLTA) is proposed to improve the pseudo range measuring accuracy. Since carrier frequency is much larger than code rate, the effect of carrier Doppler frequency error on the code phase measurement accuracy is little. Moreover, considering the carrier Doppler frequency estimation error is really small, it is ignored and assumed to be zero in the following.

3.1. The Optimal Correlation Space Based on EWPRE

Based on the generalized open-loop signal tracking method described earlier, the accuracy of the code phase measurement heavily relies on the code search grid size. If the grid size is too large, the uncertainty in the grid will distort the accuracy. On the contrary, if the grid size is too small, the accuracy may be greatly destroyed by the false peak detection. Since the length of the PRN code chip in distance is really large such as GPS L1CA signal whose code chip is as long as 300 m, the grid size should be much smaller than the code chip duration to guarantee the accuracy. If the peak with the most accurate measurement is called most accurate peak, the correlation outputs beside the most accurate peak contain partial signal power. Thus, the powers of these correlation outputs are easier to be larger than the most accurate peak and false peak detection will probably occur. It is obvious that the smaller the grid is, the more signal power the adjacent grids contains and the more likely the false detection occurs.

Therefore, the balance between the accuracy degradation induced by the uncertainty in the grid and false peak detection should be gained in the choice of the code search grid size. In order to obtain the optimal search grid size for the code phase measurement accuracy improvement, we propose the equivalent weight pseudo range error (EWPRE) to evaluate the measurement error and gain the balance. Both the errors brought by the uncertainty in the grid and false peak detection are involved so as to synthesize the effects of the two factors on the measurement accuracy.

In the design of EWPRE, we firstly evaluate the code phase measurement errors when the most accurate peak or false peak is detected. For the measurements that are directly extracted from the parameters of the search grid, as it is a biased estimation with the jitter limited within the grid, the mean error is utilized for the error assessment. When the most accurate peak in the signal image is detected, the code phase measurement error is derived from the uncertainty of the real code phase in the search grid. In this case, the real signal code phase obeys the uniform distribution in the search grid expressed as below [25]

$$f(c_e) = 1/\tau_b, -\tau_b/2 \leq c_e \leq \tau_b/2 \quad (11)$$

Thus, the mean measurement error is

$$c_{me}(\tau_b) = \int_{-\tau_b/2}^{\tau_b/2} |c_e| \cdot f(c_e) dc_e = \tau_b/4 \quad (12)$$

According to the equation, it is clear that, when the most accurate peak is detected, the measurement error is proportional to the grid size and wide search grid will induce large code phase measuring error.

When the false peak is detected, besides the error induced by the uncertainty in the search grid, the deviation of the search grid against that with the most accurate measurement is involved in the measuring error. If the n t h grid relative to the grid with the most accurate measurement is falsely detected, the mean measurement error is

$$e_n(\tau_b) = \int_{n\tau_b-\tau_b/2}^{n\tau_b+\tau_b/2} |x| \cdot \frac{1}{\tau_b} dx = n\tau_b \quad (13)$$

It is obvious that the measurement error brought by false peak detection is much larger than that in the grid. Hereafter, we will develop the weight coefficients of the measurement errors when the most accurate peak or false peak is detected in the EWPRE. In order to synthetically evaluate the effects of the two kinds of errors on the measurement, the weight coefficients should be the probability that the errors occur. Usually, in open loop tracking, the peak with the largest power is regarded as the most accurate peak and the measurement will be extracted from the grid of that peak. Although prior information has been used to judge whether the peak is the most accurate one, it cannot distinguish the false peak whose corresponding code search grid is close to the most accurate peak since measurements extracted from these grids are all probable. Only power can be utilized and false detection will occur when the correlated power of the adjacent grids is larger than the most accurate peak. Therefore, the probability that the error occurs is that the false peak has power larger than the most accurate peak.

Based on the detection variable expressed in Equation (9), the correlation outputs of two grids in code dimension with the same Doppler frequency are considered. One of them is the most accurate peak, the other one is the false peak. There are two situations for the probability calculation. When the space between the girds of the two peaks in code dimension is larger than one code chip, the correlated powers of the most accurate peak and the false peak are independent from each other and only the most accurate peak contains signal power. If P_r and P_f are the correlated power of the most accurate peak and the false peak, according to Equation (9), P_r obeys non-central χ^2 (chi-square) distribution with two degrees of freedom and P_f obeys central χ^2 distribution with two degrees of freedom. The probability density functions (PDF) of them are

$$f_p(x) = \frac{1}{2\sigma_n^2} exp(-\frac{x+\alpha^2}{2\sigma_n^2}) I_0(\frac{\alpha}{\sigma_n^2}\sqrt{x}) \tag{14}$$

$$f_a(x) = \frac{1}{2\sigma_n^2} exp(-\frac{x}{2\sigma_n^2}) \tag{15}$$

where α is the non-central parameter and it is equal to P which is the energy of the most accurate peak. $I_0(\cdot)$ is the zero-order modified Bessel function of the first kind [26]. The joint PDF is

$$f_u(P_r, P_f) = f_p(P_r) \cdot f_a(P_f) \tag{16}$$

The probability false detection occurs is that when $P_f > P_r$ and it is expressed as

$$F(P_f > P_r) = \iint_{P_f > P_r} f_u(\gamma, \beta) d\gamma d\beta$$
$$= exp(-\frac{\alpha^2}{4\sigma_n^2}) \cdot \int_0^{+\infty} \frac{1}{2\sigma_n^2} exp(-\frac{2\beta + \alpha^2/2}{2\sigma_n^2}) I_0(\frac{\alpha/\sqrt{2}}{\sigma_n^2}\sqrt{2\beta}) d\beta$$
$$= \frac{1}{2} exp(-\frac{\alpha^2}{4\sigma_n^2})$$

(17)

When the space between the girds of the most accurate peak and the false peak in code dimension is smaller than one code chip, P_r and P_f are not independent from each other and both of them contain signal power. Let I_r and Q_r be the in-phase and quadrature branch correlation outputs of the most accurate peak and I_f and Q_f be the in-phase and quadrature branch correlation outputs of the false peak, supposing the Doppler frequency error is zero, their expressions are as below according to Equations (6) and (7)

$$I_r = \sqrt{P_r} cos\Delta\varphi_1 = \sqrt{P} R(\tau - \tau_d) cos\Delta\varphi_1 + n_{I,r} \quad (18)$$

$$Q_r = \sqrt{P_r} sin\Delta\varphi_1 = \sqrt{P} R(\tau - \tau_d) sin\Delta\varphi_1 + n_{Q,r}$$

(19)

$$I_f = \sqrt{P_f} cos\Delta\varphi_2 = \sqrt{P} R(\tau - \tau_d - \Delta\tau_b) cos\Delta\varphi_2 + n_{I,f}$$

(20)

$$Q_f = \sqrt{P_f} sin\Delta\varphi_2 = \sqrt{P} R(\tau - \tau_d - \Delta\tau_b) sin\Delta\varphi_2 + n_{Q,f}$$

(21)

where $\Delta\tau_b$ is the code phase space between the two peaks and $|\Delta\tau_b| < 1$ *code chip.* $\Delta\varphi_1$ and $\Delta\varphi_2$ are the carrier phase residuals. $n_{I,r}$, $n_{Q,r}$, $n_{I,f}$ and $n_{Q,f}$ are the noise components. They are AWGN and the variances of them are all σ_n^2 according to Equation (8). Because of the orthogonal carrier phase relationship, $n_{I,r}$ and $n_{I,f}$ are independent of $n_{Q,r}$ and $n_{Q,f}$. The covariance between $n_{I,r}$ and $n_{I,f}$ is

$$cov(n_{I,r}, n_{I,f}; \Delta\tau_b) = \frac{1}{N_c^2} E[n_w^2(t_n)(1 + cos(4\pi(f_{IF} + f_d)t_n))] E[\sum_{n=0}^{N_c-1} c(t_n - \tau_d)c(t_n - \tau_d - \Delta\tau_b)]$$
$$= \sigma_n^2 \cdot R(\Delta\tau_b) \qquad (\Delta\tau_b < 1\ code\ chip) \tag{22}$$

$$cov(n_{Q,r}, n_{Q,f}; \Delta\tau_b) = \frac{1}{N_c^2} E[n_w^2(t_n)(1 - cos(4\pi(f_{IF} + f_d)t_n))] E[\sum_{n=0}^{N_c-1} c(t_n - \tau_d)c(t_n - \tau_d - \Delta\tau_b)]$$
$$= \sigma_n^2 \cdot R(\Delta\tau_b) \qquad (\Delta\tau_b < 1\ code\ chip) \tag{23}$$

Thus, I_r, I_f, Q_r and Q f are all Gaussian random variables. According to the definition of the variables [I_r, I_f, Q_r, Q_f] , the covariance matrix K is given by

$$\mathbf{K} = \begin{bmatrix} \sigma_n^2 & \sigma_n^2 \cdot R(\Delta\tau_b) & 0 & 0 \\ \sigma_n^2 \cdot R(\Delta\tau_b) & \sigma_n^2 & 0 & 0 \\ 0 & 0 & \sigma_n^2 & \sigma_n^2 \cdot R(\Delta\tau_b) \\ 0 & 0 & \sigma_n^2 \cdot R(\Delta\tau_b) & \sigma_n^2 \end{bmatrix} \tag{24}$$

Supposing $\mathbf{P}_{r,f} = [I_r, I_f, Q_r, Q_f]^T$, the joint PDF is

$$f(I_r, I_f, Q_r, Q_f; \Delta\tau_b) = \frac{1}{(2\pi)^2\sqrt{|\mathbf{K}|}} exp(-\frac{1}{2}(\mathbf{P}_{r,f} - \mu)\mathbf{K}^{-1}(\mathbf{P}_{r,f} - \mu)^T) \tag{25}$$

where $\mu = [\mu_{I_r}, \mu_{I_f}, \mu_{Q_r}, \mu_{Q_f}]^T$ and it is the mean value vector of $P_{r,f}$. By computing the inverse matrix of K, the expression of the joint PDF is

$$f(I_r, I_f, Q_r, Q_f; \Delta\tau_b) = \frac{1}{(2\pi)^2 \cdot \sigma_n^4(1 - R^2(\Delta\tau_b))} exp(-\frac{1}{2\sigma_n^2(1 - R^2(\Delta\tau_b))}$$
$$[(I_r - \mu_{r,I})^2 + (I_f - \mu_{f,I})^2 - 2R(\Delta\tau_b)(I_r - \mu_{r,I})(I_f - \mu_{f,I})]$$
$$+ (Q_r - \mu_{r,Q})^2 + (Q_f - \mu_{f,Q})^2 - 2R(\Delta\tau_b)(Q_r - up\mu_{r,Q})(Q_f - \mu_{f,Q}]) \tag{26}$$

Submitting $I_r = \sqrt{P_r}cos(\Delta\varphi_1)$, $Q_r = \sqrt{P_r}sin(\Delta\varphi_1)$, $I_f = \sqrt{P_f}cos(\Delta\varphi_2)$ and $Q_f = \sqrt{P_f}sin(\Delta\varphi_2)$ into Equation (26), after some algebra manipulations, the expression of the joint PDF of P r and P f is

$$f(P_r, P_f; \Delta\tau_b) = \frac{1}{(2\pi)^2 \cdot \sigma_n^4(1 - R^2(\Delta\tau_b))} exp(-\frac{1}{2\sigma_n^2}(P_r + P_f + \mu_{I_r}^2 + \mu_{I_f}^2 + \mu_{Q_r}^2 + \mu_{Q_f}^2$$
$$- 2R(\Delta\tau_b)(\mu_{I_r}\mu_{I_f} + \mu_{Q_r}\mu_{Q_f}))) \times \int_0^{2\pi}\int_0^{2\pi} exp(\frac{\sqrt{P_r \cdot P_f}}{\sigma_n^2} R(\Delta\tau_b)cos(\Delta\varphi_1 - \Delta\varphi_2)$$
$$+ \frac{\sqrt{P_r}}{\sigma_n^2}(\mu_{f,I}R(\Delta\tau_b) - \mu_{r,I})cos(\Delta\varphi_1) + \frac{\sqrt{P_f}}{\sigma_n^2}(\mu_{r,I}R(\Delta\tau_b) - \mu_{f,I})cos(\Delta\varphi_2)$$
$$+ \frac{\sqrt{P_r}}{\sigma_n^2}(\mu_{f,Q}R(\Delta\tau_b) - \mu_{r,Q})sin(\Delta\varphi_1) + \frac{\sqrt{P_f}}{\sigma_n^2}(\mu_{r,Q}R(\Delta\tau_b) - \mu_{f,Q})sin(\Delta\varphi_2))d\Delta\varphi_1 d\Delta\varphi_2 \tag{27}$$

When $\Delta\ \tau_b$ varies from 0 to 1 chip, the probability that false detection occurs is

$$F(P_f > P_r; \Delta\tau_b) = \int_0^{+\infty}[\int_0^{P_f} f(P_r, P_f)dP_r]dP_f \tag{28}$$

Taking the coherent integration length of 300 ms as an example, the false detection probability for different $\Delta_{\tau b}$ and different carrier-to-noise density ratio (C/N_0) is shown in Figure 2. It is obvious the false peak detection probability is related to the space between the most accurate peak and the false peak. Since more signal power is contained in the false peak when the space is small, the probability increases as the space between the two peaks decreases. While, when the space is too small, the probability become lower because of the significant positive correlation between the powers of the two peaks according to Equations (22) and (23). Hence, when the grid size is small, the false peak detection will probably occur especially at the point with the largest false detection probability. However, large grid size contains large inherent measurement error. In addition, when the correlation space exceeds 1 code chip, the false peak detection probability is extremely small with respect to those within the code chip duration so that it is not considered in the following analysis. In consideration that open-loop tracking extracts code phase measurement from that of the grid where the detected peak is, the results further validate that the search grids with the size too large or two small will both degrades the measurement accuracy.

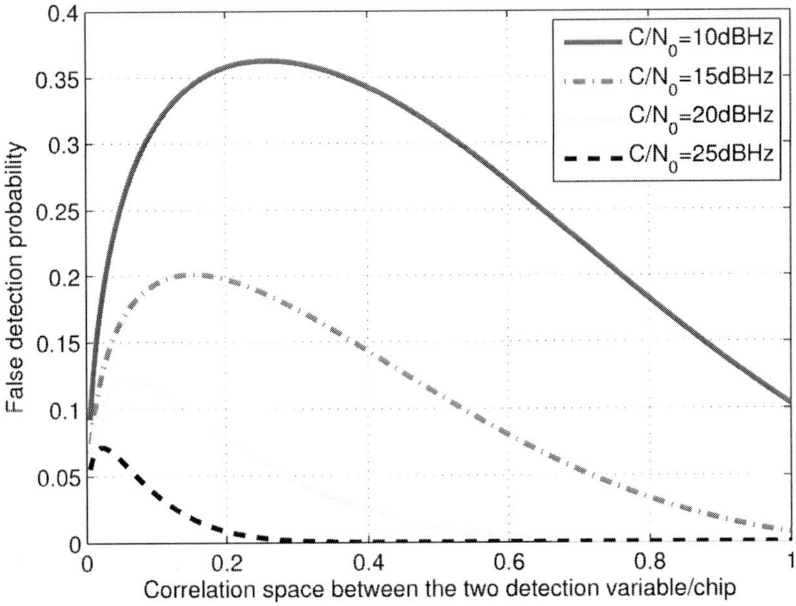

Figure 2. False detection probability of different correlation space and C/N_0.

Furthermore, it also shows that the false detection probability becomes high as (C/N_0) decreases. When C/N_0 is high, the most accurate peak with the maximum signal power is easily detected from the false peak. While, when C / N 0 is low, both the two peaks are greatly disturbed by noise so that they are difficult to be distinguished and false detection will probably occur. In addition, the code phase spaces with the largest false detection probability are various among different C/N_0 due to the positive correlation between the two peaks besides the signal power. The lower the C/N_0 is, the larger the corresponding space with the highest false detection probability is. Therefore, the signal power should be considered in the choice of code search grid size.

Considering that the false peak detection probabilities can indicate the possibility of the occurrence of the measurement errors induced by the false detection at different points, the weight coefficients are designed based on the probability. The weight coefficient for the n t h grid relative to grid with the most accurate measurement is expressed as below

$$W_n(\tau_b) = \begin{cases} \frac{1}{T_p}\int_0^{\frac{\tau_b}{2}} \mathrm{d}x + \frac{1}{T_p}\int_{\frac{\tau_b}{2}}^{T_p} (1 - F(P_f > P_r; x))\mathrm{d}x & n = 0 \\ \frac{1}{T_p}\int_{\frac{2n-1}{2}\tau_b}^{\frac{2n+1}{2}\tau_b} F(P_f > P_r; x)\mathrm{d}x & n = 1, 2, ... \end{cases}$$

(29)

where $n = 0$ corresponds to the grid with the most accurate measurement. T p is the code chip duration. The weighed coefficients are the integration of the probability that the detection result is in the false grid or in the right grid. Accordingly, it evaluates the likelihood of the measurement errors corresponding to the detection of the correct or false peak. Based on the weight coefficients, the EWPRE is

$$S(\tau_b) = W_0(\tau_b) \cdot e_{me}(\tau_b) + \sum_{n=1}^{T_p/\tau_b} [W_n(\tau_b) \cdot e_n(\tau_b)]$$

(30)

It is the weighed sum of the code phase measurement errors when the detection results vary from the grid of the most accurate peak to that one code chip away. The measurement error induced by the false detection in other grids and that when the most accurate peak is detected are both involved and weighed to accumulate together. With the help of the weight coefficients, when the false detection probability is large, the measurement error induced by the false peak detection will be dominant in the weighted error. On the contrary, the measurement error within the grid will be more significant. As a result, it synthesizes the effects of the uncertainty in the search grid and false peak detection on the code phase measurement accuracy. The value of EWPRE represents the mean equivalent relative measurement error among different grid sizes so that it can evaluate the measurement accuracy difference of these grid sizes even though it is not the estimation of the real mean measurement error. It is further verified by simulation in Section 4.1.

Consequently, the optimal code search grid size of open loop tracking method for a given C/N_0 is the one with the minimum EWPRE. After some algebraic transformation, the EWPRE is simplified as below. When the coherent integration length is 300 ms, the EWPRE of different grid sizes for different C/N_0 is shown in Figure 3.

$$S(\tau_b) = \frac{\tau_b^2}{8T_p} + \frac{1}{T_p}\sum_{n=1}^{T_p/\tau_b}\left[(e_n - \tau_b/4)\cdot\int_{\frac{2n-1}{2}\tau_b}^{\frac{2n+1}{2}\tau_b} F(P_f > P_r; x)\mathrm{d}x\right] \tag{31}$$

Taking the C / N 0 of 10 dBHz, 15 dBHz, 20 dBHz and 25 dBHz as examples, the EWPRE of different grid sizes is detailedly drawn in Figure 4.

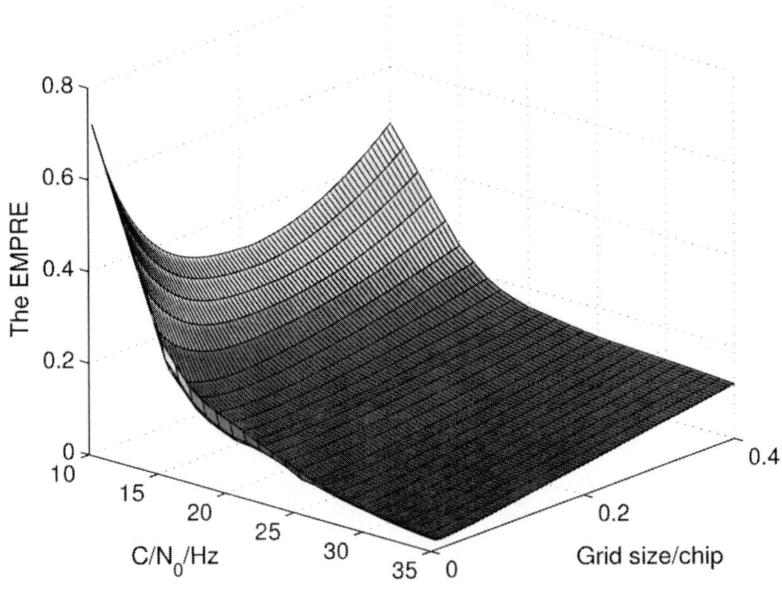

Figure 3. The EWPRE of different grid sizes and C/N_0 with 300 ms coherent integration.

One can see that there is a grid size that minimizes the EWPRE for each C/N_0 and that shall be the grid size with best accuracy. Taking the coherent integration length of 300 ms as an example, the optimal code search grid sizes of different C/N_0 are shown in Figure 5. It indicates that when C/N_0 is low, the grid size is enlarged to increase the power difference between adjacent grids so as to avoid the error brought by the false peak detection which is much more serious than the error within the grid. When the C/N_0 is high, as the detection performance is perfect, the optimal grid size is really small to reduce the measurement error in the grid. Therefore, the correlation space between the batch correlators of open loop tracking which is the code search grid size should be adjusted to the optimal size shown in Figure 5 as C/N_0 varies to improve the measurement accuracy. Considering small grid size demands large numbers of correlators which means huge computational costs, the method to calculate the measurement with less correlators but high accuracy is studied in the next subsection. The verification of the optimal grid size is conducted in Section 4 with the help of Monte Carlo simulation and test.

3.2. The Measurement Calculation Strategy

In order to reduce the correlators needed and keep the accuracy at the same time, early-minus-late discriminator (EMLD) is introduced [8]. However, no previous study has analyzed the accuracy of the method. In this section, the accuracy of the measurement derived from the discriminator is evaluated. Based on the accuracy comparison between the code phase calculated with the EMLD method and that directly extracted from the local reference signal of the detected grid, together with the computational cost assessment results, the measurement calculation strategy is developed.

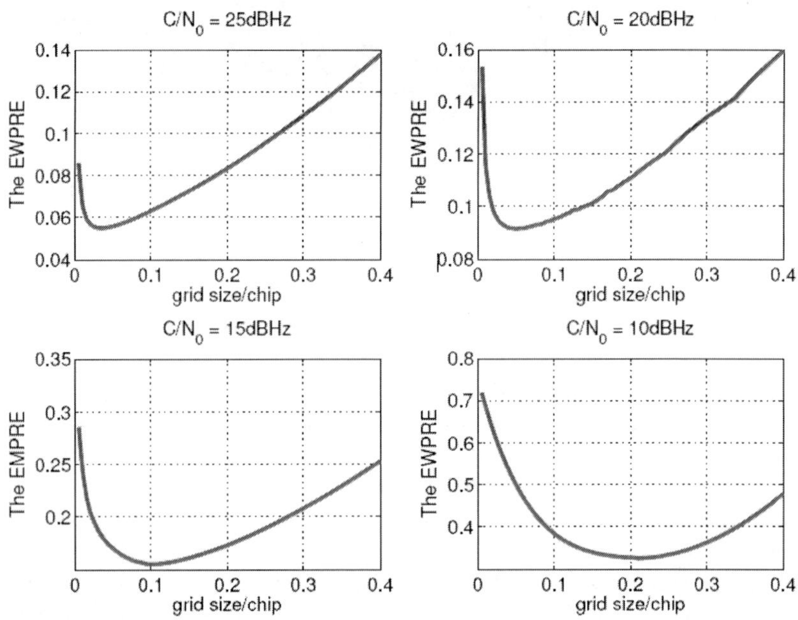

Figure 4. The detailed EWPRE of different grid size with 300 ms coherent integration.

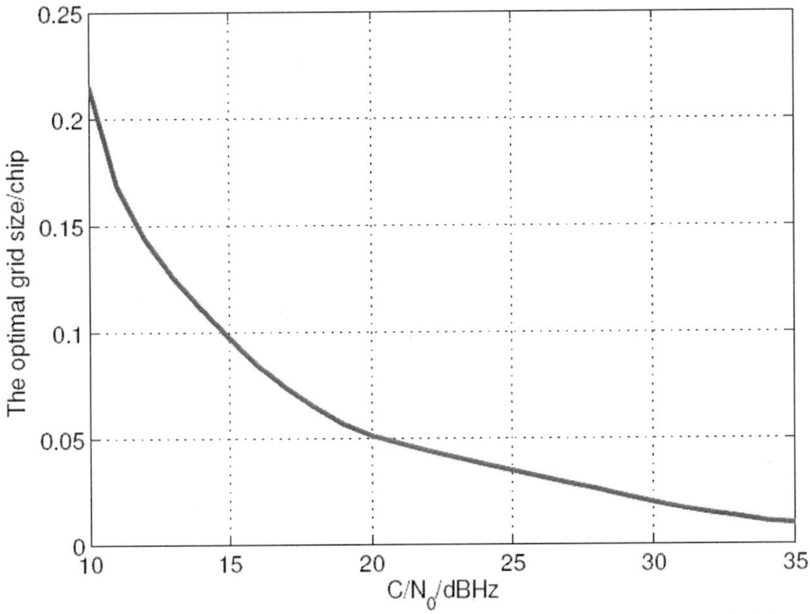

Figure 5. Optimal code search grid size of different C / N 0 with 300 ms coherent integration.

The EMLD based code phase estimation begins after the peak detection. The correlation outputs of the two code search grids adjoining the peak are used to construct the discriminator. If the two grids are represented by *E* and *L* corresponding to that with the code phase earlier or later than the peak respectively, the in-phase and quadrature correlation outputs of the two grids are expressed as below

$$\begin{aligned} E_I &= \sqrt{P}R(\tau - k\tau_b - \tau_b)cos(\Delta\varphi) + n_{I,E} \\ E_Q &= \sqrt{P}R(\tau - k\tau_b - \tau_b)sin(\Delta\varphi) + n_{Q,E} \\ L_I &= \sqrt{P}R(\tau - k\tau_b + \tau_b)cos(\Delta\varphi) + n_{I,L} \\ L_Q &= \sqrt{P}R(\tau - k\tau_b + \tau_b)sin(\Delta\varphi) + n_{Q,L} \end{aligned} \quad (32)$$

where $n_{I,E}$, $n_{Q,E}$, $n_{I,L}$ and $n_{Q,L}$ are the noise components in the correlation results. *k* is the grid index of the detected peak. The final code phase estimation result τ f is

$$\tau_f = k\tau_b + \frac{1}{1-2\tau_b}\frac{(E_I^2 + E_Q^2) - (L_I^2 + L_Q^2)}{(E_I^2 + E_Q^2) + (L_I^2 + L_Q^2)} \tag{33}$$

In this way, regardless of the noise, the discriminator can completely compensate the uncertainty of the code phase in the detected grid [8]. If the most accurate peak is detected, the code phase estimation is totally unbiased. However, noise will distort the code phase measurement so that the error standard deviation is utilized for the evaluation. Assuming that it is perfectly normalized discriminator and the front-end bandwidth is infinite, the variance of the measurement calculated from the discriminator is as below (details of derivation are shown in the appendix)

$$\sigma_{EML}^2 = \frac{\tau_b}{2 \cdot C/N_0 \cdot T_c}(1 + \frac{1}{C/N_0 \cdot T_c(1-\tau_b)}) \tag{34}$$

At the same time, for the code phase that directly extracted from the grid parameters, when the most accurate peak is detected, the variance of the measurement is

$$\sigma_c^2 = \int_{-\tau_d/2}^{\tau_d/2} c_e^2 \cdot f(c_e)dc_e = \tau_b^2/12 \tag{35}$$

In the following, we use the abbreviations below to represent the open loop tracking with the two different code phase measurement calculation methods.

-DIS: The open-loop tracking with EMLD dependent code phase estimation method

-DET: The open-loop tracking that directly calculate the code phase from the grid parameters.

Assuming that the peak detection is correct, the error standard deviations of measurements calculated by DIS and DET with different grid sizes are drawn in Figure 6 in which the coherent integration length is 300 ms. It shows that, when C/N_0 is high, the accuracy of DIS is much better than that of DET with the same grid size. Furthermore, although the accuracy of DET with optimal grid size is better than the discriminator based method with the grid size of 0.1 chip when C/N_0 is high, the improvement is a little and it needs a huge number of correlators. However, as the C/N_0 decreases, the accuracy of DIS is destroyed and it becomes worse than DET. Moreover, if false peak detection occurs, the accuracy of DIS will be further distorted. Thus, the EMLD method cannot be utilized all the time as that in [19] and a C/N_0 threshold is needed to switch the code phase calculation methods.

Taking the coherent integration length of 300 ms as an example, if the τ_b of DIS is 0.1 chip and DET has the optimal grid size shown in Figure 5, when C/N_0 is larger than 23 dBHz which is the standard deviation cross point of the measurements derived from the two methods, the correlators needed to search the area of 1 chip and the corresponding measurement error standard deviation are listed in Table 1.

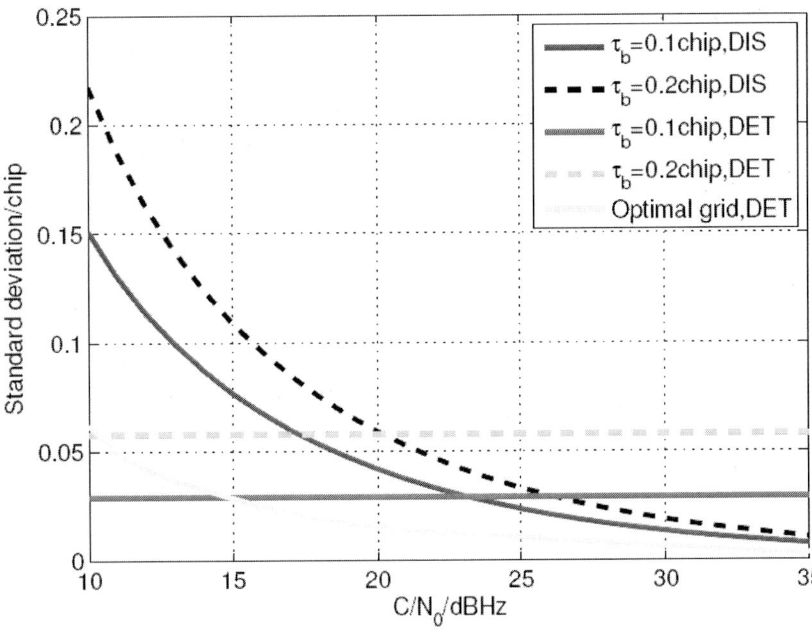

Figure 6. The code phase error standard deviation of the two measurement calculation methods without consideration of false peak detection.

Table 1. The statistic of the correlator needed and the mean accuracy.

	C/N_0 (dBHz)	23	24	25	26	27	28	29	30
DET	correlator number	24.3	26.2	28.5	30.9	34.7	38.6	44.0	50.3
	standard deviation (chip)	0.0119	0.0110	0.0101	0.0094	0.0083	0.0075	0.0066	0.0057
DIS	correlator number	10	10	10	10	10	10	10	10
	standard deviation (chip)	0.0292	0.0260	0.0231	0.0206	0.0183	0.0163	0.0145	0.0129

In the table, the measurement accuracy is calculated under the assumption that the peak detection is correct. Since the optimal grid size is smaller than 0.1 chip when C/N_0 is larger than 23 dBHz, the false peak detection probability of the open loop with the optimal grid size is a little larger than that with the grid size of 0.1 chip so that the code phase measuring error is more coincident. As a result, the measurement accuracy of DET and DIS in these cases are more or less the same which is further verified in Section 4.1. In consideration that DIS can hugely decrease the correlators needed, it is preferred when C/N_0 is high. When C/N_0 reduces, the measurement accuracy of DIS is quickly decreased due to the noise and it is much poorer than the DET with the optimal grid size. For the more, as C/N_0 decreases, with the grid size of 0.1 chip, the false detection probability become large and the accuracy will be further distorted. In fact, when C/N_0 is lower than 23 dBHz, the measurement accuracy of DIS will quickly degrade which is validated by the simulation results in Section 4.1. Therefore, we propose the measurement calculation strategy that switches the two code phase calculation methods at the C/N_0 of 23 dBHz, namely the standard deviation cross point of the measurements computed with the two methods. In the strategy, when C/N_0 is higher than 23 dBHz, the discriminator is used and the grid size is 0.1 chip. When C/N_0 is lower than 23 dBHz, measurement is calculated directly based on the parameters of grid with the optimal size.

3.3. Adaptive Correlation Space Adjusted Method

In view of the analysis above, the optimal code search grid size is different when the received signal power varies. For the purposes to improve the code phase measurement accuracy of open loop tracking in urban area, the adaptive correlation space adjusted open loop tracking approach (ACSA-OLTA) is proposed. In ACSA-OLTA, the correlation space between the batch correlators is adjusted to the optimal code search grid size calculated from the EWPRE according to C/N_0 which displays the signal power [20]. The EMLD based measurement calculation method is utilized at the proper C/N_0 to further improve the accuracy and reduce the number of correlator needed at the same time. In addition, the search grid size in frequency dimension is set to a fixed value. When the grid size in code dimension varies according to the C/N_0, the size in frequency dimension is unchanged. The Doppler frequency measuring result of the satellite signal is that of the local reference signal of the detected grid.

Taking the coherent integration of 300 ms as an example, the search grid size in frequency dimension is 5 Hz. The code search grid sizes for different C/N_0 are set as those shown in Figure 5. The process of the method is summarized as below

Step 1: Initialize the code search grid size as 0.2 chip, Doppler frequency grid size as 5 Hz. Then, carry out the coherent integration in the batch correlators.

Step 2: After the coherent integration, detect the signal in the image of the correlator outputs. Estimate the C/N_0 of the incoming signal with the help of the correlation result of the detected peak.

Step 3: Adjust the correlation space between adjacent correlators according to the estimated C/N_0 . When C / N 0 is larger than 23 dBHz, it is set to 0.1 chip. Or else, it is set as Figure 5. The signal code phase and Doppler frequency are directly extracted from the grid and fed back to the local signal generators.

Step 4: Correlate the incoming signal with the new local replicas to form the signal image.

Step 5: After removing the multipath peaks and other improper peaks according to the prior information, the peak with the maximum power in the image is selected.

Step 6: The code phase is calculated by EMLD when C/N_0 is higher than 23 dBHz. Otherwise, the code phase is that of the grid. The estimated parameters are fed back to the local signal generators.

Step 7: Estimate the C / N 0 with the correlation output of the detected peak. Then, adjust the correlation space asFigure 5 if C/N_0 is lower than 23 dBHz, or else it is set to 0.1 chip. Return to Step 4.

For other proper coherent integration lengths, the process is probably the same. The only differences are the switch point of the measurement calculation methods and the optimal grid sizes. Since all of these parameters can be calculated based on the methods above, they are not detailed for simplicity.

In the method, at the beginning, the signal power is unknown and the correlation space has to be set to an empirical value. In order to keep the correctness of peak detection, the code search grid size is initialized as 0.2 chip which is relatively large. In this way, the most accurate peak can be detected easily in which most signal power is contained. The C/N_0 will be perfectly estimated with the corresponding correlation outputs. As the code phase of the detected peak is always aligned with the incoming signal when the correlation space is varying, almost all signal power will be involved in the peak all the time and adjusting the space has tiny effects on C/N_0 estimation. The details of C/N_0 estimation have been given in [21,27] which can provide accurate C/N_0 estimation so that it is not described here for simplicity.

With the help of the C / N 0 estimation, the correlation space of the batch correlators in open loop tracking is adjusted to the optimal one calculated from EWPRE. The measurement calculation methods are also changed according to the C/N_0. In this way, as the satellite signal power is continuously varying in the real application area due to the sheltering, scattering, etc, the proposed ACSA-OLTA can greatly improve the code phase measurement accuracy of open loop tracking. The performance of ACSA-OLTA is evaluated by tests in the next section.

4. PERFORMANCE EVALUATION

In this section, we firstly evaluate the code phase measurement accuracy of open-loop tracking method with the code search grid sizes shown in Figure 5 for different C/N_0 and verify the grid size calculated based on EWPRE is optimal. Afterwards, the performance of the proposed ASCA-OLTA is assessed with the GPS L1CA signal data collected in open air area by off-line processing. Since positioning accuracy is directly related to the pseudo range calculated from the code phase measurement, by utilizing the data collected in the open air area, the positioning accuracy of the open loop tracking methods with different inherent detection grid sizes are compared with the proposed method when C/N_0 varies. The method that uses EMLD is also considered here. Finally, with the help of the data collected from the urban area, the total performance of these methods in urban area is evaluated.

4.1. Accuracy of the Code Phase Measurement

In order to verify that the grid sizes gained from EWPRE bring the optimal code phase measuring accuracy for the corresponding C/N_0 in the presence of AGWN, Monte Carlo simulation is carried out to evaluate the measurement accuracy of open loop tracking method. In the simulation, the measurement is directly extracted from the grid. The mean measurement error of the open loop tracking is computed to testify that the EWPRE can exactly indicate the difference of the measuring accuracy among different grid sizes. The expression of the mean measurement error is

$$E_m = \frac{1}{N_t} \sum_{n=1}^{N_t} |\tau_{t,n} - \hat{\tau_{t,n}}| \tag{36}$$

where N_t is the total number of the tests which is 5000 in the simulation. $\tau_{t,n}$ is the real code phase of the nth test and $\hat{\tau}_{t,n}$ is that derived from the open loop tracking. Then, the code phase measurement error standard deviations of open loop tracking with the optimal grid size and other grid sizes are calculated for different C/N_0 . The simulation parameters are shown in Table 2. The front-end filtering, quantization and other non-idealities are not considered for analytical simplicity. The Doppler frequency error is assumed to be zero since it almost has nothing to do with the code phase measurement accuracy.

Taking the C/N_0 of 10 dBHz, 15 dBHz, 20 dBHz and 25 dBHz as examples, the simulation results of the mean code phase measurement error of open loop tracking method with the code search grid sizes listed in Table 2 is shown in Figure 7. It is clear that the shape of the mean error curves in the figure are more or less the same as those in Figure 4 and the grid sizes with the minimum mean measurement error are the same as those with minimum EWPRE. For other C/N_0 , the simulation results demonstrate that the grid size calculated from EWPRE brings the best accuracy as well. As a result, it verified that the grid size derived from EWPRE is optimal.

Table 2. The parameters of the code phase accuracy simulation.

Parameters	Value
Signal	GPS L5 pilot signal
Signal C/N_0	10 dBHz~35 dBHz
Signal C/N_0 step	1 dB
Signal sampling rate	62 MHz
Code search grid size	0.01 chip~0.4 chip
Code search grid size step	0.01 chip
Coherent integration length	300 ms
Doppler frequency error	0 Hz
Code phase	Uniformly distributed in the range of 2 adjacent chip

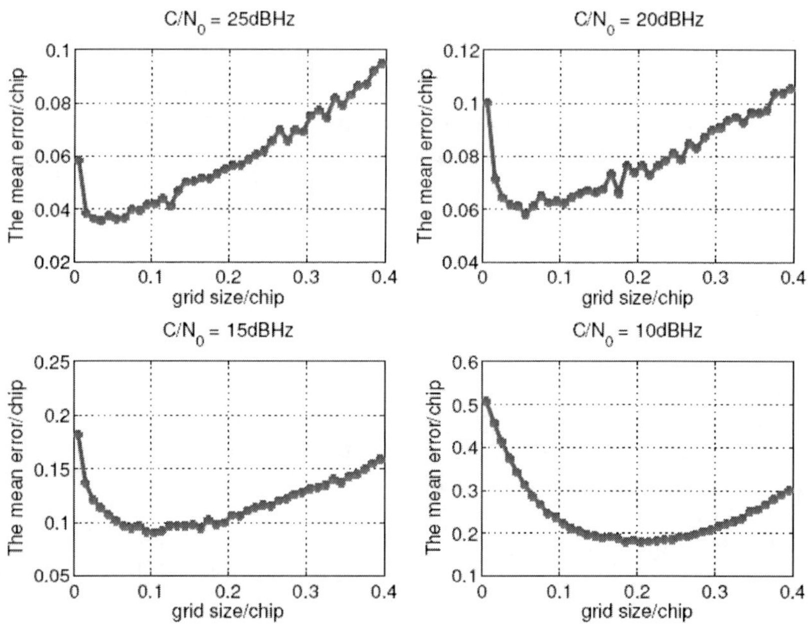

Figure 7. The mean code phase measurement error.

Figure 8 illustrates the simulation result of the code phase measurement error standard deviation for the open loop tracking method with the optimal code search grid size that is calculated in Section 3.1. The standard deviations of the code phase measurement error of open loop tracking with the grid sizes of 0.2 chip, 0.1 chip, 0.05 chip and 0.02 chip are drawn for comparison. It confirmed that the optimal code search grid sizes show higher measurement accuracy than the other grid sizes in the condition of the same C/N_0. Therefore, by changing the grid size to the optimal ones according to C/N_0, the accuracy of the code phase measurement can be greatly improved. In addition, the EMLD based open loop tracking method with the grid size of 0.1 chip is simulated as well. The error standard deviation of the code phase measurements computed by the EMLD is also drawn in the figure. It demonstrates that the accuracy of the method is more or less the same as that directly extracted from the grid with optimal size when C/N_0 is high. However, it will be quickly degraded when C/N_0 is lower than 20 dBHz which is much poorer than the measurement directly extracted from the grid.

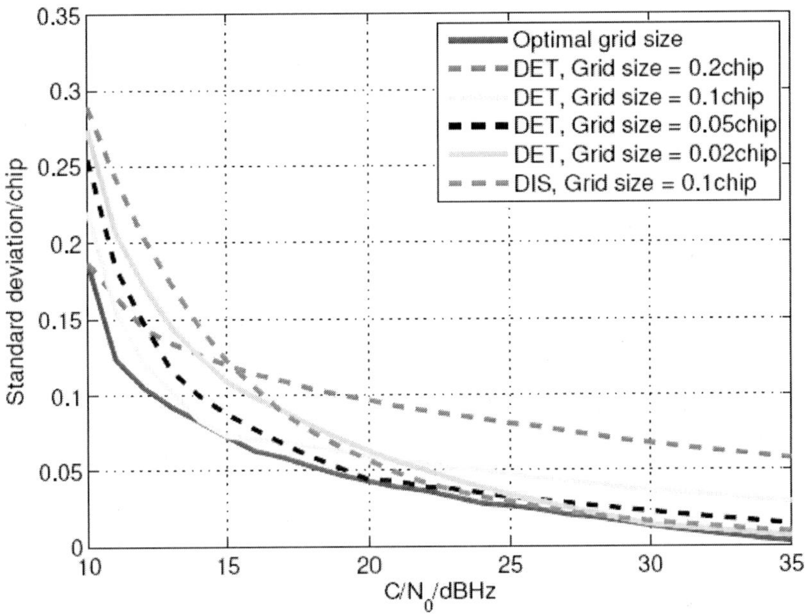

Figure 8. The standard deviation of the code phase measuring error for different C/N_0.

4.2. Positioning Accuracy Test

In this section, the ACSA-OLTA based positioning accuracy is evaluated by the off-line processing with the help of the GPS satellite signals. Since few satellites send new GNSS signals, the GPS L1CA signal is used in the test. In order to clearly confirm the experiment condition, also to correctly obtain the navigation message in the signal to eliminate the phase inversion and lengthen the coherent integration length as the pilot signal, all of the origin satellite signals with a relatively high C / N 0 are collected in the open air in the Beijing Institute of Technology campus. The constellation andC / N 0 are showed in Figure 9. The C / N 0 is gained from a NovAtel navigation receiver. The Geometric Dilution of Precision (GDOP), which is a geometry factor, representing the amplification of the standard deviation of the measurement errors onto the solution [21], is less than 2.0 during the collection.

In the first place of the test, the navigation message is collected by processing the original satellite data. The evaluation is carried out after the message in the GPS L1CA signal is demodulated based on the collected one. The coherent integration length of open-loop tracking in the test is 300 ms. The frequency search step is 5 Hz. For other feasible values, the comparison results are more or less the same so that they are not listed for simplicity. In this case, as the tests of the open loop tracking methods with different measurement extraction strategies are based on the same data, at any time during the test, the satellites that can be detected and used for positioning are all the same among these methods. Hence, the satellite constellation, ephemeris error, atmosphere error and C / N 0 are all identical in the test. The pseudo range based Least square (LS) positioning algorithm is used to test the accuracy for the reason that there is no filter in the process of LS and it can better reveal the variation of the measurement [21]. Accordingly, the positioning accuracy can remarkably indicate the code phase measuring performance.

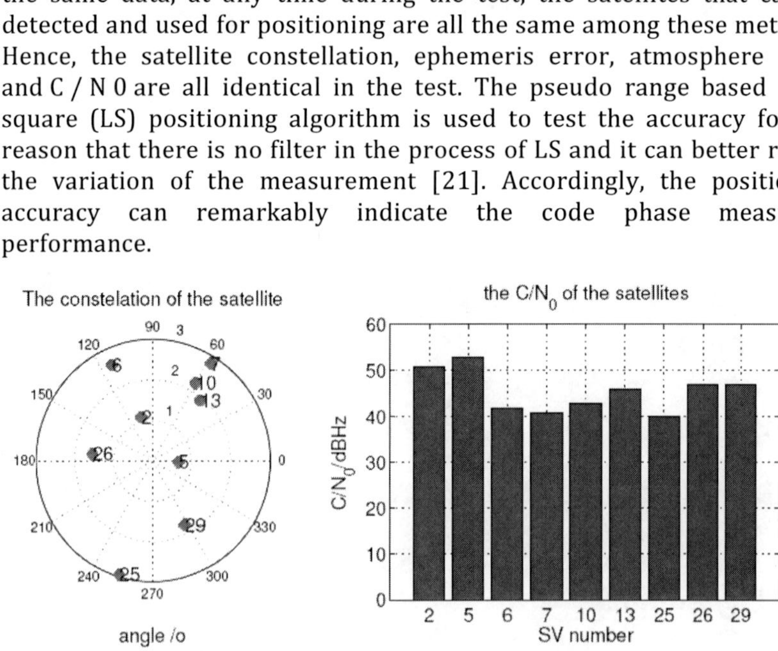

Figure 9. Constellation and C / N 0 of the satellites.

In consideration of the fact that satellite signal power may be attenuated by sheltering, inappropriate antenna direction,*etc.* in an urban area, extra AGWN is added to change the C / N 0 of the received signal. Three tests are conducted in which the signal is attenuated by 10 dB, 20 dB and 30 dB. When it is 10 dB attenuated, for all the satellites, ACSA-OLTA computes the pseudo range with the discriminator based method. As a consequence, the three-Dimension (3-D) positioning error of DET with the fixed code search grid sizes of 0.1 chip and 0.05 chip is computed to compare with that of ACSA-OLTA. When C / N 0 is 20 dB and 30 dB attenuated, ACSA-OLTA changes the measurement extraction strategy and the 3-D positioning error of ACSA-OLTA is compared with DET and DIS in which the grid sizes are both 0.1 code chip. The 3-D position error cumulative distribution of the three tests is showed in Figure 10. The position error cumulative distribution is a statistic result of the position error. It calculates the percentage of position results with the 3-D error less than the given value in all positioning results. The expression of the position error cumulative distribution (PECD) is

$$E_{PECD}(E_g) = \frac{1}{N_p} CNT(\{E_n; E_n < E_g, n = 1, 2, \ldots, N_p\}) \quad (37)$$

where E g is the given position error and E n is the 3-D error of the n t h position result. N p is the total number of the position results. C N T ({ x }) is the number of the elements in the set { x } .

When C / N 0 is 10 dB attenuated, the lowest C / N 0 of these satellites is 30 dBHz. For ASCA-OLTA and other open loop tracking methods with the fixed grid sizes, all of the satellites shown in Figure 9 can be detected and used for position. It is obvious that the EMLD applied by ASCA-OLTA brings higher positioning accuracy than the open loop methods which directly extract the measurement from the grid with the sizes of 0.1 chip and 0.05 chip. Since the C / N 0 is relatively high, the false detection hardly occurs and the uncertainty in the code search grid can be totally eliminated by the discriminator. As a result, the positioning accuracy of ASCA-OLTA is perfect. While, for the methods that directly extracting the code phase from the grid, the inherent uncertainty will probably distort the measurement accuracy so that the positioning accuracy is disturbed although the C / N 0 is high. As shown in the figure, the larger the grid size is, the worse the position accuracy is.

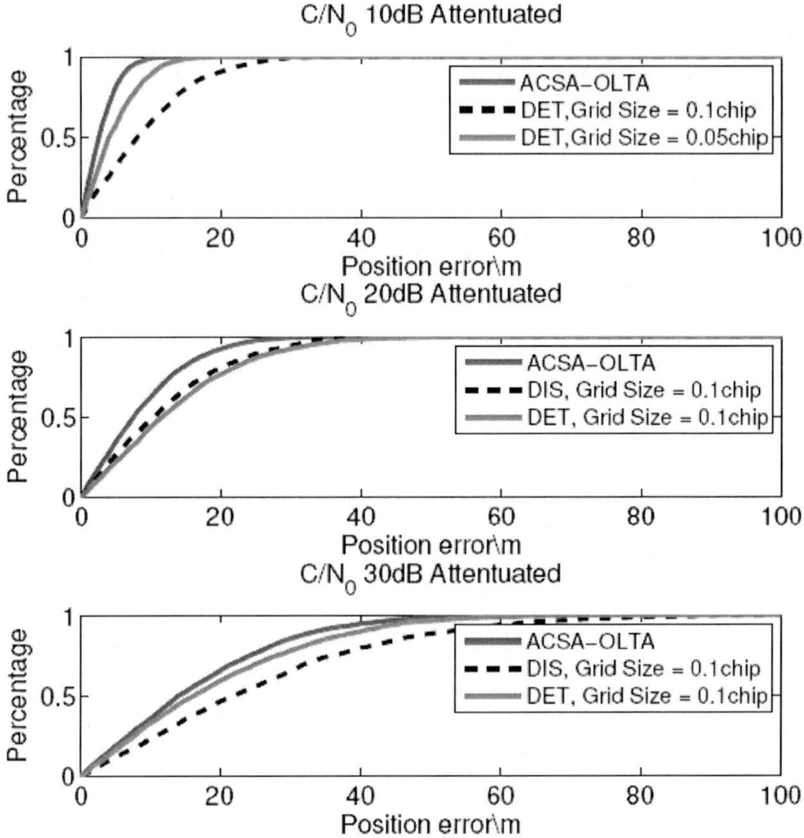

Figure 10. 3-D position error cumulative distribution of open loop tracking.

When C / N 0 is 20 dB attenuated, the lowest C / N 0 of these satellites is 20 dBHz and all of the satellites can still be detected. In this case, according to the signal C / N 0 , the grid sizes of ACSA-OLTA for some of the satellites are narrowed down to the optimal grid size and the code phase measurement is directly extracted from the grid parameters. The results show that the positioning accuracy of the EMLD based open loop tracking method and the method directly extracting the measurement from the grid with the size of 0.1 chip are probably the same and both of them are poorer than ASCA-OLTA. Since the C / N 0 of the satellites ranges from 20 dBHz to 30 dBHz, both of the two methods have satellites with accuracy highly disturbed pseudo range measurements. While

ASCA-OLTA can adaptively choose the optimal strategy and properly narrowing down the grid size based on the optimal ones so that the measurement accuracy can be greatly improved and the positioning accuracy is outstanding.

When C / N 0 is 30 dB attenuated, the signal power of all the satellites is really low and the positioning accuracy is degraded. During the test, all of the satellites except SV.25 can be detected all the time and the GDOP is still less than 2.0. In this case, the grid size of ACSA-OLTA is broadened to keep the correctness of the peak detection and it shows superiority over the open loop tracking method whose measurement is gained from the grid with the fixed size of 0.1 chip. The positioning error of the EMLD based method with the grid size of 0.1 chip is much larger than that of the ACSA-OLTA and it is even worse than the measurement directly gained from the grid. As a result, it is concluded that, when theC / N 0 is low, the noise will highly distort the accuracy of the discriminator based method. In addition, by appropriately broadening the detection grid size as the ACSA-OLTA does, the accuracy will be improved in the condition of low C / N0 .

4.3. Test in Urban Area

An experiment was conducted to further evaluate the performance of the proposed ACSA-OLTA in urban area according to the method in [28]. The same as the position accuracy test in Section 4.2, the experiment is carried out by off-line processing. The data utilized is GPS L1CA signal that is collected in Zhongguancun, Beijing, China where is a typical urban scenario. There are a lot of skyscrapers aside the road and the satellite signal is easily sheltered and attenuated so that it is a difficult environment for satellite navigation. The navigation message is gained by AGPS so that coherent integration length is not limited. The total duration of the test is about 30 min and the ephemeris gotten at the beginning is in effect all the time. The vehicular velocity varied from 0–40 km/h with random stops due to the traffic lights and the total distance traveled is about 5 km.

The test equipment is composed of a digital intermediate frequency signal collector which can down convert the satellite to intermediate frequency and store it in real time. Besides that, the SPAN-CPT Single Enclosure GNSS/INS Receiver is utilized to obtain the reference solution. It is an integrated system powered by NovAtel's world class OEM6 technology which is comprised of Fiber Optic Gyros (FOG) and Micro

Electrical Mechanical Systems (MEMS) accelerometers. It works in double difference mode using phase and Doppler measurements. The baseline separation (relative to a base station located on the Weigongcun) varied between 3–4 km. The reference solution accuracy in these conditions is summarized in Table 3. A power divider is utilized after the antenna and the two outputs are connected to the data collector and the SPAN-CPT receiver respectively to guarantee that the reference solutions and the collected data are correspondence. The equipments and trajectory of the test is showed in Figure 11.

Table 3. The reference solution accuracy.

Information	Accuracy
Position	0.1 m
Velocity	cm/level
Attitude	<1 deg

(a) (b)

Figure 11. (**a**) Equipments of the test; (**b**) Trajectory of the test (Gained from Google Earth).

To facilitate the data analysis, the trajectory is divided into two segments which are marked by the yellow line and red line in Figure 11b. The yellow line is Segment 1 and the red line is Segment 2. They represents different operational scenarios that are considered and examined separately. The satellite number, GDOP and the lowest C / N 0 of all the satellite signal received during the test are drawn in Figure 12, where Segment 1 is drawn in blue and Segment 2 is in red.

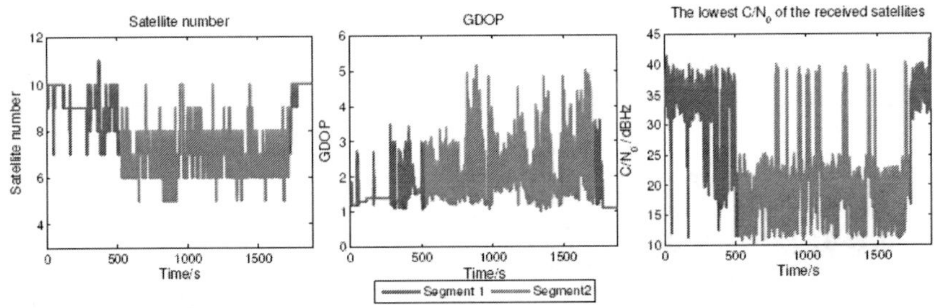

Figure 12. Variation of the satellite number, GDOP and C / N 0 in the test.

Segment 1 is on a wide road where the satellite visibility is good and the operational conditions can be considered semi-open sky. In this area, the satellite is difficult to be sheltered and attenuated. In most time, the satellite number is larger than 8, the GDOP is smaller than 3 and the C / N 0 is above 30 dBHz. Segment 2 represents a very difficult scenario. As shown in the figure, although the sensitivity of open loop tracking is outstanding, there are sudden drops in the satellite number for the reason that the buildings can totally shelter the satellite signals propagating from one direction. The GDOP will fall down as well because of the poor constellation. There is also a long time when the C / N 0 is continuous low which is caused by the window plafond in the air. All of them will distort the positioning accuracy.

The primary objective of the experiments is to evaluate the code phase measurement accuracy of the proposed ACSA-OLTA and the traditional open loop tracking methods with fixed grid size and measurement calculation approach in urban area. Hence, the positioning algorithm is still LS which has no process of filter and is able to better indicate the measurement accuracy. Since the evaluation focuses on the effects of the code search grid size and measurement extraction method on the positioning accuracy of open loop tracking, other parameters such as coherent integration length and frequency search step are all identical among the methods tested. The main parameters of ACSA-OLTA are listed in Table 4. Besides that, the C / N 0 for ACSA-OLTA to adjust the correlation space is accurate to 1dB. Since the C / N 0 of NovAtel receiver is estimated from the correlator whose code phase is best aligned with the incoming signal, which is the same as that estimated in the process of ACSA-OLTA, the test results by using the two sources of C / N 0 will be almost the same. Thus, the C / N 0 of NovAtel receiver is utilized as reference for simplicity. The parameters of the traditional methods with fixed grid sizes and measurement calculation approaches considered for comparison are also listed in Table 4. The abbreviations of DET and DIS represent the open loop tracking methods with the code phase measurement computed from the EMLD and directly extracted from the grid respectively.

Table 4. The parameters of the methods tested.

Method	Grid Size	Frequency Step	Integration Length	Positioning Rate
ACSA-OLTA	Adaptively adjusted based on C/N_0 as Figure 5	5 Hz	300 ms	1 Hz
DIS	0.1 chip	5 Hz	300 ms	1 Hz
DET	0.2 chip	5 Hz	300 ms	1 Hz
	0.1 chip	5 Hz	300 ms	1 Hz
	0.05 chip	5 Hz	300 ms	1 Hz
	0.02 chip	5 Hz	300 ms	1 Hz

Figure 13 illustrates the positioning results of the proposed ACSA-OLTA together with DIS with the grid size of 0.1 chip, DET with the grid sizes of 0.2 chip, 0.1 chip, 0.05 chip and 0.02 chip in Segment 1. The solutions of DET with the grid sizes of 0.1 chip and 0.2 chip shows large errors during the test. Narrowing down grid sizes shows evident improvements relative to those with larger grid sizes. As the C / N 0 is high, the detection performance of open loop tracking system is perfect. In this scenario, for DET, the most accuracy peak is easily detected and the measurement will be improved when the grid size is small. Whereas the large uncertainty in wide grid will deteriorate the accuracy. In addition, DIS provides better performance and it is similar to DET with the grid size of 0.02 and 0.05 chip. All of the results are coincide with the analysis in Section 3. Finally, the trajectory obtained by ACSA-OLTA, in which most of the code phase measurements are calculated by the EMLD with the grid size of 0.1 chip, shows perfect positioning accuracy with only small disagreements with the reference trajectory.

In order to further clarify the contribution of the proposed ACSA-OPTA, the position error in the segment is evaluated by the position error cumulative distribution expressed in Equation (37). The result is drawn in Figure 14. As can be seen from the figure, the accuracy of DET with the grid size of 0.02 chip and DIS with the grid size of 0.1 chip is close and 90% position results have the error less than 10 m. DIS is preferable due to the low computational costs and reduction of the correlators. When the grid size is 0.05 chip, the accuracy of DET is a little degraded and 80% position errors are smaller than 10 m. However, when the grid size of DET extends to 0.1 chip, 40% positioning results have error larger than 10 m. For DET with the grid size of 0.2 chip, the maximum error reaches about 40 m and only 40% positioning results have error less than 10 m. ACSA-OPTA is better than all the other methods tested because there is still some sudden power reduction during the segment. ACSA-OPTA can adaptively adjust the grid size and measurement extraction method to keep accuracy.

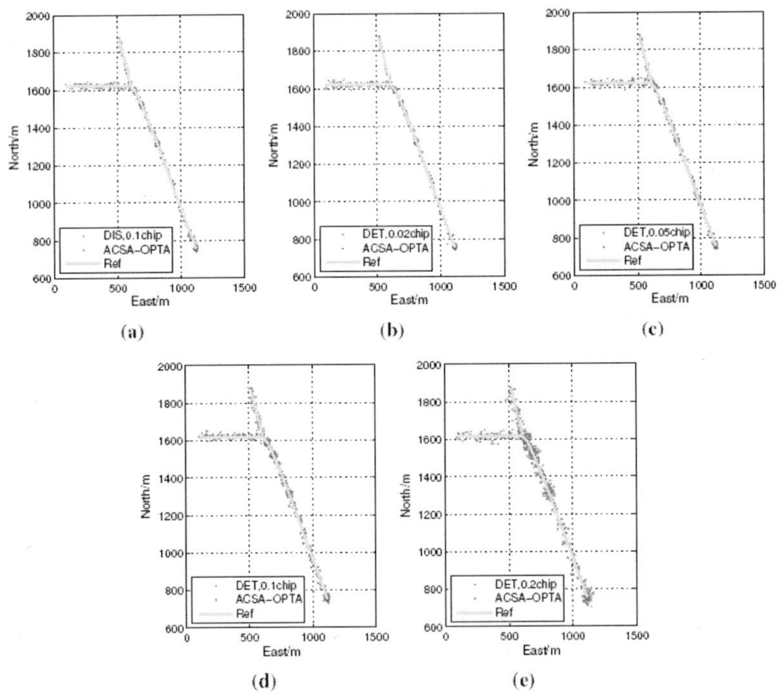

Figure 13. Positioning results of open loop tracking in Segment 1.

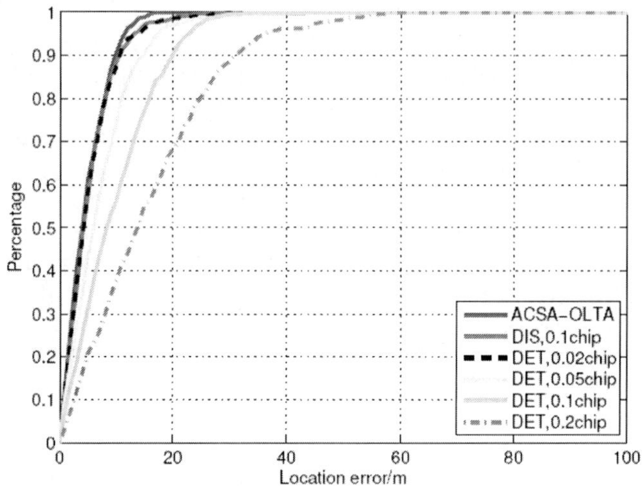

Figure 14. Position error cumulative distribution of the positioning results in Segment 1.

The trajectories obtained with the open loop tracking with different configurations for Segment 2 are shown in Figure 15. The position accuracy of DET with small grid sizes is highly degraded. Owing to the poor detection performance caused by the low C / N 0 , false peak detection frequently occurs which will extremely distort the position results especially when the grid size is small. The maximum error reaches about 80 m. For DIS, the position error is also larger since the huge noise in the signal destroys the estimation results. As the grid size is broadened to 0.1 chip for DET, the positioning performance becomes better. However, when the grid size of DET is 0.2 chip, the error is a little larger due to the incremental uncertainty within the grid. ACSA-OPTA shows remarkable improvements relative to the other configurations. As C / N 0 is varying intensely, the ACSA-OPTA can adaptively adjust the grid size to the optimal one so that it can enhance the measurement accuracy in all cases.

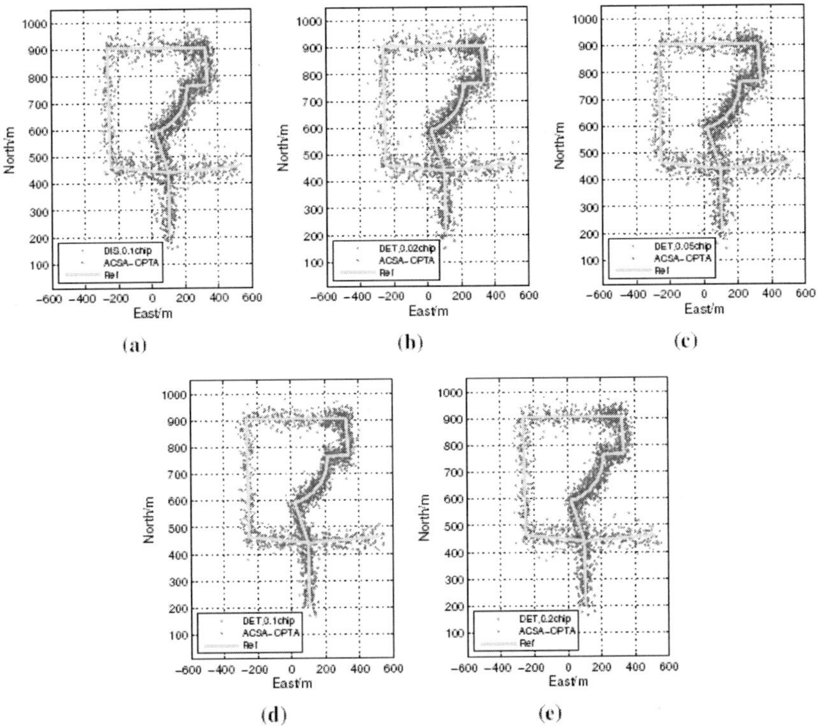

Figure 15. Positioning results of open loop tracking in Segment 2.

Figure 16 illustrates the position error cumulative distribution of the results in Segment 2. It shows that the performance of ACSA-OLTA is outstanding compared with the traditional open loop tracking methods whose configurations of grid sizes and measurement calculation method are fixed. 70% of the positioning results have error smaller than 20 m and less than 10% positioning results have errors larger than 40 m. In addition, the positioning errors of DET with the grid size of 0.02 chip and 0.05 chip and DIS with the size of 0.1 chip are extremely increased. More than 30% position results have error larger than 40 m. On the other hand, large numbers of positioning results computed from DET with the grid size of 0.1 chip have the errors in the range from 15 m to 40 m due to the fact that it can avoid the false detection to some extent but the uncertainty will distort the accuracy as well. Continuously enlarging the grid size to 0.2 chip, the performance of DET is degraded as well and more of the position results have the error larger than 20 m. Although it can eliminate the false peak detection in most time, the uncertainty in the grid is really large so that the accuracy is also distorted.

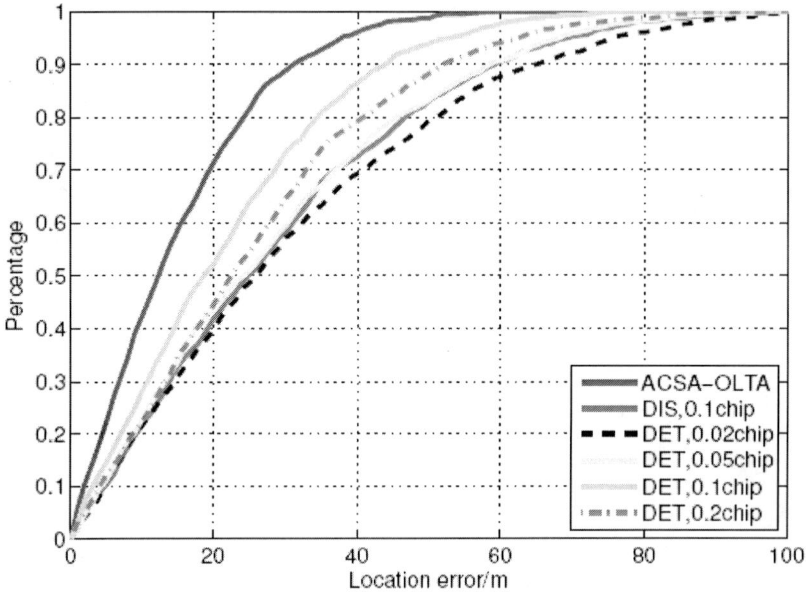

Figure 16. Position error cumulative distribution of the positioning results in Segment 2.

5. CONCLUSIONS

This work looks at the pseudo range measurement of open loop tracking for GNSS and proposes the Adaptive Correlation Space Adjusted Open Loop Tracking Approach (ACSA-OLTA) to improve the measurement accuracy of open loop tracking for vehicle positioning in urban area. A novel expression to obtain the code search grid size which can bring the best measuring accuracy, named Equivalent Weighed Pseudo Range Error is presented in Section 3.1. Using the new expression, the optimal code search grid sizes for different C/N_0 are computed. Following a thorough analysis, the accuracy of the code phase measurement calculated from the early-minus-late discriminator and that directly extracted from the grid is evaluated for different grid sizes when C/N_0 varies. The computational costs are also assessed. Based on the analysis results, the measurement calculation strategy is developed.

ACSA-OLTA adaptively adjusts the correlation space to the optimal code search grid size and switches the measurement calculation methods as the developed strategy according to C / N 0 .

With the help of the Monte Carlo simulation, the mean code phase measuring errors for different C / N 0 and grid sizes are evaluated. The results verify that the proposed EWPRE can indicate the measuring error level for open loop tracking with different grid sizes. The optimal code search grid size obtained according to EWPRE for the corresponding C / N 0 is valid. Data was collected in the campus of Beijing Institute of Technology and Zhongguancun, Beijing and processed using various configurations. For the data analyzed herein, based on the positioning results computed with the Least Square method, which is directly related to the code phase measurement accuracy, the main conclusions are as follows:

- In open air area where the C / N 0 of most satellites are larger than 30 dBHz, the errors of most positioning results calculated from the code phase measurements that are gained from the EMLD is smaller than 10 m. It outperforms those directly extracted from the grid with the sizes of 0.05 chip, 0.1 chip and 0.2 chip and it is similar to that with the grid size of 0.02 chip. The results validate the analysis in Section 3.2 that the early-minus-late discriminator can keep the measurement accuracy but reduce the correlator utilized when C / N 0 is high. It also indicates that narrowing down grid size can help to improve the accuracy in this case which is the same as that EWPRE shows.

- In urban canyon area where the C / N 0 of a number of satellites is lower than 20 dBHz, the performance of the EMLD based measurement calculation method is greatly degraded. At the same time, the accuracy of the measurements directly extracted from the grids with size of 0.02 chip and 0.05 chip are also deteriorated. On the contrary, the accuracy of the measurements directly extracted from grids with larger grid sizes is better. When the grid sizes are 0.1 chip and 0.2 chip, the positioning errors are much less than those with smaller grid sizes. It verifies that properly enlarging the search grid size can improve the measurement accuracy when C / N 0 is low as the EWPRE shows in Section 3.1.

- In both the two scenarios, ACSA-OLTA can greatly improve the measurement accuracy compared with the traditional open loop tracking method with fixed search grid sizes and measurement calculation methods. The results of the test in the paper show that when C / N 0 is high, the EMLD is utilized and 90% positioning results have the error smaller than 10 m. When C / N 0 is low, the search grid size is adjusted and 70% positioning results have the error smaller than 20 m. The results also validate that the grid sizes gained from EWPRE are preferable.

Based on these results, ACSA-OLTA does improve the measurement accuracy of open loop tracking and enhance the positioning performance of the open loop tracking based GNSS receiver for vehicular application in urban areas.

ACKNOWLEDGMENTS

This work is supported by the National Natural Science of China (Grant No. 61101128) and 111 Project of China under Grant B14010.

AUTHOR CONTRIBUTIONS

Hang Ruan conceived the work, proposed the approach, carried out the simulation, performed the experiment, analyzed the experiment results and wrote the manuscript. Jian Li helped to conduct the experiment. Lei Zhang revised the manuscript. Teng Long proposed the research direction.

REFERENCES

1. Rebeyrol, E.; Julien, O.; Macabiau, C.; Ries, L.; Delatour, A.; Lestarquit, L. Galileo civil signal modulations. *GPS Solut.* **2007**, *11*, 159–171.
2. Spilker, J.J.; Van Dierendonck, A.J. Proposed new civil GPS signal at 1176.45 MHz. In Proceedings of the U.S. ION GPS Conference, Nashville, TN, USA, 16 September 1999.
3. Fontana, R.D.; Cheung, W.; Novak, P.; Stansell, T.A. The new L2 civil signal. In Proceedings of the U.S. ION GPS Conference, Salt Lake City, UT, USA, 11–14 September 2001.
4. Razavi, A.; Gebre-Egziabher, D.; Akos, D.M. Carrier Loop Architectures for Tracking Weak GPS Signals. *IEEE Trans. Aero. Electron. Syst.* **2008**, *44*, 697–710.

5. De Haag, M.U. An Investigation into the Application of Block Processing Techniques for the Global Positioning System. Ph.D. Dissertation, The Ohio University, Athens, OH, USA, 1999.

6. Krasner, N.F. Method for Open Loop Tracking GPS Signals. U.S. Patent No. 6,633,255 B2, 14 October 2003.

7. Van Graas, F.; Soloviev, A.; De Haag, M.U.; Gunawardena, S.; Braasch, M. Comparison of two approaches for GNSS receiver algorithms: Batch processing and sequential processing considerations. In Proceedings of the ION GNSS, Long Beach, CA, USA, 13–16 September 2005.

8. Van Graas, F.; Soloviev, A.; De Haag, M.U.; Gunawardena, S. Closed-Loop Sequential Signal Processing and Open-Loop Batch Processing Approaches for GNSS Receiver Design. *IEEE J. Sel. Top. Sign. Proces.* **2009**, *3*, 571–586.

9. Misra, P.; Enge, P. *Global Positioning System: Signal, Measurements and Performance*, 2nd ed.; Ganga-Jamuna: Boston, MA, USA, 2006; pp. 381–384.

10. Soloviev, A.; Van Graas, F. Use of Deeply Integrated GPS/INS Architecture and Laser Scanners for the Identification of Multipath Reflections in Urban Environments. *IEEE J. Sel. Top. Sign. Proces.* **2009**, *3*, 786–797.

11. Ali, K.; Chen, X.; Dovis, F.; De Castro, D.; Fernandez, A.J. Multipath Estimation in Urban Environments from Joint GNSS Receivers and LiDAR Sensors. *Sensors* **2012**, *12*, 14592–14603.

12. Sokolovskiy, S.V. Tracking tropospheric radio occultation signals from low earth orbit. *Radio Sci.* **2001**, *36*, 483–498.

13. Chiang, K.Q.Z.; Psiaki, M.L. Kalman Filter Tracking of Limb Scan Signal using a Bank of Correlators. *IEEE Trans. Aero. Electron. Syst.* **2013**, *49*, 118–133.

14. Jacob, C.; Manfred, S.; Magnus, B.; Anders, C. Apparatus and method for performing open loop tracking of a signal. European Patent No. 1,243,940 B1, 31 May 2006.

15. Luo, P.; Petovello, M.G. Collaborative Tracking of Weak GPS Signals Using an Open-loop Structure. In Proceedings of the ION ITM 2011, San Diego, CA, USA, 24–26 January 2011.

16. Lin, T.; O'Driscoll, C.; Lachapelle, G. Development of a Context-Aware Vector-Based High-Sensitivity GNSS Software Receiver. In

Proceedings of the ION ITM 2011, San Diego, CA, USA, 24–26 January 2011.

17. Xie, P.; Petovello, M.G.; Basnayake, C. Multipath Signal Assessment in the High Sensitivity Receivers for Vehicular Applications. In Proceedings of the ION GNSS 2011, Portland, OR, USA, 3 September 2011.

18. Xie, P.; Petovello, M.G. Improving High Sensitivity Receiver Performance in Multipath Environment for Vehicular Applications. In Proceedings of the ION GNSS 2012, Nashville, TN, USA, 17–21 September 2012.

19. Yan, K.L.; Guo, W.F.; Zhang, H.P.; Niu, X.J. Open-loop GNSS (Global Navigation Satellite System) Signal Carrier Tracking Method and System. Chinese Patent No. 103,499,824B, 8 January 2014.

20. Parkinson, B.; Spilker, J., Jr. *Global Positioning System: Theory and Applications Volume I*, 1st ed.; American Institute of Aeronautics and A stronautics, Inc.: Washington, DC, USA, 1996; pp. 90–91.

21. Kaplan, E.; Hegarty, C. *Understanding GPS: Principles and Applications*, 2nd ed.; Artech House: Norwood, MA, USA, 2006; pp. 113–142, 219–231 and 301–334.

22. GPS Joint Program Office. *Interface Specification NAVSTAR GPS Space Segment/Navigation L5 User Interfaces*; Technical Report IS-GPS-705; GPS Joint Program Office: El Segundo, CA, USA, 2005.

23. Diggelen, F. Generating Assisted Data. In *A-GPS: Assisted GPS, GNSS and SBAS*; Artech House: Boston, MA, USA, 2009.

24. Coenen, A.; Van Nee, D.J.R. Novel fast GPS/GLONASS code-acquisition technique using low update-rate FFT.*Electron. Lett.* **1992**, *28*, 863–865.

25. Qaisr, S.U.; Dempster, A.G. Assessment of the GPS L2C Code Structure for Efficient Signal Acquisition. *IEEE Trans. Aero. Electron. Syst.* **2012**, *48*, 1889–1902.

26. Proakis, J. *Digital Communication*, 4th ed.; McGraw-Hill: New York, NY, USA, 2000.

27. Satyanarayana, S.; Borio, D.; Lachapelle, G. C/N_0 estimation: Design criteria and reliability analysis under global navigation satellite system (GNSS) weak signal scenarios. *IET Radar Sonar Navig.* **2012**, *6*, 81–89.

28. Angrisano, A.; Petovello, M.; Pugliano, G. Benefits of Combined GPS/GLONASS with Low-Cost MEMS IMUs for Vehicular Urban Navigation. *Sensors* **2012**, *12*, 5134–5158.

10
CHAPTER

GNSS SPACE-TIME INTERFERENCE MITIGATION AND ATTITUDE DETERMINATION IN THE PRESENCE OF INTERFERENCE SIGNALS

Saeed Daneshmand, Ali Jafarnia Jahromi, Ali Broumandan and Gérard Lachapelle

PLAN Group, Department of Geomatics Engineering, Schulich School of Engineering, University of Calgary, 2500 University Drive, N.W., Calgary, AB T2N 1N4, Canada

ABSTRACT

The use of Space-Time Processing (STP) in Global Navigation Satellite System (GNSS) applications is gaining significant attention due to its effectiveness for both narrowband and wideband interference suppression. However, the resulting distortion and bias on the cross correlation functions due to space-time filtering is a major limitation of this technique. Employing the steering vector of the GNSS signals in the filter structure can significantly reduce the distortion on cross correlation functions and lead to more accurate pseudorange measurements. This paper proposes a two-stage interference mitigation approach in which the first stage estimates an interference-free subspace before the acquisition and tracking phases and projects all received signals into this subspace. The next stage estimates array attitude parameters based on detecting and employing GNSS signals that are less distorted due to the projection process. Attitude parameters enable the receiver to estimate the steering vector of each satellite signal and use it in the novel distortionless STP filter to significantly reduce distortion and maximize Signal-to-Noise Ratio (SNR). GPS signals were collected using a six-element antenna array under open sky conditions to first calibrate the antenna array. Simulated interfering signals were then added to the digitized samples in software to verify the applicability of the proposed receiver structure and assess its performance for several interference scenarios.

Keywords: antenna array receiver; global navigation satellite system (GNSS); interference mitigation; space-time processing

1. INTRODUCTION

GNSS is used worldwide, however, the performance of location-based services provided by receivers can still be compromised by interfering signals. There is an ever increasing attention to safe and secure applications. As an example, countless time tagging and synchronization systems in the telecom and electrical power grid industries relying primarily on GNSS signals are vulnerable to in-band electronic interference because of being extremely weak signals broadcasted over wireless channels. Therefore, even low-power interference can easily jam receivers within a radius of several km. Interference can dramatically degrade the performance of receivers or completely deny position and time services provided by these systems. The spread spectrum modulation technique applied in the structure of GNSS signals provides a certain degree of protection against narrowband interfering signals; however, the spreading gain alone is not sufficient to suppress high power or wideband interference signals.

Over the years, GNSS interference suppression methods based on time and frequency processing have been widely studied in the literature (e.g., [1,2]). Although they are effective for suppressing continuous wave (CW) or multi-tone jammers, their performance degrades when they deal with wideband interference signals (e.g., Gaussian jammers) or when interfering signals change rapidly in time and frequency such as swept continuous wave interference [3].

On the contrary, interference mitigation techniques utilizing an antenna array can effectively detect and suppress both narrowband and wideband interfering signals regardless of their time and frequency characteristics. One of the earliest space-based processing methods has been referred to as the Capon beamformer, also called minimum variance distortionless response (MVDR) beamformer [4,5]. The MVDR beamformer has a distortionless response for the desired signal while suppressing all signals arriving from other directions. Antenna array processing in GNSS applications has been mostly centered on interference suppression [6,7,8,9,10]. Reference [10] drew the attention on utilizing minimum power distortionless response (MPDR) beamforming for GPS applications to reject interference signals whose power is significantly higher than that of the GPS signals. MPDR and MVDR beamformers have been considered popular methods for a variety of signal processing applications such as radar, wireless communications, and speech enhancement.

Despite the effectiveness of antenna array-based methods, they suffer from hardware complexity. In spatial processing, the number of antennas determines the number of interfering signals that can be suppressed. Limitations on the number of the antennas and size of the array can be considered the main practical challenge of these methods. To deal with this problem, techniques employing both time/frequency and spatial domain processing have been of great interest since, in contrast to time/frequency based methods, they are able to deal with both wideband and narrowband interference. These techniques are generally referred to as Space-Time Adaptive Processing (STAP) or Space-Frequency Adaptive Processing (SFAP) techniques. STAP and SFAP approaches in GNSS applications have been studied in the literature for several years [11,12,13,14,15]. These methods combine spatial and temporal filters to suppress more radio frequency interfering signals by increasing the Degree of Freedom (DoF) of the array without physically increasing the antenna array size. The term "adaptive" means that the array follows changes in environment and constantly adapts its own pattern by means of a feedback control. The main focus here is on space-time processing and the study of adaptive methods is not considered. Nevertheless, the proposed approach can be easily extended to the adaptive cases as well.

Besides the superior advantages of space-time filtering, specific considerations should be taken into account in designing such filters in order to prevent induced biases in pseudorange measurements and achieve accurate and precise time and position solutions [16,17]. The output of the space-time filter is basically a direction-frequency dependent response. Even if the filter completely nullifies interfering signals, the non-linearity behavior of its frequency response may result in biased measurements, distortion or broadness of the cross correlation functions during receiver acquisition and tracking stages. This may not be tolerable especially for high precision GNSS applications. The effects of this distortion on GNSS signals were recently studied [16,17,18,19].

To reduce this distortion, one effective approach is to incorporate the satellite signal steering vector, which contains all the spatial information of the incoming signal, in the structure of the space-time filter as a constrained optimization problem [19,20,21,22,23]. In fact, these methods are extended versions of MVDR and MPDR beamformers for space-time processing. Although employing the satellite steering vectors in the beamformer structure avoids unintentional signal attenuation, the resulting cross correlation functions after beamforming may still be distorted, which in turn produces biases on pseudorange measurements and errors in the position solutions. This is due to the fact that in these techniques there is no explicit assumption on the linearity of the STP filter response. In [9], it is suggested that the space-time filter be designed to have a real frequency response (formed from a filter multiplied by its conjugate); however this was not analyzed and a practical realization of filter coefficients was not reported. There are other effective approaches to reduce the induced bias error; however, they do not guarantee a distortionless response for GNSS signals [16,18].

Although employing the satellite signal steering vector has been widely employed in the STP processing, limited papers addresses the steering vector estimation in the presence of interference. The steering vector conveys the spatial information that can be also employed for various applications such as multipath mitigation, SNR maximization and Angle of Arrival (AOA) estimation. Steering vector estimation in a jammed environment for attitude determination was studied in [24]; in this paper it is assumed that the spatial covariance matrix is positive definite and invertible which may not be the case in all inference scenarios where the covariance matrix becomes ill-conditioned.

The steering vector can be obtained by either measuring phase differences of the received signals at the antenna array elements, or by calculating it from array platform attitude parameters and satellite azimuth and elevation angles. The first approach needs to acquire and track the received signals, some of which may not be available in challenging environments. The second approach can calculate the steering vector of all satellites regardless of the signal quality and availability but requires attitude parameters. Herein both methods are employed in the proposed structure of the receiver such that steering vectors, measured from the available satellite signals, are used to estimate attitude parameters and consequently the steering vector of all satellite signals without imposing any assumption on the spatial covariance matrix.

The other part of this research focuses on space-time filter design for GNSS interference mitigation. Due to the simplicity in implementation, Space-Time Processing (STP) filters in GNSS applications are mostly implemented before the despreading process (*i.e.*, correlation and Doppler removal). However, since satellite steering vectors are not employed in the structure of the filter, some satellite signals may be attenuated or distorted, which will adversely affect the performance of the receiver. Therefore, herein a two-stage receiver structure is proposed. The first stage is implemented before the despreading process. In this stage, by estimating a projection matrix from the Singular Value Decomposition (SVD) of the pre-despreading space-time covariance matrix, the received signals are projected into the interference-free subspace. Estimating the signal steering vector from the projected subspace is an underdetermined problem and it is shown that extracting the spatial information from the projected signals may not be possible or can be partially done for some signals. In fact, some of the array DOF information is used to remove interfering signals. Therefore, in the second stage, attitude parameters are estimated considering those steering vectors that could be estimated. By using attitude parameters and satellite azimuth and elevation angles from ephemeris data, the steering vector of all satellite signals can be then accurately calculated and employed in designing a space-time filter. A novel approach for designing the space-time filter is also proposed not only to nullify the interference signals but also to increase the C/N_0 and to avoid biases and distortions on cross correlation functions.

In order to verify the effectiveness of the proposed method for steering vector estimation in the presence of interference and assess its performance, a set of real GPS L1 signals was collected and simulated interfering signals were added to the digitized samples in software. A tactical-grade IMU was used as reference to evaluate the accuracy of satellite steering vectors and heading estimates. Moreover, to evaluate the performance of the proposed space-time filter, it was compared to the well-known space-only MPDR and space-time MPDR beamformer methods.

2. SIGNAL MODEL

Without loss of generality and for the sake of simplicity, only one GNSS signal is considered in the formulations below. Complex baseband representation of the received signal vector at an arbitrary N-element array configuration for the satellite signal and K interference signals can be written as:

$$\underset{N\times1}{\mathbf{r}} = \underset{N\times1}{\mathbf{a}}\, s + \underset{N\times K}{\mathbf{B}}\, \underset{K\times1}{\mathbf{v}} + \underset{N\times1}{\boldsymbol{\eta}} \tag{1}$$

where $\mathbf{B}$ is a matrix whose columns indicate interfering signal steering vectors and $\boldsymbol{\eta}$ is a complex additive white Gaussian noise vector. s represents the GNSS signal waveform and $\mathbf{v}$ is a vector specifying K interfering signal waveforms. In Equation (1), $\mathbf{a}$ is an $N \times 1$ vector representing the steering vector (or array manifold vector) of the satellite signal defined as:

$$\mathbf{a} \triangleq \begin{bmatrix} e^{j\frac{2\pi}{\lambda}(\mathbf{d})^T \mathbf{z}_1} \\ e^{j\frac{2\pi}{\lambda}(\mathbf{d})^T \mathbf{z}_2} \\ \vdots \\ e^{j\frac{2\pi}{\lambda}(\mathbf{d})^T \mathbf{z}_N} \end{bmatrix} \tag{2}$$

in which λ is the wavelength of the signal and $\mathbf{z}_n$, $n = 1, 2, ..., N$ is a 3×1 unit vector pointing to the nth antenna element and $\mathbf{d}$ is a 3×1 unit vector pointing to the satellite direction in the body frame coordinate system and Tstands for the transpose operation. Figure 1 shows that the standard implementation of the STP filter in which each antenna is followed by a temporal filter or a Tapped Delay Line (TDL) with the typical delay time of a sampling duration denoted by T_s.

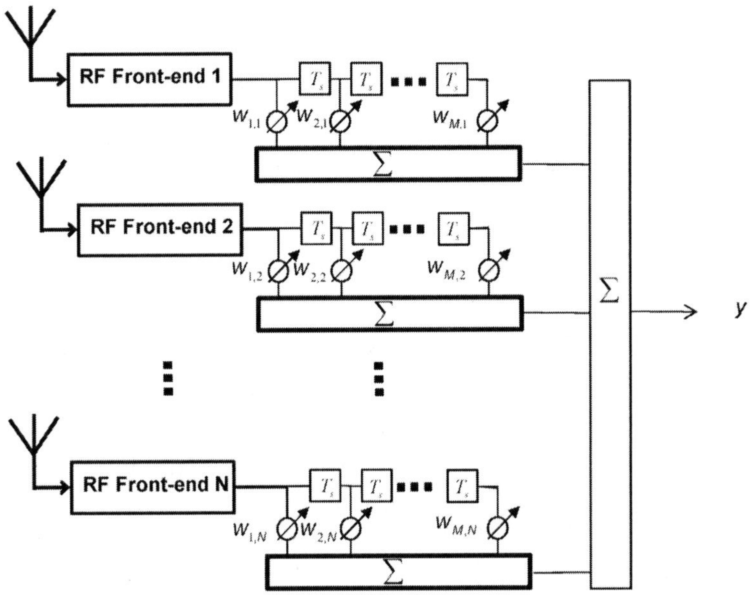

Figure 1. Generic structure of a space-time filter.

An antenna array with N elements and TDLs with $M-1$ taps leaves MN unknown filter coefficients which should be determined. For each time snapshot, MN received samples form a $MN \times 1$ vector can be written as:

$$\vec{\mathbf{r}} = \begin{bmatrix} r_{1,1} & r_{1,2} & \cdots & r_{1,N} & r_{2,1} & r_{2,2} & \cdots & r_{2,N} & \cdots & r_{M1} & r_{M,2} & \cdots & r_{M,N} \end{bmatrix}^T \quad (3)$$

in which $r_{m,n}$ is the mth delayed sample at the nth antenna element. Filter coefficients corresponding to these samples are defined as:

$$\vec{\mathbf{w}} = \begin{bmatrix} w_{1,1} & w_{1,2} & \cdots & w_{1,N} & w_{2,1} & w_{2,2} & \cdots & w_{2,N} & \cdots & w_{M,1} & w_{M,2} & \cdots & w_{M,N} \end{bmatrix}^H \quad (4)$$

Hence, the space-time filter output is obtained as:

$$y = \vec{\mathbf{w}}^H \vec{\mathbf{r}} \tag{5}$$

in which H denotes the conjugate transpose. In order to suppress high power interference, power of the filter output should be minimized as:

$$\min_{\|\vec{\mathbf{w}}\|=1} E\left\{\|y\|^2\right\} = \vec{\mathbf{w}}^H \mathbf{R}_{\vec{\mathbf{r}}} \vec{\mathbf{w}} \tag{6}$$

where $E\{\}$ represents the statistical expectation and $R_{\vec{r}}$ is the spatial-temporal covariance or correlation matrix defined as:

$$\underset{MN \times MN}{\mathbf{R}_{\vec{\mathbf{r}}}} = E\left\{\vec{\mathbf{r}}\vec{\mathbf{r}}^H\right\} \tag{7}$$

In the following section, a projection matrix into the interference-free subspace is calculated based on this correlation matrix in order to mitigate interference.

3. PROPOSED RECEIVER STRUCTURE

The proposed receiver structure incorporates two interference suppression modes namely blind and distortionless and is shown in Figure 2.

The blind STP block calculates a projection matrix to the interference-free subspace from spatial temporal down-converted samples. Each row of the projection matrix provides a set of space-time filter coefficients. Assigning all these sets to the corresponding number of the space-time filters results in several interference-free outputs. Afterwards, only the satellite signals at one of these outputs are acquired and tracked, and the locally generated codes are used to despread the signals at other outputs. The steering vector estimator block employs the output of the blind STP block and pre-determined array calibration information to estimate the projected component of the steering vector of available satellite signals into the interference-free subspace. This enables the attitude determination block to calculate the array attitude parameters. By having attitude parameters, satellite azimuth and elevation angles, the steering vector of all satellites before projection (even if these are not initially acquired and tracked) can be accurately estimated. This information is then employed in the distortionless STP block. This block assigns different filter coefficients for different satellite signals to maintain the main lobe of the beam pattern in the direction of each signal and to decrease the induced distortions and biases on pseudorange measurements. In the following subsections, four processing blocks namely “Blind STP”, “Distortionless STP”, “Steering vector estimator” and “Attitude determination” are described. For the sake of simplicity in problem formulation, the organization of these subsections is based on the processing level in the receiver rather than the data flow in the receiver structure.

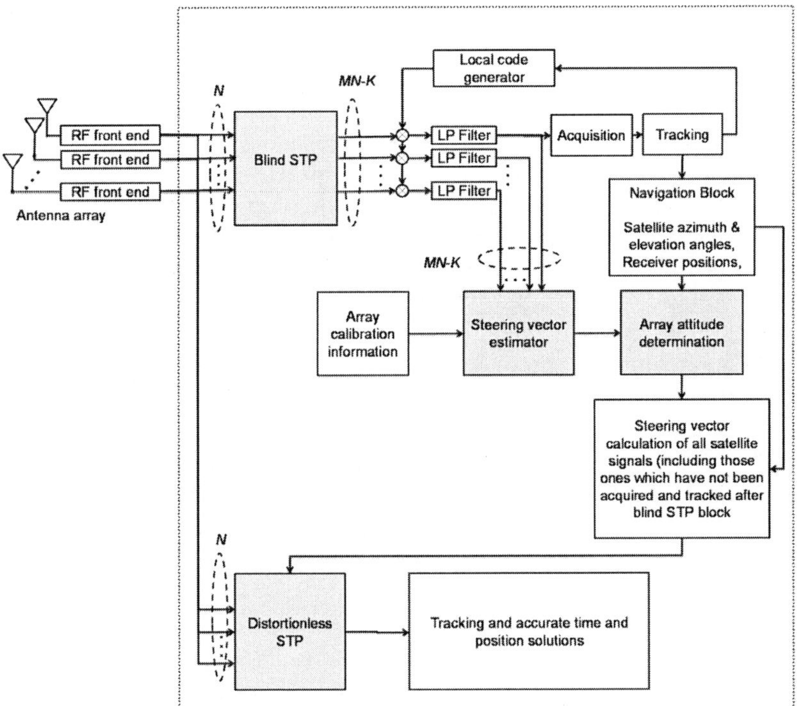

Figure 2. Structure of the proposed receiver.

3.1. Blind STP

Here the projection to an interference-free subspace is referred to as a blind method since satellite signal steering vectors are not considered in the filter design. In order to be destructive at correlator outputs, the power of the interference signal should be significantly higher than that of the GNSS and noise signals. This makes the interference subspace easily distinguished and estimated. A projection matrix into the interference-free subspace can be estimated by performing an Eigenvalue Decomposition (EVD) or Singular Value Decomposition (SVD) of Rr⃗ as:

$$\mathbf{R}_{\bar{\mathrm{r}}} = \begin{bmatrix} \underset{MN\times K}{\mathbf{U}_{Int}} & \underset{MN\times(MN-K)}{\mathbf{U}_{Null}} \end{bmatrix} \begin{bmatrix} \underset{K\times K}{\mathbf{\Lambda}_{Int}} & 0 \\ 0 & \underset{(MN-K)\times(MN-K)}{\mathbf{\Lambda}_{Null}} \end{bmatrix} \begin{bmatrix} \mathbf{U}_{Int}{}^{H} \\ \mathbf{U}_{Null}{}^{H} \end{bmatrix} \quad (8)$$

where $\mathbf{U}_{int}$ and $\mathbf{U}_{null}$ are eigenvector matrices of interference and the noise-plus-GNSS signal subspaces, respectively, and $\mathbf{\Lambda}_{int}$ and $\mathbf{\Lambda}_{Null}$ are corresponding eigenvalue matrices. In Equation (8), K indicates the rank of the interference subspace (without loss of generality, it is assumed that interfering signals are uncorrelated). Hence, a projection matrix into the reduced-rank interference-free subspace can be calculated as UHNull which is formed from $MN - K$ eigenvectors corresponding to the $MN - K$ smallest eigenvalues. In fact the filter gain vector w⃗, which minimizes the filter output power in Equation (6), belongs to this interference-free subspace. Applying this projection matrix to the received signal vector suppresses the interference; however, some satellite signals may become distorted or attenuated in the process.

3.2. Distortionless STP

The output of the blind space-time filter is basically a direction-frequency dependent response. Even if the filter completely nullifies interfering signals, the non-linearity behavior of its frequency response may result in distorted correlation functions, degraded acquisition and tracking performance, and biases on the GNSS measurements. In addition, since the spatial information of the satellite signals is not considered in the blind beamforming process, some signals may be unintentionally attenuated due to the nulls in the beam pattern. This may not be tolerable especially for high precision applications.

A novel approach for designing a distortionless space-time filter for interference mitigation is therefore proposed such that the filter impulse response becomes symmetric and linear in phase, and therefore the same delay is added to all pseudoranges. The proposed distortionless filter is based on subspace decomposition and incorporating the knowledge of the steering vectors. This section first introduces a possible structure for distortionless space-time filter and then estimates corresponding filter coefficients not only to mitigate interference signals but also to satisfy the distortionless condition.

For each satellite signal, the filter coefficients should be assigned differently, which is commensurate to the steering vector of the signal; however the structure of the filter remains the same. Therefore, without loss of generality, one GNSS signal is considered in the formulations. Assume that the power spectrum of the signal after despreading is defined as $p(f)$ and $\mathbf{h}(f)$ is the $N \times 1$ frequency response vector of the space-time filter. The correlation function for the signal after space-time filtering, despreading and Doppler removal can be written as [9]:

$$R(\tau, a) = \int_{-\infty}^{\infty} \mathbf{h}^H(f)\mathbf{a}p(f)e^{-j2\pi f\tau}df \quad (9)$$

where **a** is the steering vector defined in Equation (2). $p(f)$ is a symmetric function of frequency and therefore in the absence of $\mathbf{h}^H(f)\mathbf{a}$, the correlation function has a peak at τ. In order to avoid bias and to have a symmetric correlation function, $\mathbf{h}^H(f)\mathbf{a}$ should have a symmetric conjugate impulse response. This results in a linear phase space-time filter and consequently a symmetrically broadened correlation function for each satellite signal with a bias that is similar for all signals. Therefore, all pseudorange measurements are affected in the same way and position performance is not affected. The only performance degradation ance may be in tracking due to the broadness of the correlation functions. In this paper, the following structure for the distortionless STP is proposed:

$$\mathbf{h}(f) = \sum_{i=1}^{M} \mathbf{A}\mathbf{P}_{(i)}^{H}\mathbf{u}e^{j2\pi i f T_s} \quad (10)$$

where **A** is a diagonal matrix whose elements are equivalent to those of the steering vector and **u** is a vector weighting the rows of the projection matrix and P(i),i=1,2,..M can be obtained by partitioning the projection matrix of interest **P** as:

$$\mathbf{P} = \begin{bmatrix} \mathbf{P}_{(1)}\mathbf{A}^H & \mathbf{P}_{(2)}\mathbf{A}^H & \cdots & \mathbf{P}_{(M)}\mathbf{A}^H \end{bmatrix} \quad (11)$$

P and **u** should be determined to hold the following conditions:

(1)

Considering Equation (11), **P** should suppress the interfering signals and has the following structure:

$$\begin{cases} M\ odd & \begin{cases} \mathbf{P}_{(i)} = \left(\mathbf{P}_{(M-i+1)}\right)^* & ,\ i = 1,2,\ldots,\dfrac{M-1}{2} \\ \mathbf{P}_{(\frac{M+1}{2})} \quad is \quad real & \end{cases} \\ M\ even & \mathbf{P}_{(i)} = \left(\mathbf{P}_{(M-i+1)}\right)^* \qquad ,i = 1,2\ldots,\dfrac{M}{2} \end{cases} \quad (12)$$

where ()* denotes conjugate operation.

(2)Vector **u** is determined such as to maximize the correlation function and consequently maximize the SNR of the projected satellite signal.

This structure coerces the filter coefficients for each TDL to be conjugate symmetric and therefore to be a linear phase. Moreover, by applying matrix **A** in the projection matrix, the filter compensates for spatial phase differences of the received signal among antenna elements and consequently the resulting space-time filter from the combination of the linear phase TDLs remains linear phase and the cross correlation function is symmetrically broadened. In order to design the projection matrix in the form of Equation (12), the filter output is calculated as:

$$y = \vec{\mathbf{w}}^H \mathcal{A}^H \vec{\mathbf{r}} \quad (13)$$

in which the following condition on the filter coefficients is applied:

$$w_{m,n} = \left(w_{M-m+1,n}\right)^*,\ m = 1,\ldots,\frac{M}{2} \quad (14)$$

and A is a block diagonal matrix defined as:

$$\underset{NM\times NM}{\mathcal{A}} = \begin{bmatrix} \mathbf{A} & \mathbf{0} & \cdots & \underset{N\times N}{\mathbf{0}} \\ \mathbf{0} & \mathbf{A} & \cdots & \mathbf{0} \\ \vdots & \vdots & \ddots & \vdots \\ \mathbf{0} & \mathbf{0} & \cdots & \mathbf{A} \end{bmatrix} \tag{15}$$

In Equation (14), *M* is assumed even. The process can be easily repeated with a few modifications when *M* is odd. Considering Equation (14), half of the filter coefficients are related to each other and only half of the coefficients should be determined. In fact the DoF is halved to constrain the filter structure. The filter output in Equation (13) can be spilt into the sum of two terms as:

$$y = \vec{\mathbf{w}}_s^H \, \bar{\mathcal{A}}^H \, \vec{\mathbf{r}}_1 + \left(\vec{\mathbf{w}}_s^H \, \bar{\mathcal{A}} \, \vec{\mathbf{r}}_2 \right)^H \tag{16}$$

where:

$$\begin{aligned}
\vec{\mathbf{r}}_1 &= \begin{bmatrix} r_{1,1} & r_{1,2} & \cdots & r_{1,N} & r_{2,1} & r_{2,2} & \cdots & r_{2,N} & \cdots & r_{\frac{M}{2},1} & r_{\frac{M}{2},2} & \cdots & r_{\frac{M}{2},N} \end{bmatrix}^T \\
\vec{\mathbf{r}}_2 &= \begin{bmatrix} r^*_{M,1} & r^*_{M,2} & \cdots & r^*_{M,N} & r^*_{M-1,1} & r^*_{M-1,2} & \cdots & r^*_{M-1,N} & \cdots & r^*_{\frac{M}{2}+1,1} & r^*_{\frac{M}{2}+1,2} & \cdots & r^*_{\frac{M}{2}+1,N} \end{bmatrix}^T \\
\vec{\mathbf{w}}_s &= \begin{bmatrix} w_{1,1} & w_{1,2} & \cdots & w_{1,N} & w_{2,1} & w_{2,2} & \cdots & w_{2,N} & \cdots & w_{\frac{M}{2},1} & w_{\frac{M}{2},2} & \cdots & w_{\frac{M}{2},N} \end{bmatrix}^H \\
\underset{N\frac{M}{2}\times N\frac{M}{2}}{\bar{\mathcal{A}}} &= \begin{bmatrix} \mathbf{A} & \mathbf{0} & \cdots & \underset{N\times N}{\mathbf{0}} \\ \mathbf{0} & \mathbf{A} & \cdots & \mathbf{0} \\ \vdots & \vdots & \ddots & \vdots \\ \mathbf{0} & \mathbf{0} & \cdots & \mathbf{A} \end{bmatrix}
\end{aligned} \tag{17}$$

In order to suppress interference, $\vec{w}^H_s$ should be determined so that the filter output power is minimized. To calculate a closed form solution for $\vec{w}^H_s$, instead of minimizing the total power, the sum of power of each term in Equation (16) is minimized. This sub-optimization is expressed as:

$$\min_{\|\vec{\mathbf{w}}_s\|=1} E\left\{\left\|\vec{\mathbf{w}}_s^H \overline{\mathcal{A}}^H \vec{\mathbf{r}}_1\right\|^2 + \left\|\vec{\mathbf{w}}_s^H \overline{\mathcal{A}} \vec{\mathbf{r}}_2\right\|^2\right\} \tag{18}$$

The optimization in Equation (18) can be written as:

$$\min_{\|\vec{\mathbf{w}}_s\|=1} \vec{\mathbf{w}}_s^H \left(\overline{\mathcal{A}}^H \mathbf{R}_{\vec{\mathbf{r}}_1} \overline{\mathcal{A}} + \overline{\mathcal{A}} \mathbf{R}_{\vec{\mathbf{r}}_2\vec{\mathbf{r}}_1} \overline{\mathcal{A}}^H\right) \vec{\mathbf{w}}_s \tag{19}$$

in which:

$$\begin{aligned} \underset{\frac{M}{2}N\times\frac{M}{2}N}{\mathbf{R}_{\vec{\mathbf{r}}_1}} &= E\left\{\vec{\mathbf{r}}_1 \vec{\mathbf{r}}_1^H\right\} \\ \underset{\frac{M}{2}N\times\frac{M}{2}N}{\mathbf{R}_{\vec{\mathbf{r}}_2}} &= E\left\{\vec{\mathbf{r}}_2 \vec{\mathbf{r}}_2^H\right\} \end{aligned} \tag{20}$$

From Equation (19), it can be concluded that the projection matrix to the interference-free subspace (w⃗ sbelongs to this subspace) can be calculated as the following SVD problem:

$$\overline{\mathcal{A}}^H \mathbf{R}_{\vec{\mathbf{r}}_1} \overline{\mathcal{A}} + \overline{\mathcal{A}} \mathbf{R}_{\vec{\mathbf{r}}_2} \overline{\mathcal{A}}^H = \begin{bmatrix} \mathbf{U} & \mathbf{V} \end{bmatrix} \begin{bmatrix} \mathbf{\Lambda}_{Int} & 0 \\ 0 & \mathbf{\Lambda}_{Null} \end{bmatrix} \begin{bmatrix} \mathbf{U}^H \\ \mathbf{V}^H \end{bmatrix} \tag{21}$$

Hence a projection matrix in the form of **P** in Equation (12) can be obtained as:

$$\mathbf{P} = \begin{bmatrix} \mathbf{V}_{(1)}^H \mathbf{A}^H & \mathbf{V}_{(2)}^H \mathbf{A}^H & \cdots & \mathbf{V}_{(\frac{M}{2})}^H \mathbf{A}^H & \mathbf{V}_{(\frac{M}{2})}^T \mathbf{A}^H & \cdots & \mathbf{V}_{(2)}^T \mathbf{A}^H & \mathbf{V}_{(1)}^T \mathbf{A}^H \end{bmatrix} \tag{22}$$

where $\mathbf{V}_{(i)}, i = 1, 2, .. \frac{M}{2}$ can be obtained by partitioning the eigenvector matrix **V** in Equation (21) associated to the interference-free subspace as:

$$\underset{\frac{MN}{2}\times\left(\frac{MN}{2}-K\right)}{\mathbf{V}} = \begin{bmatrix} \underset{N\times\left(\frac{MN}{2}-K\right)}{\mathbf{V}_{(1)}} \\ \mathbf{V}_{(2)} \\ \vdots \\ \mathbf{V}_{(\frac{M}{2})} \end{bmatrix} \tag{23}$$

In the last step, **u** in Equation (10) is obtained to maximize the SNR for the projected signal. The problem of interest is finding the optimal combination of the columns of the projection matrix to satisfy the following relation:

$$\mathbf{a} = \mathbf{A}\mathbf{P}_{(i)}^{H}\mathbf{u}, \ \ i = 1, 2, \ldots, M \tag{24}$$

or equivalently:

$$\vec{\mathbf{a}} = \mathbf{P}^{H}\mathbf{u} \tag{25}$$

where a→ is defined as:

$$\underset{MN\times 1}{\vec{\mathbf{a}}} = \begin{bmatrix} \mathbf{a} \\ \mathbf{a} \\ \vdots \\ \mathbf{a} \end{bmatrix} \tag{26}$$

Taking Equations (9) and (10) into account, using Equation (24) to obtain **u** results in a constructive combination of the projected signal. From Equation (25), **u** can be obtained as:

$$\mathbf{u} = \mathbf{P}\vec{\mathbf{a}} \tag{27}$$

u can be also determined by allocating power to certain taps to partially control the shape of the resulting correlation function. This can be realized by changing the elements of $\vec{a}$ in Equation (25).

3.3. Steering Vector Estimator

In the previous section, it was assumed that the steering vectors are known. This section addresses the steering vector estimation from the output of the blind STP block. Herein, it is assumed that the array is calibrated which means constant uncertainties such as unequal cable lengths and coupling coefficients between antennas; also the antenna elements imperfections and their dependency on the signal AOA are assumed to be compensated [25]. Therefore for a calibrated array, the simplified baseband representation of the received signal in Equation (1) after despreading and projection can be written as:

$$\underset{(MN-K)\times 1}{\mathbf{y}} = \mathbf{U}_{Null}{}^{H}\vec{\mathbf{a}} + \mathbf{n} \tag{28}$$

in which $\vec{a}$ is defined in Equation (26) and **n** is the noise vector in the correlator outputs. During the tracking stage, the phase of the signal at one of the blind STP block outputs (reference output) is kept approximately zero and the phase of the signals at the other outputs are measured relative to that of the reference output (see Figure 2). For simplicity, amplitudes are also normalized with respect to the signal at the reference output. Therefore, the normalized signal vector that is actually measured from the multi-antenna receiver, denoted by $\tilde{y}$, is related to **y**as:

$$\tilde{\mathbf{y}} = \frac{\mathbf{y}}{\boldsymbol{\delta}^{T}\mathbf{U}_{Null}{}^{H}\vec{\mathbf{a}}} \tag{29}$$

in which $\underset{(MN-K)\times 1}{\boldsymbol{\delta}} = [1, 0, \cdots 0]^{T}$ (without loss of generality, the first output of the blind STP block is chosen as the reference for despreading the signals of other outputs). Given Equations (29) and (28), $\vec{a}$ can be estimated using the following minimization problem:

$$\min_{\substack{\vec{a}\\ \|\vec{a}\|=1}} E\left\{\left\|\tilde{\mathbf{y}}\boldsymbol{\delta}^T \mathbf{U}_{Null}{}^H \vec{\mathbf{a}} - \mathbf{U}_{Null}{}^H \vec{\mathbf{a}}\right\|^2\right\} \tag{30}$$

in which the constraint avoids the trivial solution that is $\vec{a}=0$. After some matrix manipulations, one can verify that $\vec{a}$ can be estimated from the following eigenvalue decomposition problem as:

$$\min_{\substack{\vec{a}\\ \|\vec{a}\|=1}} \vec{\mathbf{a}}^H \mathbf{U}_{Null} \mathbf{D} \mathbf{U}_{Null}{}^H \vec{\mathbf{a}} \tag{31}$$

where **D** is defined as:

$$\mathbf{D} = E\left\{\left(\tilde{\mathbf{y}}\boldsymbol{\delta}^T - \mathbf{I}_{(MN-K)}\right)^H \left(\tilde{\mathbf{y}}\boldsymbol{\delta}^T - \mathbf{I}_{(MN-K)}\right)\right\} \tag{32}$$

and **I** is the MN−K×MN−K identity matrix. Assume that $\vec{a}$ is written with respect to the parallel ($a_{\|}$) and orthogonal ($a_{\perp}$) components in the projected interference-free subspace as:

$$\vec{\mathbf{a}} = \mathbf{U}_{Null}\mathbf{a}_{\|} + \mathbf{U}_{Int}\mathbf{a}_{\perp} \tag{33}$$

By substituting Equation (33) in Equation (31), the optimization problem becomes:

$$\min_{\substack{\mathbf{a}_{\|}\\ \|\mathbf{a}_{\|}\|=1}} \quad \mathbf{a}_{\|}{}^H \mathbf{D} \mathbf{a}_{\|} \tag{34}$$

D has a rank deficiency of 1 and hence by SVD of **D**, $a_{\|}$ is obtained as the eigenvector corresponding to the smallest eigenvalue. As expected, $a_{\perp}$ cannot be estimated from the optimization problem since it is orthogonal to the projection matrix. For some satellite signals, $\|a_{\|}\|$ can be significant whereas for some of them, it is a small value. For better estimation of attitude parameters, only estimated steering vectors with larger values of ||a||| are employed. To measure how close the estimated steering vector (a^) is to the true steering vector, the following criterion is adopted:

$$Cr = \left\| \hat{\mathbf{a}}.\hat{\mathbf{a}}^{*} - \mathbf{a}.\mathbf{a}^{*} \right\| = \left\| \hat{\mathbf{a}}.\hat{\mathbf{a}}^{*} - \mathbf{1} \right\| \tag{35}$$

where (·) denotes the Hadamard product. This criterion simply determines how much the structure of the estimated a^ is similar to the structure of a typical steering vector defined in Equation (2). The value of *Cr*decreases if the structure of a^ becomes similar to the one defined in Equation (2) and becomes zero if a^ has the exact same structure.

3.4. Attitude Determination

Precise attitude determination is essential in many applications. Generally, differential carrier phase measurements from several receivers with precisely determined baselines or/and INS are employed for attitude determination. Since the proposed receiver is already equipped with an antenna array, herein beamforming using a calibrated antenna array is employed for attitude determination [26]. Moreover, in contrast to the standard attitude determination based on carrier phase measurements, in this method one GNSS signal along with an adequate number of antenna elements is enough for attitude determination. This is one of the main advantages of array-based attitude determination techniques, especially in interference environments where the number of available signals is still low after blind STP interference mitigation. It should be considered that the performance of the proposed method could become affected by multipath propagation. Herein, multipath propagation was not taken into account in the mathematical model of the proposed method. In [26], it was shown that for heading estimation in open sky conditions an approximate agreement of 1° with a SPAN GPS-INS system serving as reference was achieved; the experiment is repeated in interference environments here by applying the proposed receiver structure.

Considering Equation (2), $\mathbf{d}_l$ (a vector pointing to the *l*th satellite in the body frame coordinate system) can be derived as:

$$\mathbf{d}_l = \frac{K}{2\pi}(\mathbf{Z}^H\mathbf{Z})^{-1}\mathbf{Z}^H \angle\hat{\mathbf{a}}_l,\ l = 1, 2, \dots, L \tag{36}$$

in which a^l is a vector indicating the phase of the estimated steering vector elements of the *l*th satellite, L is the number of available satellites and **Z** is the array configuration matrix defined as:

$$\mathbf{Z} = \begin{bmatrix} \mathbf{z}_1 & \mathbf{z}_2 & \cdots & \mathbf{z}_N \end{bmatrix}^T \qquad (37)$$

Moreover, from the approximate position of the receiver and ephemeris information, this vector can be also expressed in the East-North-Up (ENU) coordinate system. Unit vectors pointing to all satellites expressed in the ENU and body frame coordinate systems are related to each other through:

$$\begin{aligned} \mathbf{E}^{ENU} &= \mathbf{R}_B^{ENU}\mathbf{E}^B, \\ \mathbf{E}^B &= \begin{bmatrix} \mathbf{d}_1 & \mathbf{d}_2 & \cdots & \mathbf{d}_L \end{bmatrix} \end{aligned} \qquad (38)$$

where EENU is a matrix consists of satellite pointing vectors in the ENU coordinate systems and RENUB is the transformation matrix from the body frame to the ENU coordinate system [27]. It is convenient to express this transformation by using an Euler angles parameterization that has a direct physical interpretation. In this way, RENUB is formed based on three angles r, p and h referring to the roll, pitch and heading (yaw) angles and can be expressed as:

$$\mathbf{R}_B^{ENU} = \begin{bmatrix} \cos(h)\cos(r)-\sin(r)\sin(p)\sin(h) & -\sin(h)\cos(p) & \cos(h)\sin(r)+\sin(h)\sin(p)\cos(r) \\ \sin(h)\cos(r)+\sin(r)\sin(p)\cos(h) & \cos(h)\cos(p) & \sin(h)\sin(r)-\cos(h)\sin(p)\cos(r) \\ -\sin(r)\cos(p) & \sin(p) & \cos(p)\cos(r) \end{bmatrix} \qquad (39)$$

For a given EB and EENU, estimating the transformation matrix RENUB from Equation (38) is equivalent to finding a solution to a least squares estimation problem. Therefore, by having attitude parameters, satellite azimuth and elevation angles, the steering vector of each satellite signal before blind projection can be accurately estimated even if it is not acquired or tracked after the blind STP filter. It is worth mentioning that due to the long distance between satellites and the receiver, azimuth and elevation angles can be estimated with enough accuracy even when degraded positioning performance occurs due to blind STP filtering.

4. EXPERIMENTAL RESULTS

Due to frequency regulations, outdoor radio frequency (RF) power transmission in the GNSS frequency bands is prohibited. Therefore, special considerations have to be taken into account while testing the performance of anti-interference techniques. Some previous work has suggested combining interference signals to GNSS signals through wires. However, for an array antenna, this type of test requires many combiners, cables and connectors and moreover control on the angle of arrival of GNSS and interferer signals would be difficult. Herein for testing and evaluating the performance of the proposed method, interference has been generated in software and added to the digitized GNSS samples.

Figure 3. Data collection scenario and setup.

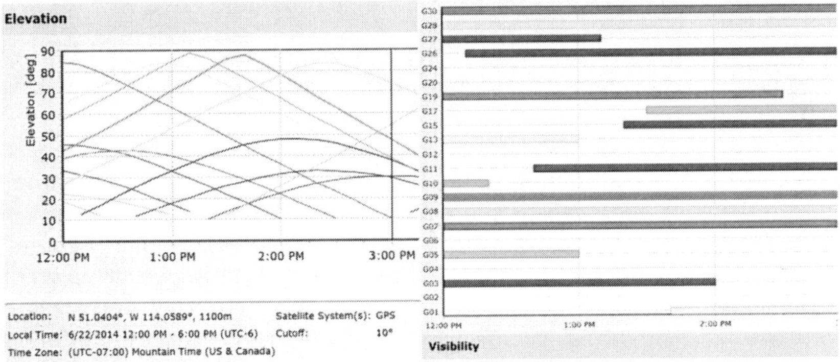

Figure 4. Availability and elevation angles of satellites during test (http://www.trimble.com).

The test set up and trajectory are shown in Figure 3; GPS L1 C/A signals were collected using an array of six antennas. The data collection was performed with an open view of satellites. A circular trajectory was driven for the calibration process to cover all azimuth angles. In order to cover a wide range of elevation angles, the circular trajectory was repeated several times over a two-hour interval. The data collection interval was long enough to allow satellites to move significantly in the sky to cover most elevation angles. Satellite visibility and elevation during the data collection are shown in Figure 4. Satellite Pseudo Random Noise (PRN) codes, elevation and azimuth angles used for the proposed space-time interference mitigation method are also shown in Table 1. The antenna array was mounted on the top of a vehicle and the six antenna elements were connected to a phase coherent six-channel Fraunhofer/TeleOrbit RF front-end. The received signals were then sampled, down converted and stored for post processing. Moreover, a NovAtel SPAN™ LCI system, which includes a NovAtel SPAN® enabled GNSS/INS receiver (SPAN SE) and a tactical grade IMU LCI was used as a reference to provide reference heading values and positions for comparison purpose. IMU measurements were sent to the receiver where a coupled GNSS/INS position, velocity and attitude solution was generated. Raw GPS data was also collected under Line of Sight (LOS) conditions using another receiver as a base station to provide differential positioning. The data collected by SPAN and the base station file were then fed to the NovAtel Inertial Explorer®post-processing software to produce accurate reference orientation angles (the estimated standard deviations for position in each direction are below 5 cm and the estimated standard deviations for attitude parameters are below 0.1°).

The precise antenna array calibration method proposed in [25] was employed to calibrate the antenna array. In this method, a two-stage optimization for precise calibration is used in the form of two EVD problems. In the first stage, constant uncertainties are estimated whereas in the second stage the dependency of each antenna element gain and phase patterns to the received signal AOA is considered for refined calibration. An open source MATLAB-based single antenna software receiver [28] was modified as a multi-antenna receiver where the acquisition, tracking and position solution parts of the original software were modified. The interference mitigation units were added to the receiver and the structure of the receiver was changed from a "single-antenna tracking" to a "multi-antenna multi-delay tracking" receiver employing both spatial and temporal processing.

Table 1. PRNs used during test and corresponding azimuth and elevation angles and C/N_0 after each interference mitigation stage.

PRN	Azimuth (degrees)	Elevation (degrees)	C/N_0 Blind STP (dB-Hz)	C/N_0 Distortionless STP (dB-Hz)
1	342	19	39.1	44.1
3	45	6	37.7	46.4
7	321	31	42.6	44.7
8	328	55	50.5	49.1
9	308	72	54.1	54.6
11	4	32	35.1	43.5
15	139	25	45.7	44.5
17	256	30	37.8	49.9
19	46	18	41.1	49.8
26	179	48	—	50.2
28	159	81	39.0	50.4
30	319	62	53.6	52.3

In the first test, one CW signal as an interference signal at GPS L1 centre frequency is added to the received signals. The elevation and azimuth of the interference signal are 7.5° and 120° and the interference-to-noise density ratio (I/N_0) is 90 dB-Hz. The number of the TDL taps is 4. Figure 5 shows the array gain pattern after applying the blind projection matrix. As shown, a deep null is placed in the direction of the interference; however, since the steering vectors of the satellite signals have not been employed in the space-time filter structure, some of the desired signals are also unintentionally attenuated or even nulled out.

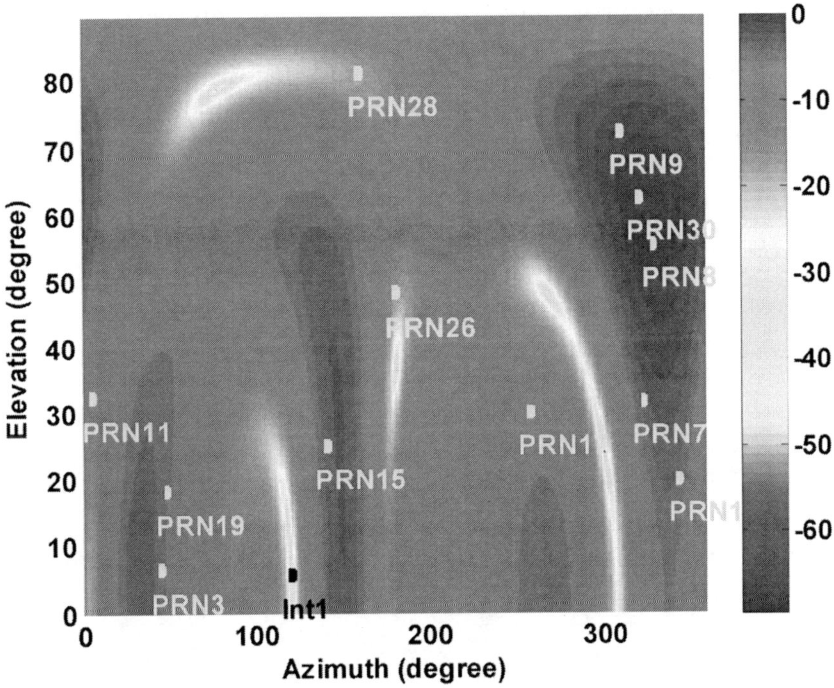

Figure 5. Normalized antenna array gain pattern at the interference frequency.

In the second stage of the proposed receiver and after employing the steering vectors, the STP filter coefficients can be determined to not only nullify the interference but also steer the main lobe of the array gain pattern into the direction of the desired signal as shown in Figure 6. Considering Equation (9), for a space-time filtering, the gain pattern (in dB) is calculated as $10\log^{\left(\left|\mathbf{h}^{H}(f)\mathbf{a}\right|^{2}\right)}$ where hH(f)a is a response of the filter to the impinging signal with the steering vector a and frequency *f*. In fact, this gain pattern determines the space-time filter gain at a specific frequency, azimuth and elevation angles.

In order to obtain an actual sense of the improvement achieved, Figure 7 compares C/N_0 values between blind and distortionless STP filtering for the same 20 s of received satellite signals. Average C/N_0 values are also shown in Table 1. Since PRN 8, 9 30 are located close to the main lobe of the gain pattern in the blind filter, after applying the distortionless filter, their C/N_0 values are slightly decreased or not considerably changed. This is due to the fact that the DoF is halved to maintain linearity of the phase response for the distortionless filter. However, the average C/N_0 of all PRNs is increased approximately by 5 dB-Hz (up to 12 dB-Hz for PRNs 28 and 17). In addition, PRN 26 was significantly attenuated and denied after blind filtering but it could be acquired and tracked by employing the distortionless STP filtering.

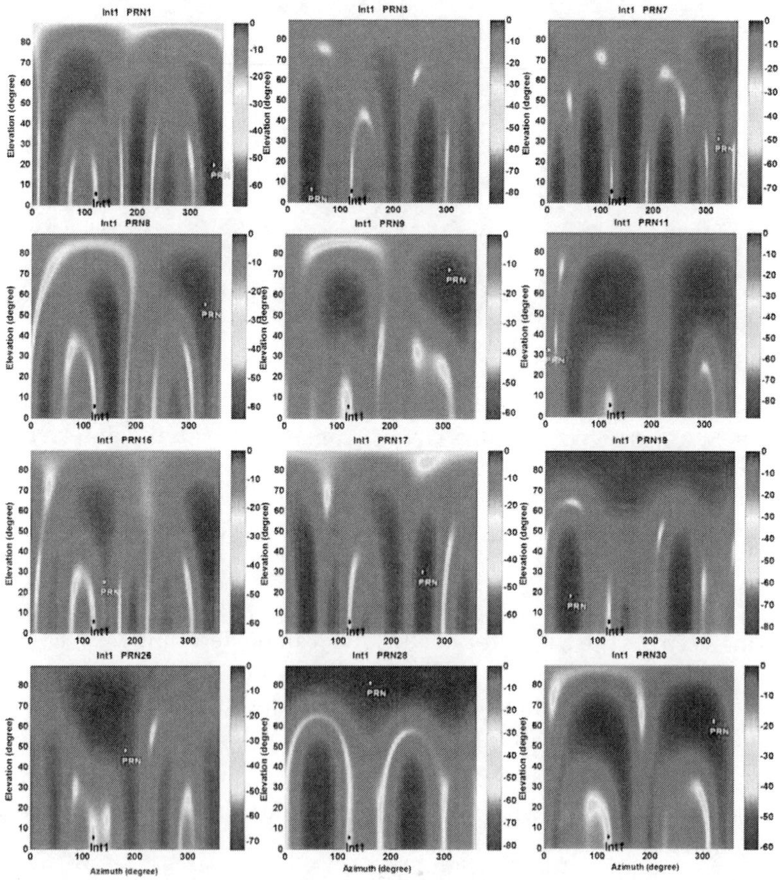

Figure 6. Normalized antenna array gain patterns at the interference frequency for the proposed distortionless STP filtering.

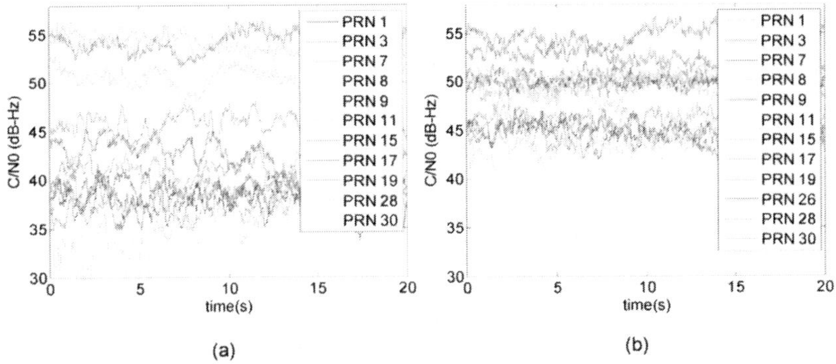

Figure 7. (**a**) Measured C/N_0 after employing the blind STP filtering (**b**) and after employing the proposed distortionless STP filtering.

The performance of the attitude determination block is now reported. Recall that the SPAN LCI system was used as a reference to provide external and independent heading and attitude parameters. Assuming a horizontal motion and considering Equations (38) and (39), the heading angle can be estimated using the following simplified relation:

$$\mathbf{E}^{ENU} = \begin{bmatrix} \cos(h) & \sin(h) & 0 \\ -\sin(h) & \cos(h) & 0 \\ 0 & 0 & 1 \end{bmatrix} \mathbf{E}^{B} \tag{40}$$

Based on the criterion in Equation (35), PRNs 30, 9 and 8 are chosen for heading determination. In fact, as shown in Figure 5, these PRNs are located close to the main lobe of the array gain pattern. Therefore, employing these signals leads to more accurate estimates of steering vectors and consequently heading angles. In this test, heading angles were calculated within a 40 s interval in static mode.

Figure 8a displays two error types. The first error (StV) compares the estimated steering vector from the proposed receiver (a^) with that measured from the SPAN system (a) and the second error (StVProj) compares a^with the parallel projected component of the measured steering vector by the SPAN system (a‖). These errors are calculated as:

$$StV\ \ error = \frac{\left\| \hat{\mathbf{a}} - \mathbf{a} \right\|}{\left\| \mathbf{a} \right\|}$$
$$StV\,\mathrm{Proj}\ \ error = \frac{\left\| \hat{\mathbf{a}} - \mathbf{a}_{\|} \right\|}{\left\| \mathbf{a}_{\|} \right\|} \tag{41}$$

Results show that the StVProj errors are less than the StV errors, which verifies the fact that a^ is an estimate of the parallel component a‖, and the orthogonal component a⊥ cannot be estimated (see Equations (33) and (34)).Figure 8a also shows the value of *Cr* calculated in Equation (35), which reveals that the structure of estimated steering vectors is close to that of a true steering vector as defined in Equation (2). Figure 8b compares the estimated heading angles for each of these three PRNs with those obtained with the SPAN system. Figure 9shows the errors of the estimated heading angles considering all these PRNs, showing an approximate agreement of 1.5° for heading estimates. In practice and in a dynamic operation environment, a receiver should constantly select among satellite signals with lower *Cr* values for estimating attitude parameters. Selecting proper satellite signals also depends on their elevation angles. In fact, satellites with higher elevation angles show poorer accuracy for heading estimation such that the satellite located at the receiver's zenith cannot be used to sense the horizontal motion. However, the errors due to multipath and noise are higher for satellites with low elevation angles. Therefore, mid elevation satellites could be a good option for extracting attitude parameters.

As mentioned before, the blind STP may distort the correlation functions and produce biases in the pseudorange measurements. Moreover, some satellites may be unintentionally nullified, which in turn reduces the Dilution of Precision (DOP). These biases and resulting attenuations degrade the position solution accuracy.Table 2 lists errors in the ENU coordinate system for three interference scenarios for distortionless and blind filters.

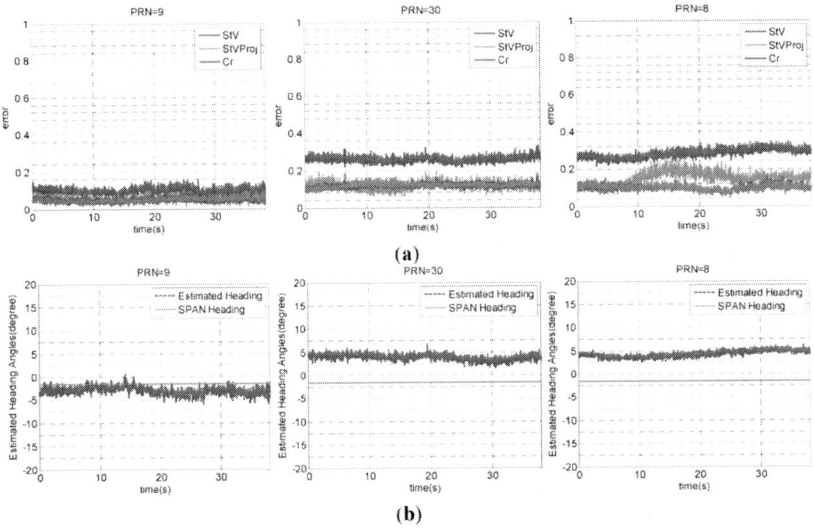

Figure 8. (**a**) Comparison of estimated steering vectors (**b**) and estimated heading angles for PRN 9, 30 and 8.

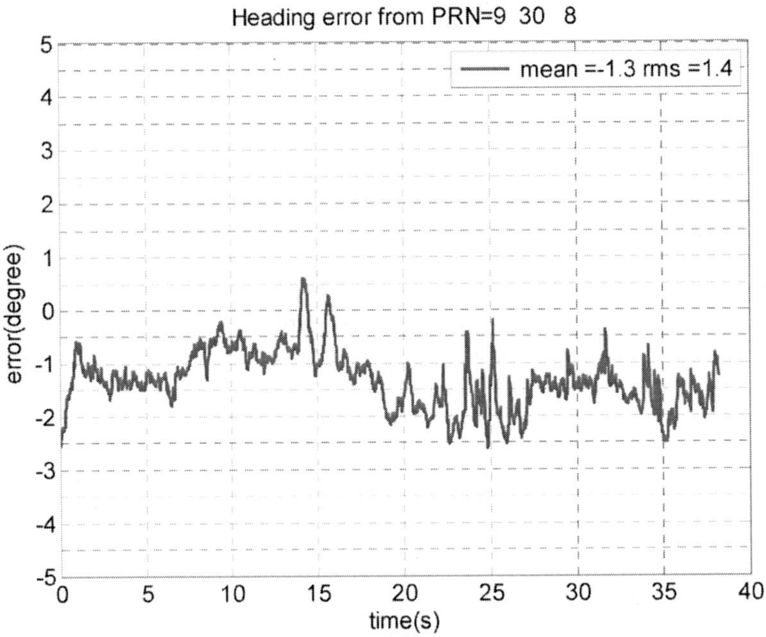

Figure 9. Error in heading angles obtained from PRNs 9, 30 and 8.

Table 2. Results for different interference scenarios.

20 s of data in the static mode		Scenario 1 One CW Interfernce	Scenario 2 One CW & One Wideband Interfernce	Scenario 3 Six CW Interfernce
		$I/N_0 = 90$ dB-Hz TDL = 4	$I/N_0 = 90$ dB-Hz TDL = 4	$I/N_0 = 90$ dB-Hz TDL = 6
Blind STP ENU error (m)	E (mean,rms)	(−0.8,0.9)	(−4.0,4.8)	~(500,500)
	N (mean,rms)	(−0.7,1.3)	(−8.1,9.6)	~(1000,1000)
	U (mean,rms)	(−3.9,4.1)	(−8.0,8.2)	~(1000,1000)
MPDR ENU error (m)	E (mean,rms)	(−0.6,0.6)	(−0.8,1.0)	-
	N (mean,rms)	(−0.9,1.0)	(−1.2,1.3)	-
	U (mean,rms)	(−1.9,2.0)	(−2.5,2.7)	-
STP MPDR ENU error (m)	E (mean,rms)	(−0.8,0.9)	(−3.5,3.6)	(−70.2,75.0)
	N (mean,rms)	(−1.7,1.8)	(−4.0,4.1)	(116.9,126.5)
	U (mean,rms)	(−8.4,8.4)	(−13.5,13.6)	(−148.8,149.2)
Proposed Distortionless STP ENU error (m)	E (mean,rms)	(−0.4,0.6)	(−1.2,1.5)	(−0.1,0.3)
	N (mean,rms)	(−0.4,0.5)	(−0.6,0.7)	(−0.5,0.6)
	U (mean,rms)	(−0.2,1.0)	(−0.5,0.7)	(2.7,2.7)
Blind STP Average C/N_0 (dB-Hz)		43.3	44	41.9
MPDR Average C/N_0 (dB-Hz)		51.4	49.6	-
STP MPDR Average C/N_0 (dB-Hz)		51.4	50.0	46.2
Proposed Distortionless STP Average C/N_0 (dB-Hz)		48.3	44.2	45.7
Blind STP Number of PRNs acquired & tracked		11	8	5
MPDR Number of PRNs acquired& tracked		12	12	-
Proposed Distortionless STP & STP MPDR Number of PRNs acquired& tracked		12	12	12

The I/N_0 for each wideband or narrowband signal is 90 dB-Hz. The interference signals are spread over the GPS L1 frequency band and have incident elevation angles in the range of 2° to 10°. In this table, the number of acquired and tracked satellites after employing the proposed blind and distortionless STP filters and the resulting average C/N_0 are also shown. In the simple interference scenario (Scenario 1), the number of acquired satellite signals is slightly different between blind and distortionless STPs and the resulting improvement in position solution is due to decreasing the distortion and bias on correlation functions; in the harsh interference scenario (Scenario 3), improvement occurs also because of increasing the number of acquired satellites and DOP improvement. In general the improvement obtained with the proposed methods depends on the interference scenario, antenna array configuration, number of taps in TDLs, calibration accuracy, power and direction of GNSS and interference signals, DOP and other factors which have not been analyzed in this paper.

In order to highlight the advantages of the purposed distortionless STP filter, its performance is compared to the conventional MPDR (space only processing) and modified STP MPDR beamformers briefly introduced as follows:

MPDR Beamformer: this beamformer has been widely employed in array based applications. In GNSS applications, as long as the array is calibrated and its orientation is determined, MPDR is one of the powerful approaches available to suppress interfering signals while maintaining desired signals. The optimization problem for the MPDR beamformer can be expressed as:

$$\begin{aligned} &\underset{\mathbf{w}}{Min} \quad \mathbf{w}^H \mathbf{R} \mathbf{w} \\ &\mathbf{a}^H \mathbf{w} = 1 \end{aligned} \tag{42}$$

where R is the spatial correlation matrix, a is the steering vector defined in Equation (2) and w is the array weighting vector. The goal is to minimize the power subject to the constraint. This minimization problem can be solved by using a Lagrange multiplier approach. The optimal gain vector is obtained as [5]

$$\mathbf{w} = \mathbf{R}^{-1}\mathbf{a}\left(\mathbf{a}^H \mathbf{R}^{-1}\mathbf{a}\right)^{-1} \tag{43}$$

STP MPDR Beamformer*:* this beamformer is the extended version of the MPDR beamformer for space–time processing and also employed in several papers (e.g., [18,19,20]). The degree of freedom of the array compared to MPDR beamformer is increased. The optimization problem for the MPDR beamformer can be expressed as:

$$\begin{aligned} &\underset{\bar{\mathbf{w}}}{Min} \quad \bar{\mathbf{w}}^H \mathbf{R}_T \bar{\mathbf{w}} \\ &\mathbf{c}^H \bar{\mathbf{w}} = 1 \end{aligned} \tag{44}$$

where Rr→ and w→ are defined in Equations (4) and (6) and the vector c is defined as:

$$\underset{NK\times 1}{\mathbf{c}} = \begin{bmatrix} \mathbf{a}^T & \underset{N\times 1}{\mathbf{0}^T} & \cdots & \underset{N\times 1}{\mathbf{0}^T} \end{bmatrix}^T \tag{45}$$

In this filter, in order to force the beamformer to have a fixed group delay, only one group of tap gains with a certain delay (without loss of generality here the first group is chosen) is required to pass the satellite signal undistorted. However, the filter response is not necessarily linear in phase and cross correlation functions may be asymmetrical and distorted. This minimization problem can be solved using a Lagrange multiplier approach.

For the MPDR beamformer, the array DoF, indicating the number of unwanted signals that can be nullified, is equal to the number of antenna elements minus one. Since this beamformer is only based on spatial processing, cross correlation functions and position solutions do not experience any distortion due to time filtering. ENU error results for this beamformer, reported in Table 2, verify the fact that the MPDR beamformer can suppress the first two scenarios without generating significant ENU errors but is not able to mitigate six uncorrelated narrowband interference signals. However, the proposed distortionless STP filter not only provides extra DoF for narrowband interference mitigation but also keeps the cross correlation functions undistorted. The results also show that the proposed STP filter outperforms the STP MPDR beamformer in terms of distortions and ENU errors. Contrary to the STP MPDR beamformer, the proposed STP filter is designed not only to maximize the SNR but also to be linear in phase. In other words, the proposed STP filter is spatially and temporally distortionless. It should be noted that for the proposed STP filter the resulting average C/N_0 values are slightly lower than those of the space-only and STP MPDR beamformers because of applying a restriction on the structure of the filter. However, as shown in Table 2, the positioning performance of the proposed approach considerably outperforms that of the STP MPDR method.

5. CONCLUSIONS

An array-based GNSS receiver capable of interference mitigation and attitude determination was proposed. Two types of space-time filtering were employed in the structure of the receiver, namely blind and distortionless STP methods. The blind STP block projects the received signals into the reduced rank interference-removed subspace without any consideration of GNSS distortion, bias and attenuation. It was shown that the array attitude parameters could be estimated from the output of the blind STP block and this information were then employed to calculate the steering vector of all available satellite signals. Using the calculated steering vectors, the proposed distortionless STP block was able to assign specific filter coefficients to the signal of each satellite and shape the array gain pattern to simultaneously nullify the interference signals and put the main lobe toward the direction of each desired GNSS satellite while maintaining the filter phase response linear. Simulation and experimental results showed that the proposed distortionless STP method acquired and tracked a higher number of satellites compared to the blind STP. Moreover, the distortionless STP resulted in less distortion and bias on pseudorange measurements and consequently provided more accurate position and timing solutions (the improvement of a few metres in ENU errors in the simple interference scenario to a few hundred metres in a challenging one). Experimental results also showed that the positioning performance of the proposed approach is considerably higher than that of the STP MPDR method. The proposed STP receiver structure removes the need for a priori knowledge of the antenna array attitude since it is able to accurately estimate it using available GNSS satellites. The results reported here were limited to a few specific cases to show the applicability of the proposed receiver structure. The next step could be the design of an approach to adaptively nullify interference signals to allow the proposed receiver to operate in dynamic mode.

AUTHOR CONTRIBUTIONS

The main idea of the paper was conceived by Saeed Daneshmand. Technical discussions, experimental tests and data processing were performed by Saeed Danehsmand, Ali Jafarnia Jahromi and Ali Broumandan. The software was developed by Saeed Danehsmand and Ali Broumandan. Gérard Lachapelle supervised the work and provided vital technical feedback. All the authors contributed to the writing of the manuscript.

ACKNOWLEDGMENTS

The work reported in this paper has been partly performed under a contract of the European Space Agency in the frame of the Announcement of Opportunities on GNSS Science and Innovative Technology within the European GNSS Evolutions Programme (EGEP). The views presented in the paper represent solely the opinion of the authors. This research was also supported by iCORE/Alberta Innovates Technology Futures.

REFERENCES

1. Madhani, P.H.; Axelrad, P.; Krumvieda, K.; Thomas, J. Application of successive interference cancellation to the GPS pseudolite near-far problem. *IEEE Trans. Aerosp. Electron. Syst.* **2003**, *39*, 481–488.

2. Badke, B.; Spanias, A. Partial band interference excisions for GPS using frequency-domain exponents. In Proceedings of the IEEE International Conference on Acoustics, Speech, and Signal Processing (IEEE ICASSP 2002), Orlando, FL, USA, 13–17 May 2002; pp. 3936–3939.

3. Amin, M.G.; Zhao, L.; Lindsey, A.R. Subspace array processing for the suppression of FM jamming in GPS receivers. *IEEE Trans. Aerosp. Electron. Syst.* **2004**, *40*, 80–92.

4. Capon, J. High-Resolution Frequency-Wavenumber Spectrum Analysis. *IEEE Proc.* **1969**, *57*, 1408–1419.

5. Van Trees, H.L. *Optimum Array Processing, Detection, Estimation, and Modulation Theory Part. IV*; John Wiley & Sons: New York, NY, USA, 2002; pp. 428–699.

6. Daneshmand, S.; Broumandan, A.; Nielsen, J.; Lachapelle, G. Interference and multipath mitigation utilising a two-stage beamformer for global navigation satellite systems applications. *IET Radar Sonar Navig.* **2013**, *7*, 55–66.

7. Cuntz, M.; Konovaltsev, A.; Sgammini, M.; Hattich, C.; Kappen, G.; Meurer, M.; Hornbostel, A.; Dreher, A. Field Test: Jamming the DLR Adaptive Antenna Receiver. In Proceedings of the 24th International Technical Meeting of The Satellite Division of the Institute of Navigation (ION GNSS 2011), Portland, OR, USA, 20–23 September 2011; pp. 384–392.

8. Seco-Granados, G.J.; Fernández-Rubio, A.; Fernández-Prades, C. ML estimator and hybrid beamformer for multipath and interference mitigation in GNSS receivers. *IEEE Trans. Signal. Process.* **2003**, *53*, 1194–1208.

9. Fante, R.; Vaccaro, J.J. wideband cancellation of interference in a GPS receive array. *IEEE Trans. Aerosp. Electron. Syst.* **2000**, *36*, 549–564.

10. Zoltowski, M.D.; Gecan, A.S. Advanced adaptive null steering concepts for GPS. In Proceedings of the 1995 IEEE Conference

Record, Military Communications Conference (MILCOM'95), San Diego, CA, USA, 5–8 November 1995; pp. 1214–1218.

11. McDonald, K.F.; Costa, P.J.; Fante, R.L. Insights into Jammer Mitigation via Space-Time Adaptive Processing. In IEEE Position Location and Navigation Symposium, Atlanta, GA, USA, 22–26 April 2006; pp. 191–198.

12. Myrick, W.L.; Goldstein, J.S.; Zoltowski, M.D. Low Complexity Anti-jam Space-Time Processing for GPS. In Proceedings of the IEEE International Conference on Acoustics, Speech, and Signal Processing (IEEE ICASSP 2001), Salt Lake City, UT, USA, 7–11 May 2001; pp. 2233–2236.

13. Moelker, D.J.; van der Pol, E.; Bar-Ness, Y. Adaptive antenna arrays for interference cancellation in GPS and GLONASS receivers. In Proceedings of IEEE Position Location and Navigation Symposium, Atlanta, GA, USA, 22–26 April 1996; pp. 191–198.

14. Ramos, J.; Zoltowski, M.; Urgos, M. Robust blind adaptive array, a prototype for GPS. In Proceedings of the IEEE International Symposium on Phased Array Systems and Technology, Boston, MA, USA, 15–18 October 1996; pp. 406–410.

15. Agamata, B.N. Time Domain Adaptive Filter Implementation of a Minimum Variance Distortionless Response (MVDR) Beamformer for a Sensor Array. In Proceedings of the 4th International Technical Meeting of the Satellite Division of The Institute of Navigation (ION GPS 1991), Albuquerque, NM, USA, 11–13 September 1991; pp. 443–452.

16. O'Brien, A.J.; Gupta, I.J. Mitigation of adaptive antenna induced bias errors in GNSS receivers. *IEEE Trans. Aerosp. Electron. Syst.* **2011**, *47*, 524–538.

17. Fante, R.L.; Fitzgibbons, M.P.; McDonald, K.F. Effect of Adaptive Array Processing on GPS Signal Crosscorrelation. In Proceedings of the 17th International Technical Meeting of the Satellite Division of The Institute of Navigation (ION GNSS 2004), Long Beach, CA, USA, 21–24 September 2004; pp. 579–583.

18. De Lorenzo, D.S.; Lo, S.C.; Enge, P.K.; Rife, J. Calibrating adaptive antenna arrays for high-integrity GPS.*GPS Solut.* **2012**, *16*, 221–230.

19. Myrick, W.L.; Zoltowski, M.; Goldstein, J.S. Exploiting conjugate symmetry in power minimization based pre-processing for GPS: Reduced complexity and smoothness. In Proceedings of IEEE International Conference on Acoustics, Speech, and Signal Processing (ICASSP 2000), Istanbul, Turkey, 5–9 June 2000; pp. 2833–2836.

20. Church, C.M.; Gupta, I.J.; O'Brien, A. Adaptive Antenna Induced Biases in GNSS Receivers. In Proceedings of the 63rd Annual Meeting of The Institute of Navigation (2007), Cambridge, MA, USA, 23–25 April 2007; pp. 204–212.

21. Chen, Y.-H.; Juang, J.-C.; Seo, J.; Lo, S.; Akos, D.M.; de Lorenzo, D.S.; Enge, P. Design and implementation of real-time software radio for anti-interference GPS/WAAS sensors. *Sensors* **2012**, *12*, 13417–13440.

22. Konovaltsev, A.; de Lorenzo, D.S.; Hornbostel, A.; Enge, P. Mitigation of Continuous and Pulsed Radio Interference with GNSS Antenna Arrays. In Proceedings of the 21st International Technical Meeting of the Satellite Division of The Institute of Navigation (ION GNSS 2008), Savannah, GA, USA, 16–19 September 2008; pp. 2786–2795.

23. Daneshmand, S. GNSS Interference Mitigation Using Antenna Array Processing. Ph.D. Thesis, Department of Geomatics Engineering, University of Calgary, Calgary, AB, Canada, 2013.

24. Markel, M.; Sutton, E.; Zmuda, H. An Antenna Array-Based Approach to Attitude Determination in a Jammed Environment. In Proceedings of the 14th International Technical Meeting of the Satellite Division of The Institute of Navigation (ION GPS 2001), Salt Lake City, UT, USA, 11–14 September 2001; pp. 2914–2926.

25. Daneshmand, S.; Sokhandan, N.; Zaeri-Amirani, M.; Lachapelle, G. Precise Calibration of a GNSS Antenna Array for Adaptive Beamforming Applications. *Sensors* **2014**, *14*, 9669–9691.

26. Daneshmand, S.; Sokhandan, N.; Lachapelle, G. Precise GNSS Attitude Determination Based on Antenna Array Processing. In Proceedings of the 27th International Technical Meeting of The Satellite Division of the Institute of Navigation (ION GNSS+ 2014), Tampa, FL, USA, 8–12 September 2014; pp. 2555–2562.

27. Noureldin, A.; Karamat, T.B.; Georgy, J. *Fundamentals of Inertial Navigation, Satellite-Based Positioning and Their Integration*; Springer: New York, NY, USA, 2013.

28. Borre, K.; Akos, D.; Bertelsen, N.; Rinder, P.; Jensen, S.H. *A Software-Defined GPS and Galileo Receiver: Single-Frequency Approach*; Birkhäuser: Boston, MA, USA, 2007.

INDEX